GENERAL TOXICOLOGY

JAROSLAVA ŠVARC-GAJIĆ

Nova Science Publishers, Inc.
New York

For permission to use material from this book please contact us:
Telephone 631-231-7269; Fax 631-231-8175
Web Site: http://www.novapublishers.com

LIBRARY OF CONGRESS CATALOGING-IN-PUBLICATION DATA

Svarc-Gajiae, Jaroslava.
General toxicology / Jaroslava Svarc-Gajiae.
p. ; cm.
Includes bibliographical references and index.
ISBN 978-1-60741-022-5 (hardcover)
1. Toxicology. I. Title.
[DNLM: 1. Toxicology--Handbooks. QV 607 S968g 2009]
RA1211.S93 2009
615.9--dc22
2008054772

Published by Nova Science Publishers, Inc. ✝ New York

CONTENTS

PREFACE

In the past century the population entered into the chemical era, characterized by intense development of the chemical industry and an increased use of chemicals in different areas of human life. As a consequence, people are continuously exposed to chemicals of various origins. More than ten million chemicals were synthesized in the past decade. Of these, about one hundred thousand substances are categorized as economically justified industrial compounds. Furthermore, many natural products encountered in everyday life are very toxic. Transgenetic products further broaden the scope of toxicology and bring many controversies to scientific, juristic and government communities. Surroundings overburdened with foreign compounds threaten human life and alienate us from nature. The purpose of this book is to serve as an aid to anyone who wants to be more aware of their surroundings and of the risks and consequences of inevitable encounters with either naturally occurring toxicants or synthetic ones, introduced by chance or intentionally, and usually for financial gain.

Writing a book aimed at broadening views on food, environment and everything else that surrounds us, was highly demanding and complicated task for primarily two reasons. One reason is the challenge to clearly define a subtle borderline in inevitable exposure to toxicants distinguishing high risk situations from the exposures in which organisms manages to counteract the adverse effects. While a single person cannot globally mitigate chemical dangers, one can contribute, on an individual level, by learning to detect when that borderline is crossed. No one can afford to live in surroundings free of risks, chemical or otherwise, but the skill to recognize a real threat can be built on the basis of knowledge, curiosity and a high level of awareness.

The other challenge inherent in writing this book was to determine which topics to discuss in a general approach to toxicology. Selection in itself is a difficult task, making the development of this general approach to toxicology even more complex. Consideration of all significant aspects of toxicology would exceed the scope of this book and every judgment regarding the volume that would adequatly explain the complex toxicological problems is inappropriate.

This book is intended, therefore, as a handbook for wide array of professionals, students and general population; and is devoted to presenting general principles of toxicology and explaining the complexity of chemical behavior in the human body. Intended as a tool, not only to specialists and people dealing with the problems of intoxication, but also to the general population and to analysts dealing with problems of toxicant detection and determination, this book is envisioned also as an aid in general education, since everyone should be familiar with their body and with the everyday risks that life and surroundings bring.

Chapter I

Introduction to Toxicology

History of Toxicology

There are numerous ways to define toxicology, but the simplest and narrowest definition describes toxicology as the science of poisons and toxins, as the word itself implies. The term *toxicology* derives the Greek *toxicon*, meaning poison, and *logos*, meaning science. The science of toxicology studies the symptoms of poisoning and mechanisms of toxic effects. Besides dealing with chemical and physical properties of toxic substances and their physiological effect on the living organisms, toxicology also includes qualitative and quantitative methods of chemical analysis and develops procedures for treatment of poisoning. The term toxin, or poison, refers to highly poisonous substances produced by higher plants, animals, pathogenic bacteria or humans. To date, this scientific discipline has expanded to include wide range of related topics, such as the evaluation of risk involved in the use of pharmaceuticals, pesticides and food additives; as well as studies of occupational poisoning, exposure to environmental pollution and the effects of radiation.

Toxicology can also be defined as a medical and natural science studying the outcome of human activities and application of the knowledge of many other scientific disciplines, such as biology, chemistry, biochemistry, pathology, pharmacology, physiology, immunology, ecology and other related fields.

Modern toxicology is constantly developing, leading to the formation of new branches of study. In the past decade, numerous toxicological subdisciplines were born, such as *experimental toxicology*, which defines procedures for examining the effects of a substance both *in vivo* and *in vitro* and develops new testing methods; *clinical toxicology,* which deals with the diagnostic of the poisoning and therapies development; *forensic toxicology,* which is concerned primarily with the detection and estimation of poisons in tissues and body fluids obtained at autopsy and, occasionally, in blood, urine or gastric material of living persons. Forensic toxicology offers a means to develop scientific proof that the poisoning is a cause of particular pathological state. *Industrial toxicology* deals with the occupational exposure and poisoning prevention in the industry; *Ecotoxicology* examines the effects and fate of the toxic substances in the environment, their routes, means of transport and effects on flora, fauna and biodiversity. *Aquatic toxicology* studies the effects of manufactured chemicals and other anthropogenic and natural compounds and activities on aquatic organisms at various organizational levels, from subcellular and individual organisms, to communities and water ecosystem. *Molecular toxicology* tries to explain molecular-level mechanisms of adverse action of toxicants. *Behavioral toxicology* observes the influence of various anthropogenic and natural substances on the human nervous system and on psychological functions. *Agricultural toxicology* defines the risk of exposure to substances commonly used in agriculture, such as pesticides, and their effects on the environment. *Veterinary toxicology* diagnoses and treats poisoning in domestic and wild animals.

Recognition that various naturally occurring substances can affect different functions of humans and other living organisms is not recent. Several thousand years ago, humans recognized toxins of plant, animal and mineral origin. Shen Nung (2696 B.C.), the father of traditional Chinese medicine, died from a toxic dose of one of the herbs he used in experiments. In his manuscript (*Shen Nung Pen Ts'ao Ching*), he describes and defines more than 365 herbs, containing various physiologically active substances. The first written medical documents that mention poisons date from 1500 B.C. and were found in Egypt. These documents include 110 pages describing anatomy, physiology, medical treatments and incantations. The sacred Hindu texts, the *Vedas* (900 B.C.), elaborate on similar topics. In ancient Greece, much useful data on toxicants also was collected and written down. In the epic tales *The Iliad* and *The Odyssey*, Homer (800 B.C.) describes the use of arrows poisoned with venoms (*toxicon pharmacon*). When Socrates was convicted and sentenced to death in 399 B.C. on charges of religious heresy and corrupting minors, he ingested the alkaloid coniine. His death occurred quickly and was accompanied by convulsions and paralysis. Hippocrates, a Greek physician born on the island of Cos in 460 B.C., is known as the father of modern medicine and was regarded as the greatest physician of his time. He gave a name to "cancer," based on the Greek *karkinos* (crab), because of the visual similarity of cancerous tissue with the spreading of crab claws. Hippocrates moved medicine away from the superstition and toward science. He also originated a physician's oath of ethics that is still used today. Dioscorides, a Greek pharmacologist and physician at Nero's court, was the first to classify the toxins in plants, animals and minerals. He wrote *De Materia Medica*, which is accepted as a basis for the modern pharmacopoeia. In antic Rome, knowledge of poisons and their use was highly sophisticated. Roman king Mithridates VI (131-63 B.C.) from his youth repeatedly poisoned himself. He considered poisoning an art and systematically studied the prevention and counteraction of poisons using himself and his prisoners to test antidotes. Over the years, he consumed mixtures of more than 50 different compounds to protect himself from potential poisoning and, as a result, he failed in an attempt to commit suicide by poison. In ancient Rome, poisoning between political rivals was very common and in 82 B.C. Sulla established "Lex Cornelia de sicariis et veneficis," a law against poisoning and the selling and possession of poisons.

Natural poisons were readily available and extensively studied in the ancient world. Cleopatra, the Queen of Egypt (69-30 B.C.), experimented on prisoners and the poor with strychnine and other natural poisons. Even though many natural toxins were well examined and defined in the ancient world, the properties of some natural alkaloids, such as ergot alkaloids, remained unexplained for a long time. The earliest reference to ergotism, the effect of long-term ergot poisoning, appeared in the *Annales Xantenses* for the year 857, and subsequently, the condition was noted frequently throughout history. In the Middle Ages, the effect of ergot poisoning was known as St. Anthony's fire because of the burning sensation and gangrenous changes brought on by the consumption of wheat and rye contaminated with ergot.

One of the most important figures in the development of toxicology as a scientific discipline was Auroleus Phillipus Theophrastus Bombastus von Hohenheim, known as Paracelsus (1493-1541). He established principles that are still valid and are accepted by modern toxicology. Paracelsus was born on November 10, 1493, in Silbruk, Switzerland, the son of respectable physician. Paracelsus showed interest in medicine, alchemy and chemistry very early in his youth. At the age of sixteen he started chemistry, medicine and surgery studies at the University of Basel, where he eventually introduced new substances, such as minerals and metals, into the practice of medicine. The preparation of his alchemic medications sometimes took days or months. He would bury vessels containing medicine in the earth, or leave them in the sun for a very long time, to distill and purify the contents. Paracelsus is described as arrogant and prone to conflicts because he challenged the established learning of his time, and considered Galen`s principles to be primitive. Also, Paracelsus lectured in German, instead of Latin, which was the language of the academy at that time. One of Paracelsus's ideas was to extract the essence of animal, plant and mineral raw materials by using the essence of wine (*Quinta essentia vini)*, obtained by distillation. The distillation process entailed the extraction of an essence (*quinta essentia*) with ethanol, obtained by wine distillation. Quinta essentia of bismuth, of different salts, of pearls, flowers, camphor, seeds and other materials were common components of Paracelsus's therapies. However, his most important contribution to modern toxicology is the awareness of the significance of the relationship between toxicity and dose.

He claimed that every substance is toxic and only the dose determines whether the toxic effect is expressed or not –"Dosis sola facit venenum" or, the dosage alone makes the poison. The statement is rendered in more descriptively in German – "Alle Dinge sind Gift und nichts ist ohne Gift; allein die Dosis machts, dass ein Ding kein Gift ist." (All things are potentially poisonous and nothing is without poison. Only the dosage determines whether a thing is or is not poisonous.)

Paracelsus also introduced into therapy and healing the use of metals, such as mercury and antimony. Metal tablets had a dual effect: the first, a mechanical effect that provokes peristalsis, the second is a chemical effect, due to emetic properties of antimony oxide that forms on the surface of the pill. By same principle, Paracelsus used *calix vomitivus* (a vomitory cup or chalice), made of antimony glass to treat alcoholics. Wine held in those glasses for certain period produced antimony wine (or *vino stibiato),* containing antimony tartarate (an emetic), which causes nausea in the consumer. Subsequently, mercury was used to treat syphilis, which was described for the first time in 1530.

During the Renaissance, undoubtedly the most commonly used poisons were arsenic compounds because of their neutral smell and taste. Professional assassins, such as Rodrigo and Cesare Borgia, used arsenic as their weapon of choice and operated in Italy for political and monetary gain. Female professional poisoners, such as Hieronyma Spara, who operated in Rome, and Giulia Tophania and Catherine Monvoisin, who worked in France, helped women use arsenic to kill their wealthy husbands and become rich widows. All three women were accused, convicted and sentenced to death by burning at the stake and by strangulation. Poisonings were so frequent at the time that King Louis XIV (1638-1715) was forced to pass a royal decree forbidding apothecaries to sell arsenic and other poisonous substances except to persons sent by the king.

With the development of reliable methods for detecting arsenic, the number of arsenic poisonings significantly decreased. In 1775, Karl Wilhelm Scheele, the Swedish chemist, discovered that white arsenic (arsenic trioxide) is converted to arsenous (or arsenious) acid by exposure to chlorine. Further, the addition of metallic zinc reduces arsenous acid to arsine gas, which, when gently heated, deposits metallic arsenic on the surface of a cold vessel, proving its presence. Additional methods for arsenic detection were subsequently developed. In 1821, Sevillas used the decomposition of arsine gas to detect small quantities of arsenic in stomach contents and urine. In 1836, James Marsh, a chemist at the Royal British Arsenal in Woolwich, U.K., used the generation of arsine gas to develop the first reliable method to determine the presence of absorbed poison in body tissues and fluids, including liver, kidney or blood.

However, until the contributions of Percival Pott (1714-1788), only the acute effects of various toxic substances were recognized and examined. Pott, an English surgeon, stressed the effects of chronic exposure to toxins and was the first to discover the link between occupational carcinogenesis and scrotal cancer in chimney sweeps. He also made significant discoveries in cancer research and surgery techniques.

The development of organic chemistry in 19th century further contributed to the development of toxicology. Morphine crystals were isolated from the opium poppy *(Papaver somniferum)* by Friedrich Serturner (1783-1841) and were used to create strong and effective analgetics. In 1820, quinine was isolated from the bark of the *Cinchona* tree by pharmacists Joseph Caventou (1795-1887) and Pierre Pelletier (1788-1842). Morphine, quinine, strychnine and other alkaloids were subjects of the research of French pharmacologist Francois Magendie (1783-1855), who is known as the father of experimental pharmacology. In 1855, Friedrich Gaedcke (1828-1890) isolated cocaine from the leaves of Erythroxylon coca, from Peru and Bolivia, and by 1860 it was used as an anesthetic. Using a solution of acetic acid in ethanol, Belgian chemist Jean Servials Stas (1813–1891), extracted nicotine from the tissues of a victim of poisoning, identified the poison and solved the case. Modified by German chemist, Friedrich Julius Otto (1809-1870), Stas's method was quickly applied to isolate numerous other alkaloid poisons, including colchicine, coniine, morphine, narcotine (aka noscapine, nectodon, nospen or anarcotine) and strychnine.

Forensic toxicology originated at the beginning of the 19th century with the activity of toxicologists Mateu Orfila (1787-1853) and Robert Christison (1797-1882). With their extensive knowledge of chemistry and poisons, they independently promoted the significance of chemical evidence in crime scene investigation and brought toxicology into the courtroom. As a result, numerous criminal cases were solved. In 1812, Orfila published *Traité des Poisons*, the first systematic approach to the study of the chemical and physiological nature of poisons.

With the development of industry, human exposure to various dangerous substances has increased. Unfortunately, many social catastrophes marked by exposure to toxins have occurred in the 20th century. During the 1950s, an acetaldehyde factory in Minamata, Japan, released effluents containing heavy metals into a local bay, causing severe poisoning of the local population. Many inhabitants who consumed marine species from local waters lost their basic motor functions and became permanently paralyzed due to the irreversible binding of bioaccumulated methylmercury from toxic fish to brain tissue. In 1971, a mass poisoning occurred in Iraq when high concentrations of mercury fungicide used to preserve crops remained in grain and contaminated bread was consumed, exposing more than 50,000 people to toxic mercury: Over 6,000 people were hospitalized and 459 died.

In 1970, in Bangladesh large numbers of people suffered severe exposure to arsenic. when tube wells were drilled to provide clean drinking water in terrains naturally rich in arsenic, poisoning millions of people who drank the water. Then in 1984, accidental release of 40 tons of methyl isocyanate gas from a pesticide plant in Bhopal, India, injured hundreds of thousands of people. In 1983, Times Beach, Missouri, became a toxic ghost town during the largest civilian exposure to dioxins ever in the USA. The town was evacuated and demolished. Another case of careless dumping of toxins forced the relocation of people living in the region of the Love Canal, near Niagara Falls, New York. Subsequent investigation revealed that the town had been built over a landfill containing hazardous waste.

Modern toxicology comprises three major components: toxicants, biosystems and their effects. Toxicants are defined by their physiochemical properties, their occurrence and their interactions with other compounds. In order to define the *biosystem*, biology and medicine should be involved. To accurately explain and define the term *toxic effects* requires examination of both toxicokinetic and toxicodynamic processes. When considering toxicokinetic processes, it is important also to consider the biological processes of absorption, distribution, biotransformation and excretion of the toxicant. Effects of toxic substances on macromolecules, organelles and enzymes are biodynamic processes that also should be thoroughly examined.

Rapid industrial and chemical development introduces many new, toxicologically undefined substances into the global environment and poses new challenges for toxicology as a scientific discipline. The dynamic properties of toxicology allow this scientific discipline readily to respond to such changes as they enable rapid harmonization of the scientific knowledge and cognitions to newly generated conditions.

TOXICOLOGICAL TESTING

Toxicological testing is best performed with pure substances since examination with chemical mixtures is practically impossible and may give rise to erroneous conclusions due to possible chemical interactions. If it is necessary to determine the toxicity of a complex mixture, investigation should start by establishing the composition of the mixture and determining which components of the mixture are bioactive. Each substance then must be classified and characterized with respect to its physicochemical properties. The stability of a given substance determines how it should be administered for testing, particularly if it is inactivated in the gastrointestinal tract.

In general, the starting point for toxicological testing of new substances is testing in laboratory animals. A chemical may be administrated to animals by oral, dermal, intravenous, intramuscular or intra-peritoneal routes. The ideal experimental animal is one that metabolizes and secretes a substance in the same manner as humans do. Thus, rodents, hamsters, rabbits, cats and dogs are the most frequent human surrogates used for toxicological testing, mainly because they are economical, amenable to frequent handling and lacking in the vomiting reflex. Other animals, such as monkeys, are also used. Metabolic differences between humans and surrogate test animals can result in erroneous conclusions about the activity of a substance in humans. Use of younger animals provides the advantage of obtaining faster results due to their higher rate of metabolism, as well as their being cheaper to purchase and house. Age and gender of animals can influence the outcome of the investigation.

The purity of the substance is very important, since impurities may significantly influence toxicological findings. Before animal testing is started, a literature search must be done to collect available results of any previous testing. If the test substance is administrated to experimental animal by mixing it with food or water, it may be affected by oxidation (i.e., through contact with air), or by temperature variations or by the feeding habits of the experimental animals. Gastric intubation (gavage), an artificial means of substance introduction, has advantages with respect to dosing exactness and reduced variation. The gavage is performed after the animals have fasted in order to avoid mixing chemical test substances with stomach contents. In general, the chemical dose should not exceed 5% of the test animal's diet [1]. Housing of the animals also is very important, as the material composition of their cages, residue from wood chips, isolation and other parameters, have been shown to affect test results by affecting enzyme and hormone levels, which in turn influence biotransformation processes.

Alternative test methods also exist, including the use of laboratory-cultured single-cell organisms, isolated organs or mammalian cell lines instead of animals. The advantage of animal testing is the possibility to examine the results of different routes of exposure under controlled conditions. However, differences in genetic pool, lifespan, susceptibility, metabolic pathways and other parameters of organic life, present serious limitations when extrapolating the results of animal testing to human populations. The steps in toxicological examination are strictly preset and depend on the type of the tested substance and its exposure route. It is not uncommon to adjust a preset sequence on the basis of the results obtained.

Animal experiments are usually classified according to their duration as acute, subacute, chronic or subchronic testing. Chronic and acute toxic effects for a specific toxicant can differ both qualitatively and quantitatively. Chronic effects are evident after certain prolonged periods of exposure and are usually less severe.

For *acute* toxicity testing, one large single dose, or that dose equivalent divided into several smaller doses, is administrated during 24 hours. Repeated exposure once a day during a period of several days, or 14 or 28 days, is denoted as *subacute* toxicity testing. Experiments extending over 30 to 90 days are considered *subchronic* testing. The administration of the test substance every day during several months or years is considered *chronic* testing. Special categories of toxicological testing include reproductive testing, developmental testing and genotoxic testing. Genotoxicity testing focuses on mutagenic, carcinogenic and teratogenic effects. The characterization of the test substance begins with acute toxicity testing, and these results provide important background information and guidance for subsequent tests. The assessment is prolonged to gain information regarding genetic toxicity and toxicokinetic qualities of the substance. Negative results of genetic toxicity are usually preceded by subchronic and chronic toxicity testing.

The toxicologist selects the route of administration based on the expected (or typical) route of exposure in humans. Under normal conditions, the oral route is the most common route for human exposure. In industrial settings, inhalation of dust or vapor, as well as percutaneous (dermal) contact are typical routes of exposure. The route of administration also may influence the quantitative toxicological findings, such as in what defines a lethal dose, and also may alter the position and the slope of the dose-response curve. A dose-response curve is an x-y graph relating the magnitude of a stressor dose (substance) to the response of the organism, as illustrated in Figure 1.1. dose-response curve, below.

Acute Toxicity Testing

Acute toxicity testing is usually performed on at least two animal species. Most frequently, acute toxicity tests are performed by administrating the test substance only once to each animal. Alternatively, a series of smaller doses may be administrated during 24 h. The animals are then kept under observation for 14 days. Administration of the test substance is usually performed by a plastic or stainless-steel dosing tube which leads from mouth to stomach. Test samples may be dissolved or suspended in either water, 0.25% agar or oil, if the hydrophobic substances are encountered. The dosage volume should be constant and is usually 5 ml/kg. Observations of the animals are performed by placing them into a rink with other non-treated animals in order to compare their behavior. Number of animal deaths, time between administration and death, as well as all symptoms before death are recorded. Upon death, any changes at necroscopy, such as the color of abdominal wall, kidneys, liver, uterus, and lungs and the degree of intestinal motility, are observed. A determination is made whether the death occurs as a result of cardiac or respiratory arrest and whether the blood vessels are dilated or constricted at the moment of death. All other unusual changes are described.

The main purpose of acute toxicity testing is to determine the level of substance toxicity, which is usually achieved by determining the LD_{50} (*dosis letalis*), that is, the dose that causes death in 50% of the test population. The LD_{50} concept was introduced in 1927 by J.W. Trevan [2], who attempted to estimate the relative poisoning potency of various drugs. Trevan used this approach to compare even chemicals that act in different ways, as he determined that their LD_{50} data are comparable and can determine the relative acute toxicity even for different chemical substances. Without the lethal dosage criterion, relative toxicity, for example, of neurotoxic, cardiotoxic and nephrotoxic substances, would be extremely inaccurate and impossible to determine.

For gaseous substances, the LC_{50} value is defined as a measure of acute toxicity. In inhalation experiments, the concentration of a chemical in air that kills 50% of the test animals in a given time, usually 4 hours, is the LC_{50} value. The chemical, usually a gas or vapor, is first mixed at a known concentration in a special air chamber where the test animals are placed. When LC_{50} value is reported, it should also state the species of the test animal and the duration of the exposure, e.g., LC_{50} (rat) - 1000 ppm/4 hr. Every stated LD_{50} value also must be accompanied with a statement of the route of administration and the test animal species encountered in the study, since the values can significantly vary depending on animal species, gender, body weight, age and route of administration [1]. Other parameters, such as diet, gastrointestinal content, temperature, time of day or season, can also cause variations in results. For this reason, the acute toxicity testing should be standardized as much as possible and human error should be excluded.

The toxicity of different substances may be compared on the basis of LD_{50} values and the substances classified according to toxicity scales (Table 1.1. and Table 1.2.) [3]. The two toxicity scales most commonly used in toxicity testing are the Hodge and Sterner scale and the Goselin, Hodge and Sterner scale.

The Hodge and Sterner Scale provides a measure of substance toxicity for oral, inhalation and dermal administration based on LD_{50} and estimates a probable lethal dose for humans. The Gosselin, Smith and Hodge scale provides a measure of substance toxicity based on LD_{50} oral administration results. The numerical indicators ascribed to levels of severity in the two scales are reversed and therefore any ratings of toxicity should be referenced to the scale that is used.

Table 1.1. Toxicity Classes: Hodge and Sterner Scale

Toxicity Level	Toxicity class	Oral LD_{50} (mg/kg) (rats)	Inhalation LC_{50} (mg/l) (rats, 4 h)	Dermal LD_{50} (mg/kg) (rabbits)	Probable lethal dose for man of average weight
1	Extremely toxic	1 or less	10 or less	5 or less	1 grain (a
2	Highly toxic	1-50	10-100	5-43	4 ml (1 tsp)
3	Moderately toxic	50-500	100-1000	44-340	30 ml
4	Slightly toxic	500-5000	1000-10,000	350-2810	600 ml
5	Practically non-	5000-15 000	10 000-100 000	2820-22	1 litre
6	Relatively	15 000 or more	100 000	≥22 600	>1 litre

Table 1.2. Toxicity Classes: Gosselin, Smith and Hodge Scale

Toxicity rating	Toxicity class	LD_{50} (oral, rat)	Probable oral lethal dose for man of average weight
6	Super toxic	< 5 mg/kg	1 grain (< 7 drops)
5	Extremely toxic	5-50 mg/kg	4 ml
4	Very toxic	50 500 mg/kg	30 ml
3	Moderately toxic	0.5-5 g/kg	30-600 ml
2	Slightly toxic	5-15 g/kg	600-1200 ml
1	Practically non-toxic	> 15 g/kg	> 1200 ml

In general, if the acute toxicity, expressed as LD_{50}, is close for different animal species, the degree of acute toxicity in humans may also be expected to be close. Special calculations are used to translate animal LD_{50} values to possible lethal dose values for humans. A safety factor of 10,000 or 1,000 is usually included in such calculations to allow for variability between individuals and for the uncertainties of experimental test results. The LD_{50} should be considered only as one aspect of toxicity information. For a more thorough estimate of the immediate or acute toxicity of a chemical, additional information should be considered, such as the lowest dose that causes toxic effect (TDLO – Toxic Dose Low), the dose which does not provoke any effect in humans (NOEL – Not Observable Effect Level), as well as the rate of recovery from a toxic effect. The toxicological behavior of the compound may also vary when the substance is introduced together with other compounds, which can significantly alter the compounds toxicity.

The determination of LD_{50} values provides the guidance for further testing and is useful in defining symptoms consequent to administration, as well as for identifying clinical manifestations of acute poisoning. The information derived from acute toxicity testing also may be useful in designing long-term studies, in identifying target organs and in predicting possible diagnosis and treatments. Results of acute toxicity also provide information necessary to classify and label substances as required by regulatory authorities.

Besides defining the immediate toxicity of a substance, acute toxicity studies enable the definition of a dose – response curve. In defining dose-response relationship, one lethal dose, one ineffective dose and at least three log-doses in-between are determined. Depending on the outcome of the first administrated dose (usually 100 mg/kg), the next dose is logarithmically adjusted to either 1 g/kg or 10 mg/kg. The procedure is continued until the dose elicits some symptoms but is neither ineffective nor lethal. According to these data a dose-response curve is plotted (Figure 1.1.) to describe the LD_{50} value.

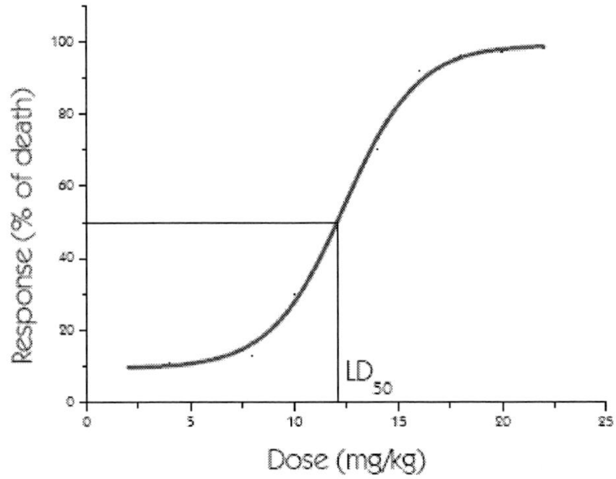

Figure 1.1. Dose-response curve.

To represent the dependence of the death ratio on dose, death incidence may be replaced by probability units (probit) [4], which are calculated on the basis of probability, which is directly dependent on standard deviation. Namely, the probability distribution of the event, i.e., the death, may be described by a Gaussian curve, and the width of the curve is defined by dispersion or standard deviation of the data. The narrower the curve, the lower the standard deviation and the less the individual variation between the animals. Thus, the dose-response curve may be linearized by using probability units instead of percentage of death events, and by showing the logarithm of the dose on "x" axis..

Within the dose-response curve, the slope of the intermediate part of the curve depends on variations in the individual susceptibility of the animals. Other factors such as an animal's age, gender, bogy weight and diet can cause the variations in the toxic effect. A flatter slope points to greater individual variations and is in reverse proportion to the dispersion (standard deviation) of the Gaussian curve, which describes the probability of the event.

Animal rights activism has led to debate regarding the need for alternative methods for toxicity testing. The outcome of this debate has resulted in recognition of a need to reduce the use of animal testing , to modify conventional testing methods, to reduce the pain and suffering in animals, and to replace animal testing with types of testing, which may be performed *in vitro* on tissue models or in cell cultures, or *in vivo* using the lower organisms.

To date, neither type of proposed alternative testing has been accepted by regulatory authorities because the accuracy of their results when extrapolate to humans is uncertain.

The British Toxicology Society (BTS) proposes an alternative to LD_{50} determination for assessing toxicity. The BTS proposed method employs a small number of animals and is based on the observation of toxic signs, instead of counting the number of dead animals. For oral testing, three fixed doses are administered - one of 5 mg/kg, one of 50 mg/kg and one of 500 mg/kg . Depending on the dose that provokes observable signs of toxicity, the substance is simply classified as very toxic, toxic, not very toxic or unclassified. Table 1.3, presents details of parameters for classifying results by this proposed method.

Table 1.3. British Toxicology Society (BTS) Proposed Method for Determination/Observation of Acute Oral Toxicity

Dose (mg/kg)	Result	Classification
5	< 90% of survived animals	Very toxic
	≥ 90% of survived animals with evident signs of toxicity	Toxic
	≥ 90% of survived animals without evident signs of toxicity	Examine the 50 mg/kg dose
50	< 90% of survived animals	Toxic. Examine the 5 mg/kg dose, if not examined
	≥ 90% of survived animals with evident signs of toxicity	Moderately toxic
	≥ 90% of survived animals without evident signs of toxicity	Examine the 500 mg/kg dose
500	< 90% of survived animals or evident signs of toxicity without lethal outcome	Moderately toxic. Examine the 50 mg/kg dose, if not examined
	Without signs of toxicity	Not classified

Subchronic Toxicity Testing

Subchronic testing provides information about some chronic effects, target organs and the possible accumulation of the substance in the tissues. Subchronic testing is performed over a period of several months to a year, and the testing parameters are determined according to results of acute testing. The study usually involves four groups of test animals, including a control group. The number of the animals in each group is usually from 10 to 20 for each gender, and usually involves two test animal species. The test substance is blended with the purified diet, ensuring the stability of the test substance. The physical appearance, behavior, body weight, food intake and excretion of the animals are monitored regularly. Biological fluids are collected for hematological and biochemical tests. Specific hepatic, eye and gastrointestinal functions are monitored and blood pressure and body temperature are measured at regular time intervals. Subchronic testing usually reveals cumulative effects of the examined substance in tissues and on metabolic systems. Upon demonstration of animal illness, or upon the completion of the study, a necroscopy is performed and organs and tissues are examined for evidence of pathological changes. The weight of major organs, such as heart, liver, thyroid, spleen, adrenals and testes is determined. Histological analysis is performed on the liver, heart, gastrointestinal tract, spleen, ovary, kidneys, pancreas, bladder, testes and other tissues. Prepared tissue slides are archived for possible review by regulatory authorities, sponsoring companies and researchers.

Chronic Toxicity Testing

The main goal of chronic toxicity testing is to elucidate the biochemical mechanisms of the toxicity, to reveal potential carcinogenic effects and to perceive adverse effects that are not clearly evident in subchronic testing. In this long-term study, a larger group of experimental animals is employed and lower doses of the test substance are applied for longer periods. The study seeks evidence of long-term effects and the doses are carefully calibrated so as not to overburden the organism's capacity for absorption, transportation, biotransformation and excretion. Since the aim of chronic toxicity study is to explain and define mechanisms by which adverse effects arise, acute toxic effects of the free substances are undesirable and are avoided by adjusting the dose. The number of test animals in a single dose-level group is usually 50, but may be higher if a higher level of confidence is desired. The increase in the number of experimental animals produces a higher probability of the event demonstration, as well as a higher the confidence level. For example, for the probability that the adverse effect will be demonstrated in one of hundred animals, the number of required animals in the test should be 295 or 670, for confidence levels of 95% and 99%, respectively [5].

Genetic Toxicity Testing

Genetically toxic, or genotoxic, compounds are mutagenic compounds that have the ability to damage the genetic material of the cell either to a lesser extent at the nucleotide level (point mutations) or to in a greater extent on chromosomal level (i.e. as chromosomal aberrations). Genetic toxicity testing is directed toward revealing either point mutations or chromosomal aberrations under the influence of certain compounds and is performed on microorganisms, mammary cell lines or animals.

Mutagenicity tests are conducted in the early stage of toxicological screening, after the acute toxicity testing is completed and the initial information about the compound is compiled. Positive results of genotoxicity tests exclude further use of the substance in food production. Damage to genetic material can be passed to offspring if the mutation occurs in germ cells, and can be a cause of many medical conditions, teratogenesis and carcinogenesis. It is also believed that mutations in somatic cells contribute to cancer formation. Theoretically, a single mutated cell can initiate cancer formation, but in a more realistic approach it is believed that a certain number of cells in proximity of the mutated cell must also undergo similar mutations in order to form tumor tissue.

In mutagenicity testing, the Ames test, a biological assay to assess the mutagenic potential of chemical compounds, is used to detect point mutations in a strain of *Salmonella typhymurium* bacterium under the influence of the test compound. The microbial *(Salmonella)* strain requires the presence of histidine for its growth. *Salmonella* is grown on nutrient agar and transferred to test tubes with and without suspected mutagen. Rat or human liver homogenate is added to each test tube in order to provide enzymatic activation of the procarcinogen. After incubation, the content of the test tubes is transferred to plates with substrate lacking in histidine. Positive growth on the plate indicates that the strain has adapted to new condition due to their mutations and reveals the mutagenic character of the test substance.

To activate the suspected mutagen by the host's enzyme system, the microbial strain is injected into the peritoneal cavity of test animals that have previously received an adequate dose of the suspected mutagen. After the incubation period, samples of peritoneal fluid containing microorganisms are applied to the contents of an agar plate deficient in nutrients. Significant subsequent growth of colonies indicates mutagenic activity in the test compound. Numerous attempts have been made to conduct this type of test using yeast or fungi (*Saccharomyces, Neurospora crassa*) instead of bacteria, but yeast/fungi-based testing failed to offer any strong advantage over conventional bacterial tests [6].

Another standard procedure for mutagenicity testing involves *in vivo* testing of eukaryotic cells, but it is also not uncommon to harvest the eukaryotic cells for *in vitro* assessment of the effects of test substances on chromosomes. The fruit fly (*Drosophila melanogaster*) is commonly used for *in vivo* assessment. The test procedure involves treating the organism with the test chemical and observing the changes in its body color, shape, eye color or thorax bristle [7].

For the detection of chromosomal aberrations, prokaryotes are unsuitable and more complex cells are required. Testing for chromosomal aberration is usually performed on cell cultures of lymphocytes, fibroblasts or hepatocytes. The cell culture is grown on glass slides in a medium containing different doses of the suspected mutagen. After an incubation period, the slides are washed in order to remove unabsorbed chemicals and are incubated again with radioactive thymidine. If mutation has occurred, a normal DNA repair system in the cells will be activated and the tritiated thymidine will be incorporated into the nucleic acid of the cells, and can be monitored. Chinese hamster ovary cells are rapidly growing and are very sensitive to cytotoxic guanine analogs (6-thyoguanine, 8-azaguanine). If mutation occurs in the ovary cells in testing, the cells will have the ability to grow in the presence of guanine analogs. Mutated cells are deficient in the hypoxantine-guanine-phosphoribosyltransferase enzyme and the cytotoxic effect is omitted [6]. Chemically induced mutation can also be demonstrated as a loss of heterozygosity of the gene responsible for the expression of thymidine kinase enzyme in mouse lymphoma cells. Discrimination between homozygous and heterozygous cell lines can be achieved by provoking cytotoxicity, specific for heterozygous cells by incorporating 5-bromo-2-deoxyuridine into the medium.

Chromosomal damage caused by a test chemical also can be observed under the microscope after appropriate staining of the chromosomes. The cell replication is usually interrupted in the metaphase stage by the addition of colchicine. Cells are first incubated with the test chemical and then are allowed to pass two DNA replication cycles in a medium containing 5-bromo-2-deoxyuridine. After the cell replication is complete, if chromosomal damage has occurred, chromosomal fragments are easily spotted under the microscope.

Teratogenic Testing

Teratogenesis, that is, the formation of congenital malformations in the fetus, can be provoked by several different mechanisms. Apart from a possible effect on genetic material of the developing fetus, a test compound may also interact with other cell materials, disturbing their replication and provoking teratogenesis. The teratogenic potential of a test compound, therefore, should be separately examined, and final conclusion regarding teratogenicity cannot be made on the basis of genotoxic tests alone.

Teratogenicity testing is usually performed using three species and three different concentrations of the test substance. The highest tested dose is one that does not express adverse effects on the mother. Teratogenicity testing also requires both a negative and a positive control group, to which the reference teratogen, such as high dose of vitamin A or trypan blue is administrated. The administration of the test substance, placebo and reference teratogen begins on the 7th day of pregnancy in rats [6]. Pregnant female rats are housed separately to avoid aggressive interaction with males and they are weighed daily. both live and dead fetuses, removed by caesarean section, are inspected for abnormalities and malformations. Mothers also undergo a complete necroscopy.

Reproductive and Developmental Toxicity Testing

Reproductive testing examines the effects of the test substance on fertility, developing fetus and the mother and her offspring after their birth. This testing also examines for potential teratogenic or mutagenic effects by monitoring the development of the fetus. At least three different doses are examined, with the highest dose a nearly maximally tolerable dose that produces no significant toxic effect on the mother. Reproductive testing is a three-generation study involving the animals that are exposed to the substance for the first time transplacentally and via mother's milk. The testing ends with testing of the offspring of the offspring. The animals are monitored from the time of the mother's conception to the time the offspring produce their own offspring. Each female generation is bred twice. The offspring from the first breeding are terminated at birth for the examination, and the offspring from the next breeding participate further in the test. In this way all levels of potential effects are studied. The sexual development, maturation and potential effects of the test substance on lactation are observed and investigated. Special attention is dedicated to all possible effects on reproduction, such as altered gonadal function, changes in mating behavior and changes in conception rate. All animals participating in reproductive testing are eventually terminated and pathologically examined [7].

Other Tests

Knowledge of dermal absorption of certain chemicals is very important in occupational exposure, where the risk of intoxication by specific compounds through a dermal route is high. Many regulatory agencies concerned about safety in the workplace demand scientific information about skin absorption. Animal testing providing this information is usually performed on rats. For testing, each rat's back is shaved and the test chemical is smeared on the skin for an exposure period of up to 24 hours. Afterward, the rat's skin is washed and the animals are housed individually in "metabolism cages" to permit the collection of their excrement for analysis. The skin, blood and excrement are analyzed, so that the rate of skin absorption can be calculated. Alternatively, various skin cultures may be used to measure the passage of a test chemical into and across the skin to a fluid reservoir. Absorption of a test chemical is measured over the time by analyzing the receptor fluid and the treated skin. Nonanimal tests offer the possibility of examining a broader range of doses.

Corrosive agents are chemicals that cause irreversible damage and destruction of the skin, often damaging several layers of the tissue. Corrosive reactions are characterized by ulcers, bleeding, bloody scabs, and discoloration. Corrosivity testing is performed on rabbits because of their sensitive skin. A test chemical is applied to the shaved skin on the rabbit's back. The application site is then covered with a gauze patch for the duration of the exposure period, normally four hours. The patch is then removed and the degree of skin damage is read and scored at specified time intervals. Untreated skin areas serve as the control. A chemical is considered to be corrosive if, by the end of the 14th day of observation, the chemical has damaged the outer layer of the skin of one or more animals, leaving visibly dead tissue. In nonanimal studies, human-derived skin cells that have been cultured to form a multi-layered model are used in the test. Alternatively, artificial protein membrane (Corrositex™) may be used to replace the skin culture. Unlike corrosive agents, which cause irreversible skin damage, irritants are chemicals that provoke reversible skin damage. Signs of irritation include the development of a rash, inflammation, swelling, scaling and abnormal tissue growth in the affected area. The test of possible chemical irritation is performed in the same manner as in the skin corrosion tests, and the substance is classified as an irritant if the observed skin damage is reversible by the end of the 14th day of observation.

Phototoxicity and photoirritation are inflammatory skin reactions caused by exposure to a chemical and subsequent exposure to the sunlight or ultraviolet radiation. This endpoint is a concern mainly with drugs and pharmaceuticals that are either ingested and then partially distributed to the skin, or applied directly to the skin in the form of a cream. Phototoxicity typically appears as exaggerated sunburn and may be accompanied by a rash, swelling and inflammation. Phototoxicity tests are performed by applying different concentrations of a test chemical to shaved animal skin on the animal's back. Half of the animals are then exposed to ultraviolet radiation for two or more hours, after which the test chemical is removed. The animals are then restrained for several days while experimenters examine their skin. If the substance is phototoxic, swelling and sores commonly appear on the treated skin. In non-animal testing, cell lines are exposed to a test chemical in both the presence and the absence of the light. Photo-cytotoxicity is then evaluated by the cell's ability to absorb neutral red dye.

Pyrogens are fever- and inflammation-causing agents that can pose serious health hazards, especially in the case of intravenous drugs or pharmaceutical products. In the testing of pyrogens, the test substance is injected into the bloodstream of experimental animals and their body temperature is monitored. As an immunological reaction, pyrogenicity involves the interaction between a substance and the cells of the immune system. Using human blood as the test medium, this non-animal method is able to fully model the interaction between the body's immune system and the test substance, thereby confirming the presence or absence of the pyrogenic effect.

REFERENCES

Barilc, FA. *Principles of Toxicology Testing*. Boca Raton; CRS Press; 2007.

Hodge, HC; Sterner, JH. Tabulation of toxicity classes. *Am. Ind. Hyg. Assoc. J.* 10, 1949, 93–98.

Jacobson-Kram, D; Keler KA. Toxicological Testing Handbook: Principles, Application and Data Interpretation. Informa Healthcare: 2006.

Omaye, ST. *Food and Nutritional Toxicology*. Boca Raton; CRS Press: 2004.

Schlede, E; Genschow, E; Spielmann, H; Stropp, G; Kayser, D. Oral acute toxic class method: A successful alternative to the oral LD50 test. *Reg. Tox. Pharm.* 42(1), 2005, 15-23.

Trevan, JW. The error of determination of toxicity. *Proc. R. Soc.* CI B. 101(712), 1927, 483–514.

Turner, RA. *Screening Methods in Pharmacology*. New York; Academic Press: 1965.

Chapter II

TOXICANT ABSORPTION

Systematic toxic effects can be demonstrated only after a toxicant has been absorbed and transferred into the blood. A local effect occurs as a result of direct interaction on a contact site with the toxicant, such as on the skin in dermal contact, or in the respiratory organs immediately after inhalation. After absorption, systematic toxic effects may be expressed on specific organs and systems, such as the gastrointestinal, respiratory or cardiovascular system. Some substances may express both local and systematic toxic effects. For example, tetraethyl lead is readily absorbed via the dermal route, leaving skin wounds and provoking a systematic effect soon after absorption, which is particularly profound in the central nervous system (CNS).

Many factors influence the dynamics of the entry of a substance into an organism and consequently, its toxicity. Among most important factors are concentration, duration of the exposure, physico-chemical properties of the compound and route of entry. Toxicant particles confront many barriers on their way to bloodstream, such as skin or mucosal membranes of the gastrointestinal or respiratory system. Each tissue level barrier is composed of numerous sub-barriers on a cellular level, and the chemical must pass cell membranes many times. The transport mechanism through lipid bilayers, i.e., cell membranes, is, therefore, very important, and is highly dependent on physico-chemical properties and on the size of the molecule, as well as on the molecular charge. The resistance to the entry of a molecule into a cell varies for neutral molecules of different sizes and for charged ions.

The most common routes of toxicant entry are dermal, and via the gastrointestinal or respiratory tract. Less frequent and probable exposure occurs intravenously and subcutaneously. When a compound is introduced intravenously, it is delivered to the bloodstream directly, skipping many of the barriers that substances normally encounter in all other route of exposure. For this reason the highest toxicity of compounds occurs with intravenous exposure. As part of the body's protective mechanism, the transport of a toxicant into certain sensitive tissues, such as the brain or a developing fetus, may be hindered by specific bodily barriers, such as the placental barrier or the blood/brain barrier. The blood/brain barrier is not completely developed in newborns, making them more susceptible to toxicants. This is the reason that morphine is 10 times more toxic to newborn rats than in adult rats. In the case of the placenta, its role is to enable the exchange of gases between mother and fetus, to excrete fetal metabolic products and to provide nutrients to the fetus. Most of the toxicants pass the placental barrier by simple diffusion, with the exception of specific nutrients that exploit specific transport mechanisms: Some toxicants are partially metabolized by placental enzymes and their toxicity is thereby reduced.

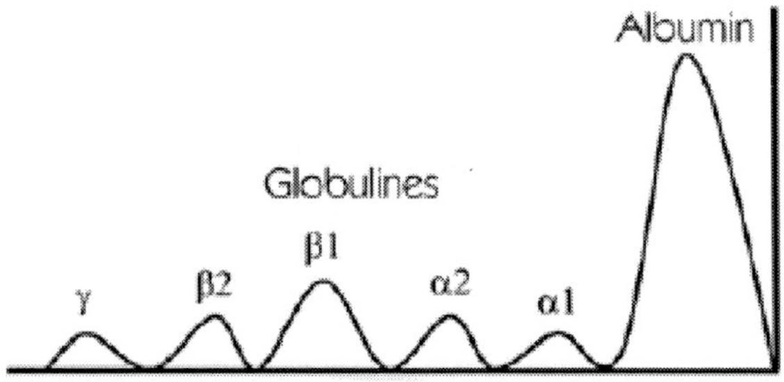

Figure 2.1. Plasma proteins separated by electrophoresis.

After the absorption and entry of a toxicant into the blood circulation, it is distributed through blood circulation to other parts of the body. Highly lipophilic substances are also transported via lymph. Plasma proteins (Figure 2.1.) also play an important role in toxicant transport and they reversibly bind newly introduced free compounds. The most abundant plasma transport protein is albumin, which binds calcium, copper and zinc ions, billirubin, vitamin C, antibiotics, histamine, barbiturates and many other substances. Transferrin, which is actually β1-globulin, binds iron and transports it throughout the body. Ceruloplasmine (α2-globulin) binds copper ion, while transporters α- and β-lipoproteins are very important for the transportation of lipophilic substances, such as steroid hormones, cholesterol, vitamins and other substances.

The portion of a toxicant that is bound to plasma proteins is not available for distribution to other tissues and organs, because only free molecules can pass capillary walls. When a portion of the free form of a toxicant exits a capillary, dissociation of the reversible protein-toxicant complex occurs and a new quantity of the free toxicant is released into the circulation. This process is determined by dynamic equilibrium. Protein bonding is not specific for certain substances, and one compound may be replaced with another with higher affinity to the protein.

After it is absorbed, a substance is distributed to all body organs and tissues. In the first stage of toxicant distribution, the most affected organs are those that are the most vascularized, such as liver, kidneys, lungs and glands. Poorly vascularized tissues, such as fat tissue or muscles, take up only a small amount of the toxicant. In the later stages of distribution, the role of circulation decreases and the retention of the compound in tissue depends mostly on chemical affinity. The concentration of lead immediately upon absorption is highest in erythrocytes, kidneys and liver. After a month, most of the remaining lead is stored into the bone tissue. Liver and kidneys have high capacity for various chemicals due to high concentration of intracellular proteins, such as ligandine and matallothioneine, which can bind the toxicants. Highly lipophilic compounds can easily enter neural tissue. Some toxicants exploit existing transport mechanisms intended for the transportation of other molecules due to the structural resemblance.

If the toxicant concentration that has entered the body is significant and exceeds the excretory capacity of the body, the compound may accumulate in specific organs or tissues where it may provoke a tissue-specific toxic effect. This type of accumulation can represent a protective storage of the toxicant excess and may explain the very long (up to decade) half-life of some toxicants.

For some substances, the highest toxicant concentration is achieved at the site of their toxic effect, while other compounds concentrate in body zones that are not closely related to the locale of their most severe toxic impact. Carbon monoxide, for example, binds to hemoglobin and disturbs oxygen transport. The accumulation site and the site of adverse influence are the same for carbon-monoxide. Similarly, the chemical pesticide paraquat binds to lung tissue, where it provokes the toxic effect. Lead, on contrary, accumulates in the bones where it is stored for a long time protecting the organism from acute lead levels. Toxicant accumulation in bone tissue is a surface phenomenon and is related to structural similarities with bone components. Pb^{2+} and Sr^{2+} ions resemble Ca^{2+} ion, while F^- ion mimics OH^- in calcium hydroxyapatite bone structure. Ion exchange occurs between extracellular fluid and bone surface. In the case of highly lipophilic xenobiotics, such as dioxins, polychlorinated biphenyls or polycyclic aromatic hydrocarbons, the deposition of the toxicant occurs in fat tissue thereby decreasing concentrations in the blood. In conditions of fasting or starvation, certain amounts of a toxicant may be mobilized from the fat depot and toxic blood levels may be reached. Not all toxicant deposition in bone tissue is locally "safe." Fluoride ions, for example, may provoke painful fluorosis, while radioactive strontium may lead to the formation of neoplasm.

It is clear that two types of deposits can be recognized. The first one provides a protective mechanism and decreases the concentration of the free form on the site of toxic action. This is the case with lead accumulation in bones or with the accumulation of highly lipophilic xenobiotics in fat tissue. Other type of toxicant deposit such as one in the kidneys, liver or brain, is a localized deposition and also the site of the most profound toxic effect. The accumulation of the toxicant in such localized storage sites leads to expression of a tissue- of organ-specific adverse effect. Cadmium damages the kidneys due to high concentrations accumulated in this organ. Similarly, hepatotoxic compounds usually have high affinity to liver tissue, where they accumulate. Relationship between the concentration of plasma toxicants and tissue toxicants is defined by dynamic equilibrium. When certain amount of the toxicant is metabolized and excreted from the blood plasma, a new portion of the toxicant is withdrawn from its depot and transferred into the blood.

Blood and Blood Elements

Blood is a crucial body fluid that transports oxygen, hormones, vitamins and minerals throughout the body. Blood cells are suspended in plasma, which represents approximately 55% of the blood and is composed mainly of water (90%), proteins and inorganic electrolytes. Plasma serves as transport medium for glucose, lipids, amino acids, hormones, carbon dioxide and oxygen, as well as for metabolic end products. Blood plasma contains many vital proteins including fibrinogen, globulins, albumin and others. Serum refers to blood plasma in which clotting factors, such as fibrin, have been removed. Frozen plasma may be stored for one year after collection. Transport, packaging and storage of dried plasma are much simpler and convenient.

All blood cells - namely erythrocytes, monocytes, macrophages, neutrophils, basophils, eosinophils, T-cells, B-cells, NK-cells (natural killer cells), megakaryocytes, as well as dendric cells - are produced from stem cells (hematopoietic cells) found in the bone marrow. Stem cells are characterized by their ability to form multiple cell types and by their ability of self-renewal, i.e., the generation of one daughter cell that remains a stem cell and one daughter cell that differentiates into another cell type. There are two types of bone marrow, red and yellow. Red blood cells, platelets and most white blood cells arise in red marrow. The color of yellow marrow is due to the much higher fat content. Both types of bone marrow contain numerous blood vessels and capillaries. Red marrow is found mainly in the flat bones, such as the hip bone, the breast bone, the skull, ribs and so on. In cases of severe blood loss, the body can convert yellow marrow to red marrow in order to increase blood cell production.

In the early 20th century, Karl Landsteiner classified blood according to two molecules present on the surface of red blood cells. The molecules were denoted as "A" and "B,",and according to their presence on the surface of erytrocytes, blood is classified as types A, B, O or AB. Type A blood contains only A molecules, type B contains B molecules and types O and AB type contain no A or B molecules, or both A and B molecules, respectively. The composition of blood also differs in respect to the presence of a certain protein, which was initially discovered in Rhesus monkeys and denoted as the Rh (Rhesus) protein, leading to further classification of blood types a Rh positive (+) or Rh negative (-).

Erythrocytes

Mammalian erythrocytes, or red blood cells, are anuclear and thus, do not contain any genetic information. They also lack other organelles and also mitochondria, and produce energy by glycolysis, accompanied by lactic acid production. The diameter of a typical human disk-shaped erythrocyte is 6-8 μm, much smaller than most other human cells. Because these cells do not contain a nucleus or organelles, they cannot produce new structures or repair proteins and enzymes, and their lifespan is limited. Erythrocytes are shaped like biconcave discs (Figure 2.2.), optimizing the exchange of oxygen. Red blood cells are flexible and their shape enables them to fit through tiny capillaries, where they release their oxygen load. A typical erythrocyte contains about 270 million hemoglobin molecules, with each molecule carrying four heme groups.

Hemoglobin is a complex molecule consisting of iron containing heme groups and four polypeptide globin chains. Heme is the pigmented iron-containing nonprotein part of a hemoglobin molecule. Red blood cells change color according to the state of the hemoglobin. When combined with oxygen, the resulting *oxyhemoglobin* is scarlet and when oxygen has been released, the resulting *deoxyhemoglobin* is a darker red.

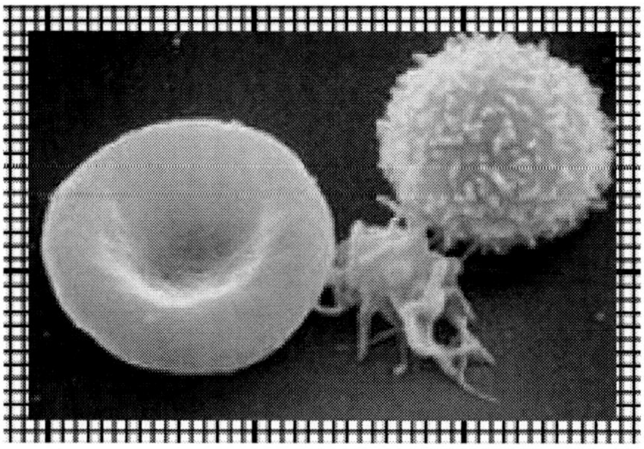

(Author Bruce Wetzel).

Figure 2.2. From left to right: erythrocyte, thrombocite and leukocyte.

Collectively red blood cells store about 3.5 grams of iron in the body. Erythrocytes are continuously being produced in the red bone marrow of large bones, at a rate of about 2 cell million per second. In the embryo, the liver is the main site of red blood cell production. Red blood cell production can be stimulated by the hormone erythropoietin, which is produced in kidneys. This hormone has been used widely for doping in sports. Erythrocytes develop from stem cells to reticulocytes (immature red blood cells) to mature erythrocytes in approximately 7 days and live about 120 days. The aging cells swell into a sphere-like shape and are engulfed by phagocytes, destroyed and their remnants are released into the blood. The main sites of destruction are the liver and the spleen. The heme constituent of hemoglobin is eventually excreted as billirubin, and the globin parts of molecules are reused in the body.

Leukocytes

White blood cells, or leukocytes, are produced in the bone marrow and are an important component of the humoral immune system. Human blood normally contains between $4x10^9$ and $11x10^9$ white blood cells per liter or about 7,000 to 25,000 white blood cells per drop of blood. In conditions such as leukemia, the number may rise to 50,000 white blood cells in a single drop of blood. White blood cells have a rather short life cycle, living from a few days to a few weeks. These cells are also found in the lymphatic system, the spleen and other body tissue. Leukocytes are classified as granulocytes and agranulocytes, depending on the presence of granules.

Granulocytes are characterized by granules in their cytoplasm which contain substances involved in the destruction of infectious agents. Wright's stain, a histologic stain, is used to determine the leukocyte type. According to their staining properties granulocytes are classified as neutrophils, basophils and eosinophils. Neutrophils, which make approximately 65% of all leukocytes, are primarily involved in defense against bacterial infections and are usually first responders. Their activity and death in large numbers form pus. Basophils are the least common of the granulocytes, representing about 0.5% to 1% of circulating leukocytes. A basophil cell that has migrated to other tissue is called a mast cell or mastocyte. Both neutrophils and basophils store histamine. When activated in the course of an allergic response, basophils secrete histamine, several proteoglycans, lipid mediators like leukotriens and several cytokines. Histamine and proteoglycans are pre-stored in the cell's granules, while other secreted substances are newly generated. All of these substances contribute to inflammation. Eosinophils represent approximately 4% of leukocytes and primarily deal with parasitic infections. Agranulocytes, which do not contain granules in their cytoplasm, are classified as lymphocytes and monocytes.

Lymphocytes

Lymphocytes, nearly colorless blood cells, play an important and integral role in the body's immune system. There are several types of lymphocytes, each having a specific role in immune response. Those designated as *B cells* (because they mature in *bursa* organs in bird species), are primarily responsible for immunological activity in relation to antibody production. . Immature B cells are produced in the bone marrow of most mammals, with the exception of rabbits whose |B cells are produced in appendix. Another type of lymphocyte, *T cells*, are so named because they mature in the *thymus* of mammalian species. Only 2% of mature T cells survive and leave the thymus. Following maturation, the T-cell lymphocytes enter the circulatory system and peripheral lymphoid organs such as the lymph nodes or spleen. Different types of B cells circulate in the blood and lymph and perform the role of immune surveillance. They do not produce antibodies until they become fully activated. Each B cell has a unique receptor protein on its surface that binds to one particular antigen. The presence of this receptor allows the distinction of B cells from other types of lymphocyte. The receptor is also a key for triggering B cell activation. Once a B cell encounters a specific antigen and binds it to the receptor, it can further receive an additional signal by interacting with a helper T cell (see description below). Activated B cell further differentiates into either a plasma cell or a memory cell. The main function of *plasma B cells* is to secrete antibodies, which assist in the destruction of microbes by binding to them and making them easier targets for phagocytes. Memory B cells circulate in the body for prolonged period and are specific to the antigen encountered during the first contact. *Memory cells* are able to live for a long time, and can respond quickly following a second exposure to the same antigen. *B-1* cells are another type of B cells which express polyspecificity, meaning that they have a low affinity for many different antigens. These cells mainly produce antibodies of the IgM type and are present predominantly in the peritoneal and pleural cavities.

A critical difference between B cells and T cells is how each type of lymphocyte recognizes an antigen. B cells recognize free antigen in the blood or lymph by using their membrane receptors. By contrast, T cells recognize an antigen in a processed form, as a peptide fragment presented by other molecules to the T cell receptor. A naive B cell, i.e., one that has not been exposed to an antigen, may be activated in a T-cell dependent or independent manner. When a B cell encounters a pathogen, it attaches its protein part. This complex is then moved to the outside of the cell membrane, where it can be recognized by a T lymphocyte, which is compatible with structures bound on the cell membrane of a B lymphocyte. If the B cell and T cell structures match, the T lymphocyte activates the B lymphocyte, which produces antibodies. In T-cell independent activation, macrophages present several antigens to the B cell, provoking the cross linking of the antibodies on the surface of the B cells.

Several different types of T cells are recognized, each with a distinct function. *Cytotoxic T* cells destroy virally infected cells and tumor cells and are also implicated in transplant rejection. They express the certain glycoprotein at their surface which is responsible for their activity. *Helper T cells* divide rapidly upon activation and secrete small proteins (cytokines) to regulate immune response. *Regulatory T* cells are crucial for signaling the end of T-cell mediated immunity. They are distinguished from other T cells by the presence of a specific intracellular molecule. Mutations in the gene encoding this molecule are a cause of serious autoimmune diseases. *Natural Killer T* cells (NK cells) are a special kind of lymphocytes that bridge between the adaptive immune system with the innate immune system. Unlike conventional T cells, which recognize peptide antigens, these specialized T cells recognize glycolipid antigens presented by molecules that are part of a complement cascade. Once activated, these specialized T cells can perform functions ascribed to both helper and cytotoxic T cells, such as cytokine production and the release of cytolytic substances.

T and B lymphocytes differ from NK (natural killer) cells in that they produce memory cells after infection. These memory cells enable rapid response to a previously encountered infectious agent quickly upon re-infection. When a naive T cell encounters an antigen, it is activated and begins to divide into many daughter cells. Some daughter cells differentiate into effector T cells, which perform the functions of that cell, such as producing cytokines (helper T cells) or evoking cell killing (cytotoxic T cells). Other cells form memory T cells, which survive in an inactive state in the host for long periods until they re-encounter the same antigen and reactivate. Both B and T cells are activated in the presence of antigen, such that B cells produce large quantity of antibodies, while T cells release cytokines or other cytotoxic substances. $\gamma\delta$ cells represent a small subset of T cells, approximately 5% of total T cells, with the highest abundance in the gut mucosa. The $\gamma\delta$ cells seem to be able to recognize whole proteins rather than peptide fragments [1].

Monocytes

Monocytes make approximately 6% of leukocytes and share the phagocytosis (ingestion of bacteria) function of neutrophils. They also present pieces of pathogens to T cells. They are produced in bone marrow from hematopoietic stem cell precursor. The role of monocytes in immunological response is to perform phagocytosis using intermediary proteins, such as antibodies, or by a complement system that coats the pathogen, as well as by binding to the microbe directly. Monocytes that migrate from the bloodstream to other tissues are called macrophages. Macrophages are responsible for protecting tissues from foreign substances but are also suspected to be the predominant cells involved in triggering arteriosclerosis.

Platelets are irregularly-shaped, colorless bodies, that are responsible for blood coagulation and are produced in the bone marrow from megakaryocytes, the largest cells found in bone marrow. Megakaryocytes release their platelets in one of two ways. They may release their platelets by rupturing and releasing their contents all at once in the marrow. Alternatively, megakaryocytes may release platelet ribbons into blood vessels. The platelet ribbons are able to continuously emit platelets into circulation. The platelets react with the fibrinogen, a fibrous coagulant in the blood, to form fibrin, and construct mesh-like structure in which other blood elements are retained in the process of clot formation. Calcium and vitamin K also must be present in the blood to support the formation of clots.

The Cell Membrane

The cell membrane has a crucial role in preserving the integrity of the cell: it forms the cytoskeleton and keeps all cell organelles in the cell's inner space. It also actively participates in the transport, by different mechanisms, of many important molecules such as glucose, amino acids, ions and others. The cell membrane also performs enzymatic activities, important in metabolic processes and immunity. Besides transporting molecules via ion pumps, ion channels or carrier proteins, the cell membrane has a key role in maintaining cell potential and in transferring information through chemical messages. Chemical messages are transferred via substances that can bind to receptors located either on the cell membrane or in the intercellular space.

Phospholipid Molecules

Phospholipid molecules, which create a three-dimensional lipid bilayer, constitute the basic structure of the cell membrane (Figure 2.3.). In phospholipid molecules, negatively charged phosphate groups are oriented toward the inner and outer membrane side, causing the core of the cell membrane to be hydrophobic; the outer layer of the cell membrane is hydrophilic.

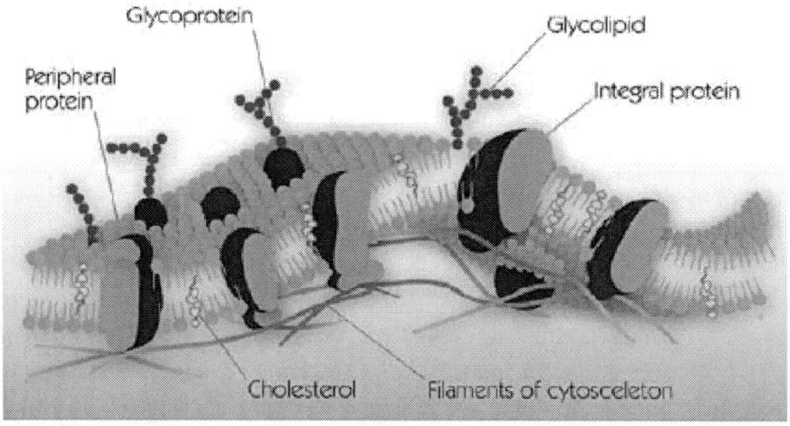

Figure 2.3. Phospholipid bilayer of the cell membrane.

The plasma membrane, a selective barrier at the boundary of every cell, is made up of approximately 30% cholesterol, 65% phospholipids (phosphatidiylcholine, phosphatidiletanolamide, sphyngomielynes) and 5% glycolipids (cerebrosides, gangliosides)

[1]. Phospholipid molecules in the cell membrane exhibit rapid lateral diffusion. The amount of cholesterol and unsaturated fatty acids determine the fluidity and permeability of the cell membrane. Neural myelin membranes are the most permeable due to their high lipid content (around 75%). Proteins are a significant part of the cell membrane structure. Individual proteins may be adjacent (*peripheral*) to the membrane or may be embedded in it (*integral*). Many integral proteins are tightly, permanently bound to the membrane and are not free to diffuse. Such proteins may be separated from the biological membranes only by using detergents, nonpolar solvents, or denaturing agents. Other integral trans-membrane proteins have a role in cell grouping in tissue formation, being cell adhesion proteins. These proteins have domains that extend both into the extracellular and intracellular space. The extracellular domain of a cell adhesion protein can bind to other molecules that might be either on the surface of an adjacent cell or a part of the extracellular matrix. Cell adhesion may be homophilic or heterothallic, depending on whether the adhesion occurs between adhesion proteins or between an adhesion protein and some other molecule. The intracellular domain of adhesion protein is bound to cytoskeleton, the framework of the cell, preventing hydrolysis by extracellular enzymes. Cell adhesion proteins, proteins located on the cell surface that are involved with binding, are also very important for the function of migratory cells, like leukocytes, and for the regulation of synaptic adhesion, as well as for learning and memory. Some genetic disorders have the defect in the gene responsible for the production of cell adhesion proteins, and are associated with increased susceptibility to carcinogenesis.

Proteins and phospholipids constantly diffusing through the cell membrane give this boundary its dynamic character, with constantly changing composition. Fusion of intracellular vesicles (small membrane-enclosed sacs) with the membrane not only excretes the contents of the vesicle, but also incorporates the components of the vesicle membrane into the cell membrane. Due to dynamic properties of cell membranes, diet can significantly influence the quality of the cells and the composition of the cell membranes.

Cell Receptors

A cell receptor's main function is to recognize and respond to a specific ligand, such as a neurotransmitter or a hormone; or to react to change in membrane potential. Cell receptors are located either on a cell membrane or on the membrane of cell organelles; but in broader sense, a receptor may be any cellular molecule that reacts to certain chemical stimuli.

Some integral membrane proteins act like transmembrane receptors and have an important role in signal transmission. Polypeptide chains can cross a lipid bilayer only once, or can span the membrane up to seven times in G-protein coupled receptors. They are usually part of a cell's plasma membrane, but can also be incorporated into the membranes of subcellular compartments and organelles. Signals are transmitted by binding a signaling molecule at one side of the membrane or sometimes binding to a pair of such molecules, initiating the response on the other side. In this way receptors play a unique and important role in cellular communication and signal transduction. Many transmembrane receptors are composed of two or more protein subunits which may dissociate upon ligand binding.

In a majority of receptors for which structural evidence exists, the transmembrane domain of the receptor is composed mostly of alpha helix. In certain receptors, such as the nicotinic acetylcholine receptor, the transmembrane domain forms a protein-lined pore or ion channel. Upon activation of an extracellular domain by binding of the appropriate ligand, the pore becomes accessible to ions, which then pass through the membrane. In other receptors, transmembrane domains are presumed to undergo a conformational change upon binding, which then initiates certain intracellular effect. The transmembrane domain may also contain a ligand binding pocket. The structure of cell receptors can be revealed by X-ray crystallography or nuclear magnetic resonance spectroscopy.

The intracellular domain of the receptor interacts with the interior of the cell or organelle, transducing the signal. Intracellular communication may be performed via specific protein-protein interactions with effector proteins or via enzymatic activity, such as tyrosine kinase or guanylyl ciklase activity. The activity of transmembrane receptors is internally regulated by phosphorylation processes.

Transmembrane receptors can be classified as *metabotropic* and *ionotropic*. Metabotropic receptors affect cells indirectly via enzymatic control of ion channels, while ionotropic receptors contain a central pore that functions as a ligand-gated ion channel. When a ligand binds to a G-protein coupled receptor, the receptor acts on the G-protein, which then stimulates the enzyme to produce a second messenger. G-protein coupled metabotropic transmembrane receptors comprise a large protein family of transmembrane receptors that sense molecules outside the cell and activate cellular responses indirectly and include the following types of receptors:

- ✧ Adenosine receptors
- ✧ Adrenergic receptors
- ✧ Angiotensin receptors
- ✧ Cannabinoid receptors
- ✧ Calcium receptors
- ✧ Cholecystokinin receptors
- ✧ Chemokine receptors
- ✧ Dopamine receptors
- ✧ Glutamate receptors
- ✧ GABA receptors
- ✧ Glucagon receptors
- ✧ Histamine receptors
- ✧ Muscarine acetylcholine receptors
- ✧ Olfactory receptors
- ✧ Opioid receptors
- ✧ Photoreceptors
- ✧ Secretin receptors
- ✧ Serotonin receptors
- ✧ Somatostatin receptors

Metabotropic transmembrane receptors that propagate the signal via intracellular activity of tyrosine kinase include the following:

Erythropoietin receptors
Insulin receptors
Cytokines receptors
Receptors for growth factors

Receptors that respond to information carried by some peptides act via activity of guanylyl ciklase. Some peripheral membrane proteins also function as receptors and may have an important role in cell communication. Another major class of receptors includes specific intracellular proteins. Such receptors can enter the cell nucleus and modulate gene expression in response to certain ligand. This group of intracellular receptors includes receptors that receive information from steroid hormones, such as estrogen, androgen and progesterone. Other hormones, such as glucocorticosteroid, mineralocorticosteroid and thyroid hormones also have their target points in intracellular space. Retinoid and vitamin D receptors are located in the cytoplasm and can readily reach the nucleus. Vitamin A, a known teratogen, acts on the level of provoking mutagenesis in fetus cell, probably by influencing gene expression by entering the nucleus upon binding to intracellular receptor. Peroxisome proliferator-activated receptors are members of the nuclear receptor superfamily of ligand-activated transcription factors, which have a central role in the storage and catabolism of fatty acids.

Ionotropic receptors are integral transmembrane proteins that function like an ion channel and whose opening is regulated by ligands. Some of these receptors include:

Nicotinic acetylcholine receptors
Glycine receptors
GABA-A and GABA-C receptors
Glutamate receptors
Serotonin receptors

In many hormonal disorders, elevated or insufficient production of a certain hormone is not always a real cause of the illness. Real cause of the illness might be a genetic disorder with hereditable receptor genes where the hormonal level may be normal but the endocrine disorder arises due to nonfunctional receptors, i.e., the receptor does not respond sufficiently to the hormone or neurotransmitter. Cell sensitivity to certain hormones or neurotransmitters may be self-regulated by increasing or decreasing the number of receptors. For example, at high insulin plasma concentrations, the number of surface receptors for insulin is gradually reduced by the endocytosis of insulin-receptor complexes and subsequent intracellular lysosomal activity. In this way, the number of sites available for the binding is regulated and the excess hormone is counteracted. Self-induced loss of insulin receptors reduces the target cell's sensitivity to elevated hormone concentration. Under certain conditions, the synthesis of new hormone receptors occurs in the endoplasmic reticulum, followed by their insertion in the cell membrane through a process of cell adaptation.

Transport Through Cell Membrane

One of the most important roles of the cell membrane is to transport the molecules, ions and excretory products from one side of the cell to the other via different mechanisms. Because it has pores, the lipid bilayer is permeable to water and to a few other small, uncharged molecules such as oxygen and carbon dioxide. However, cell membranes are poorly permeable to ions, particularly small hydrophilic molecules like glucose, and macromolecules. Such substances are transported across the membrane via different mechanisms (Figure 2.4.).

Active transport across the cell membrane utilizes the cell energy. Passive transport does not consume cell energy and may be performed as simple or facilitated diffusion. The simplest form of transport across the cell membrane is passive diffusion, which is the result of a concentration gradient. Water freely passes through the cell membrane by osmotic process. Gases like O_2, N_2 and small hydrophobic molecules, also diffuse easily through the membrane. In order to transport ions across the cell membrane, which impacts the cell membrane potential, specific transport mechanisms are implied. In most cells an unequal distribution of ions exists in the inner and outer space and this contributes to the formation of the specific cell potential.

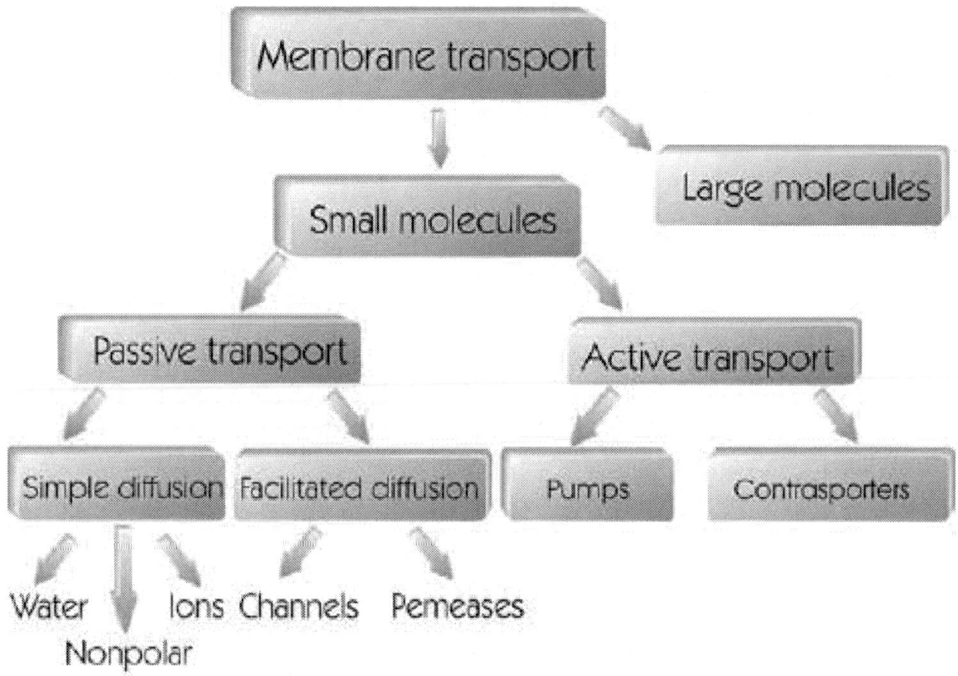

Figure 2.4. Different transport mechanism across the cell membrane.

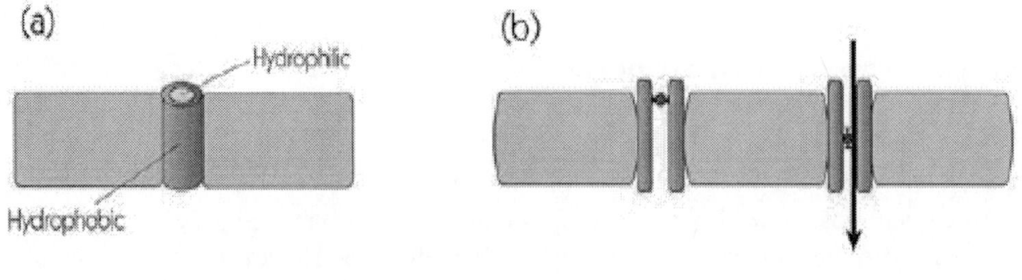

Figure 2.5. (a) Simple ion channel and (b) Gated ion channel.

In facilitated diffusion, molecules require both the concentration gradient to be transported across the cell membrane, and also a specific protein molecule to that actively take part in the transfer process. Common helper molecules are ion channels representing simple integral proteins with a hydrophobic outside surface and a hydrophilic inner space (Figure 2.5.). Transmembrane proteins create a water-filled pore, or channel, through which ions and some small hydrophilic molecules can diffuse. The channels may be open or closed according to the requirements of the cell. The kinetics of facilitated transport differs from the kinetics of simple diffusion. In simple diffusion, the rate of diffusion is proportional to the concentration of diffusing molecules. In facilitated diffusion, however, the rate is limited by the availability of helper molecules. Once all the helpers are saturated, increasing concentrations of diffusing molecules cannot further linearly increase the rate of transport, and the saturation of the process (rate of transport vs. concentration) is observed (Figure 2.6.).

More complex ion channels have gates that open for transportation in response to a chemical stimulus (phosphorilation) or an electrical stimulus. Transportation across the membrane can also be facilitated by certain carrier proteins, which often exhibit enzymatic activity. Many ion channels open or close in response to binding with a small signaling molecule or a ligand. Some ion channels are gated by extracellular ligands while others open in response to intracellular ligand. In both cases, the ligand is not the substance that is transported when the channel opens. Binding of acetylcholine at the synapse receptors opens the channels that admit Na^+ entrance and initiates a nerve impulse or a muscle contraction. Binding of gamma amino butyric acid (GABA) at synapse receptors in the central nervous system admits Cl^- ions into the cell and inhibits the transduction of a nerve impulse. Coupled transport through ion channels is performed when two substances are transported simultaneously via certain transporters. The two substances may be transported in the same direction (symport) or in opposite directions (antiport) (Figure 2.7.). Na-glucose symport takes place in the intestine, from the gut lumen into the cell lining. Ca-Na antiport takes place in cardiac muscle.

Some ion channels open or close in response to changes in the plasma membrane charge or in response to mechanical or light stimuli.

Molecules may also be transported in the direction opposite to the concentration gradient in the process of active transport, which consumes energy by ATP (adenosine triphosphate) hydrolysis (Figure 2.8.).

The energy of ATP may be used directly or indirectly. In direct, active transport, transporters bind to ATP directly and use the energy of its hydrolysis to transport the substance. Other transporters use the energy already stored in the gradient of pumped ions.

Big molecules, which cannot be transported via active or passive transport, are transferred into a cell by engulfing a substance and forming internal vesicles in a process called endocytosis. Some dyes, for example, are absorbed from the gastrointestinal tract via endocytosis. Receptor-mediated endocytosis occurs after the material to be transported has been bound to specific receptors at the cell membrane. Examples include the transport of insulin and cholesterol into animal cells. Phagocytosis is a type of endocytosis where an entire cell is engulfed. In pinocytosis, an external fluid is engulfed (Figure 2.9.). In a process opposite to endocytosis, a substance is secreted from the cell; this process is called exocytosis. In exocytosis, a transport vesicle fuses with the plasma membrane and excretes its content. This mechanism is used in the secretion of protein hormones (insulin, for example), serum proteins and extracellular matrix (collagen).

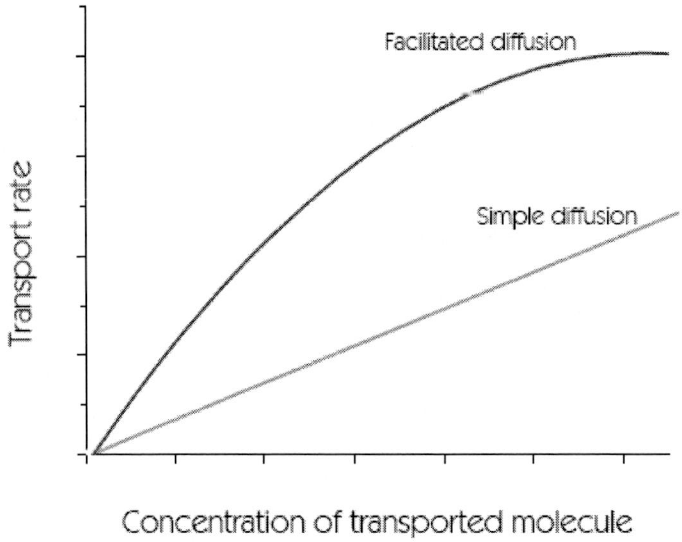

Figure 2.6. Kinetics of the transport process across the cell membrane for simple and facilitated diffusion.

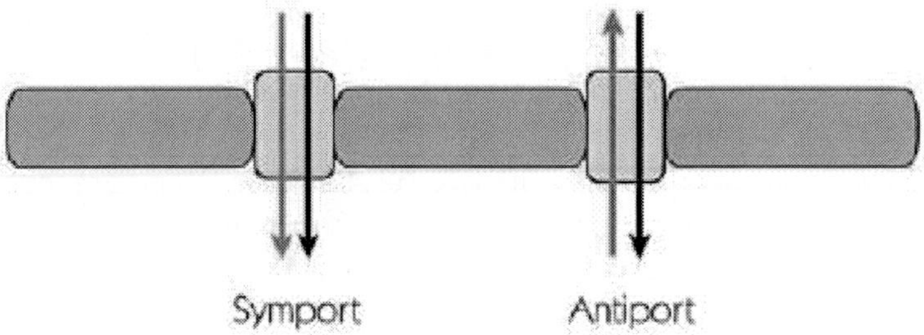

Figure 2.7. Coupled transport through ion channels.

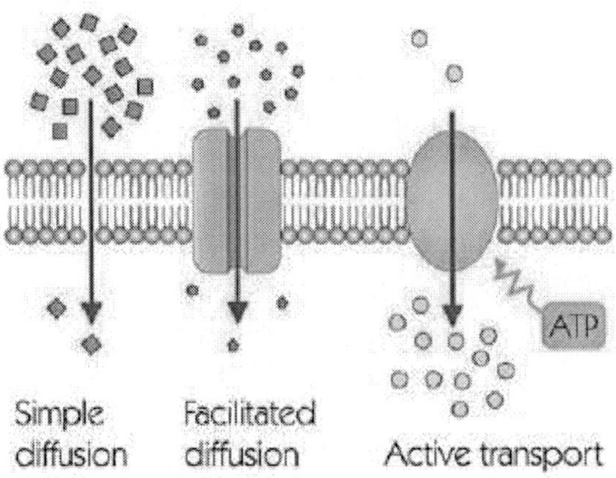

Figure 2.8. Passive and active transport.

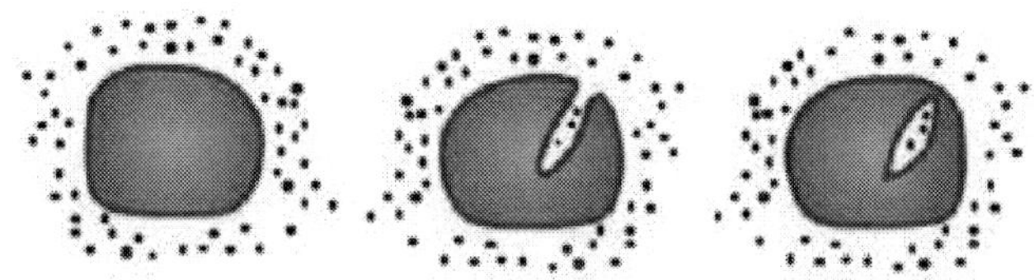

Figure 2.9. Pinocytosis.

ABSORPTION THROUGH THE GASTROINTESTINAL TRACT

The oral route of toxification is probably the most frequent. A person may be exposed to a toxicant via contaminated food or drink, or poisoning may occur due to an intentional (suicidal) poisoning or through criminal action. All orally taken xenobiotics follow the route of food absorption along the gastrointestinal tract, starting from the mouth. Absorption occurs along the entire gastrointestinal tract but is most significant in the small intestine. However, because substances are not usually retained in the stomach for a long period, absorption in the stomach is less important, even though some substances, like alcohol and aspirin, are very rapidly absorbed directly through stomach wall. Substances spend significant time in intestines, which are a coiled, muscular tube, 6-9 meters long. These large, hollow organs of the digestive system contain muscles that enable movement of substances. After food is chewed and moistened with saliva, it is swallowed and pushed down the esophagus. The esophagus is about 25 cm long and connects the throat with the stomach. In order to prevent the food entering the windpipe, the epiglottis valve closes at the opening of the windpipe. At the junction of the esophagus and stomach, a ring-like valve also may close the passage between the two organs. As food approaches the closed ring valve, the surrounding muscles relax and allow the food to pass. The food then enters the stomach, which has three mechanical tasks. The stomach stores the swallowed food and liquid, mixes the food or liquid with digestive juice produced by the stomach, and empties its contents slowly into small intestine. During a meal, the stomach gradually fills to a capacity of 1-2 liters, from an empty capacity of 50-100 milliliters. Epithelial cells of the inner surface of the stomach secrete about 2 liters of gastric juices per day. Gastric juices contain hydrochloric acid, pepsinogen and mucus. Mucus is produced by epithelial cells that make up a protective barrier between the epithelial cells and the stomach acid. Pepsin, an enzyme responsible for protein digestion, is inactivated when it comes into contact with the mucus. Bicarbonate ions, excreted in pancreatic juices, reduce the acidity of the stomach contents near the cell lining of the stomach. Secretion of gastric juices is controlled by nervous and endocrine signals. Hydrochloric acid, a component of a gastric juice, does not have a primary function in digestion, but lowers the pH of the stomach to 1.5-2.5 in order to activate pepsin. In addition to digestion, gastric juices also have a role in killing bacteria that might be introduced into the stomach with food. Several factors affect the emptying of the stomach, including the nature of the food, mainly its fat and protein content, and the degree of muscular action.

All digested nutrients are eventually absorbed through the intestinal walls. The small intestine, which is up to 6 meters long and 2-3 centimeters wide, is the primary site of digestion and absorption of nutrients. Coils, folding and finger-like structures in the intestine called villi and microvilli (Figure 2.10.) give a small intestine of 500-600 m in length a surface area of 300 m^2. Sugars and amino acids are transferred into the bloodstream via capillaries in each villus, while glycerol and fatty acids are transported into the lymphatic system. Absorption of the sugars and amino acids occurs via active transport, requiring cellular energy. In large intestine, which is 7-10 cm in diameter and approximately 3 m long, it is mostly water, vitamins and minerals that are absorbed. Secreted in the large intestine also is an alkaline mucus that protects epithelial tissue and neutralizes acids produced by bacterial metabolism. In the large intestine, bacteria such as *E. coli* produce vitamins, such as vitamin K, and folic acid, which enter systematic circulation. In addition to these substances, waste products of the digestion process also occur in the large intestine, including undigested bits of food, such as fiber, and older cells that have been shed from the mucosa.

The activity of the many glands along the digestive tract is crucial for digestion. Digestion begins in the mouth, where salivary glands excrete amilase and start to break down starches, while bicarbonate ions in saliva neutralize the acids in foods. Glands in the stomach lining produce stomach acids and enzymes responsible for protein digestion. The pancreas also produces a juice that contains a wide array of enzymes involved in carbohydrate, fat and protein digestion. Numerous glands in the walls of the intestines take part in completing the digestive process. Lactose, for example, is hydrolized by lactase produced in intestinal lining; and maltose, originating from digested carbohydrates, is further broken down by maltase.

In addition to being the largest gland in humans, the liver is the only human internal organ that can regenerate itself to a significant extent, a characteristic which may already have been known to antic Greeks, possibly due to cases where people survived liver injuries in the battles. A major portion of the biotransformation process of exogenous substances occurs in the liver. Namely, after the substance has been absorbed from intestines, it is carried by the blood to the liver, where it undergoes biotransformation through exposure to hepatic microsomal enzymes and other enzymes. Through these processes, the toxic potential of a substance is usually reduced. This is known as the first-pass effect, even though for some substances, additional toxic metabolites subsequently may be formed. In other routes of exposure, such as dermal or intravenous exposure, the substance avoids the first-pass effect and is usually more toxic. Consequently, the toxicity and behavior of xenobiotics are highly dependent on the route of entry. For this reason, snake venom is toxic only after intravenous injection. In the digestive tract, snake venom is completely inactivated by digestive enzymes.

Besides being involved in biotransformation processes, the liver has a wide array of varied functions. Plasma proteins, which are important transporters of toxicants throughout the body and a means of protecting the body, are synthesized in the liver. *Kupffer cells* are specialized macrophages located in the liver. The primary function of Kupffer cells is to recycle old red blood cells, which are broken down by phagocytic action. During this process, globin chains are reused, while iron-containing heme is further broken down into iron, which is reused, and billirubin, which is conjugated with glucuronic acid within hepatocytes and excreted into the bile. The liver also represents a storage site for energy in the form of glycogen and the site of urea production from amino acids. In carbohydrate metabolism, the liver performs *gluconeogenesis*, *glycogenolysis* and *glycogenesis*. *Gluconeogenesis* is glucose synthesis from certain amino acids, lactate or glycerol, and occurs in situations of reduced energy status, i.e., when concentrations of blood glucose are low. The formation of glucose by glycogen hydrolysis, a process called *glycogenolysis,* is the function not only of the liver, but can be performed by muscle tissue as well. The reverse process, *glycogenesis*, contributes to storage of excess glucose in the form of glycogen.

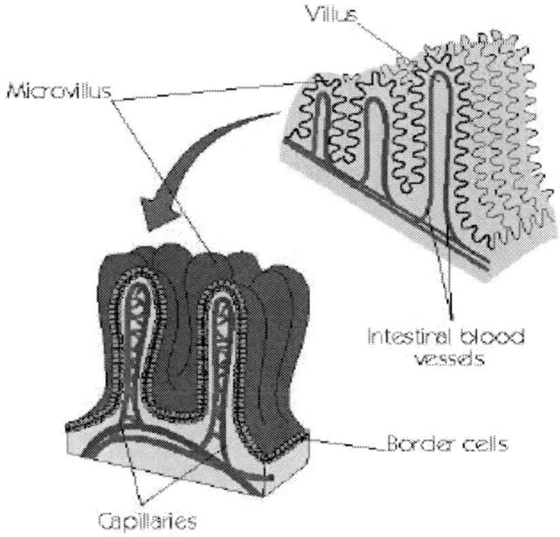

Figure 2.10. The inner surface of the small intestine.

The liver also participates in lipid metabolism, such as cholesterol synthesis and the production of triglycerides. In addition, coagulation factors fibrinogen, prothrombin, as well as protein C, protein S, antithrombin and other plasma proteins, are also produced by liver. A multitude of substances, including vitamin B12, iron and copper are also stored in this large organ, and many hormones, including insulin, are broken down by the liver.

The main hormones that control the functions of the digestive system are produced and released by cells in the stomach lining and small intestine. Three hormones control digestion: gastrin, secretin and cholecystokinin. Gastrin causes the stomach to produce acid and is also necessary for the normal growth of stomach, small intestine and colon lining. Secretin influences the pancreas to excrete a digestive juice that is rich in bicarbonate ions. Secretin also stimulates the stomach to produce pepsin, an enzyme that digests proteins, and the liver to produce bile. Cholecystokinin is released from the intestinal epithelium in response to fats, and causes the release of bile from the gall bladder and lipase from the pancreas. Ghrelin stimulates the appetite and is produced in the stomach and upper intestine in the absence of food. By contrast, the peptide YY is produced in the gastrointestinal tract in response to a meal and functions to inhibit the appetite. Both these hormones affect the brain's ability to regulate food intake.

Two types of nerves control the action of the digestive system. Extrinsic nerves come to the digestive organs from the unconscious part of the brain or from the spinal cord. They release acetylcholine and adrenaline. Acetylcholine causes the muscles of the digestive organs to contract and stimulates the stomach and pancreas to produce more digestive juices. Adrenaline relaxes the muscles of the stomach and intestines and decreases the blood flow to these organs. Intrinsic nerves make up a very dense network embedded in the walls of esophagus, stomach, small intestine and colon. The intrinsic nerves are triggered when the walls of the hollow organs are stretched by food. In response, digestive juices are produced along with many other different substances that speed up or delay the movement of the food.

Optimal acidity of the gastrointestinal tract (pH = 5-8), large resorptive surfaces and intense blood flow in intestine walls during and after ingestion of food, contribute to significant gastrointestinal absorption of the toxicants. The majority of the toxicants are resorbed by simple diffusion and in this process the degree of lipophilic character determines the rate of absorption. On the other hand, hydrophilicity of the substance contributes to better absorption due to improved contact with the resorptive surface, such as the hydrophilic mucosa. Elemental mercury passes the digestive tract practically unabsorbed due to high hydrophobicity and insufficient oxidizing conditions along the tract. Acidity in the lumen of the different compartments of the gastrointestinal tract significantly influences absorption by diffusion, because only un-ionized substances can cross the lipid bilayer. Absorption is favorable when the acidity converts molecules to un-ionized form. Consequently, the acidic conditions of the small intestine favor the absorption of the acidic substances, while alkaline conditions, such as in large intestine, support the absorption of the alkaline substances. The amount of the acidic toxicant that will be ionized, depending on the acidity of the environment, may be described by the equation:

$$pK_a - pH = \log \frac{Un\text{-}ionized}{Ionized}$$

Where pK_a represents the dissociation constant of the acidic component.

For benzoic acid with $pK_a = 4$, the stomach (pH = 2) absorption is determined by a factor:

$$4 - 2 = \log \frac{Un\text{-}ionized}{Ionized}$$

$$100 = \frac{Un\text{-}ionized}{Ionized}$$

In the stomach, the amount of un-ionized molecules, which will diffuse through cell membranes and will be readily absorbed, is 100 times greater than the ionized form, which is unable to pass cell membranes. For the same compound the absorption in the gut (pH = 6) is described:

$$4 - 6 = \log \frac{Un\text{-}ionized}{Ionized}$$

$$\frac{1}{100} = \frac{Un\text{-}ionized}{Ionized}$$

The absorption will be insignificant, because the un-ionized molecules will be present in very small concentration.

For the alkaline substances the situation is reversed, i.e. alkaline environment contributes to absorption by diffusion:

$$pK_b - pH = \log \frac{Ionized}{Un\text{-}ionized}$$

pK_b represents the dissociation constant of the alkali.

Large molecules may also be absorbed from the small intestine by other mechanisms. Examples of large molecule absorption include the poisoning with botulinum toxin or systematic allergic reaction after the absorption of certain food proteins, as well as absorption of some dyes. It is assumed that endocytosis is involved in the resorptive process of such large molecules, which can not be transferred by diffusion or active transport.

Many physico-chemical and physiological parameters influence the rate of toxicant absorption. Dissolved toxicants are more readily absorbed than solid particles, which need to be dissolved in intestinal mucosa before they can be absorbed. The dissolution rate of a toxic substance in intestinal mucosa is in proportion to the size of the particle. For smaller particles, contact surface is bigger and therefore absorption is more intense. For toxicants with a large particle size, dissolution in intestinal mucosa is slow and a large amount of the toxin is excreted before the dissolution process is complete. Organic solvents and surfactants increase the solubility and absorption rate of toxins. The amount of a toxicant that will dissolve depends also on the content of intestine. The most efficient absorption and highest blood concentration is reached when the toxicant is introduced to an empty stomach. If the stomach is full, proteins and other food components may sorb the toxicant and influence the amount absorbed. From this ensues the principle of the application of active carbon and various resins in poisoning treatment, because they sorb the toxicant, so that less of it is available for the absorption. Since the nutrients and toxicants are exploiting the same transport mechanisms, the absorption of xenobiotics also depends on the concentration of nutrients whose transport mechanism is used by the toxicant. 5-fluorouracil exploits the pyrimidine transport mechanism, while Tl, Co and Mn use iron transporters. Lead is absorbed less if the diet is rich in calcium. Digestive enzymes, bile, intestinal flora and acidity also influence the stability of a substance and its absorption. Slower peristaltic activity prolongs the retention of the toxicant in the gastrointestinal tract, while rapid peristaltic activity decreases contact time and the possibility for absorption. Bacterial enzymes take part in many hydrolytic and reductive reactions and may both activate and inactivate many toxic substances. For these reasons children are more susceptible to methemoglobinemia due to undeveloped *E. Coli,* which inhibits reductive reactions and allows more NO_3^- to oxidize the hemoglobin. Beneficial activity of the intestinal flora include reduction of aromatic nitro and amino groups, which often express carcinogenic or goitrogenic (i.e., suppressing the function of the thyroid gland) activity. In processes catalyzed by bacterial enzymes, cyanide glycosides, present in the kernel of many fruits, such as peach and apricot, may be hydrolyzed to very toxic cyanides.

ABSORPTION THROUGH THE RESPIRATORY TRACT

Exposure to toxic compounds through inhalation may results in rapid toxicant absorption, and this route of toxicant entry is often related to specific occupational exposures. Toxicants in the form of gas, vapor or solid particles may enter the respiratory system and be absorbed in a different ways, depending on their physico-chemical properties. Gases and vapors follow the same route in the lungs as air inhaled during normal breathing. The route of solid particles is somewhat different and depends significantly on particle size.

Breathing consists of two phases, inspiration and expiration. During inspiration, the diaphragm and the intercostal muscles contract and the diaphragm moves downwards increasing the volume of the thoracic cavity. Intercostal muscles pull the ribs up, expanding the rib cage and further increasing the volume of thoracic cavity. Increase in volume lowers the air pressure in the lungs below that of the atmosphere and provokes the air entrance. During expiration, the diaphragm and intercostal muscles relax. This returns the thoracic cavity to its original volume, increasing air pressure in the lungs and forcing the air out.

Air enters the body through the nose, where it is warmed and filtered. After passing the pharynx and larynx, the air moves into the bronchi, which carry air into and out of the lungs. The vocal cords are two bands of tissue that extend across the opening of the larynx. The bronchi are structures that are lined with ciliated epithelium and mucus-producing cells, both of which are very important in toxicant removal. The bronchi further branch into smaller tubes known as bronchioles. Bronchioles terminate in grape-like sacs known as alveoli. There are approximately 600 million alveoli in the lungs of adult person. The alveoli are surrounded by a network of thin-walled capillaries and are the main site of toxicant absorption (Figure 2.11.). The resorptive surface area of the alveoli is approximately 50-100 m^2 and is much larger than the skin surface area. Only about 0.2 μm separates the alveoli from the capillaries due to the extremely thin walls (1-2 μm) of both structures. Intimate contact between alveoli and capillaries, as well as mucus that moistures the alveoli walls, contribute to significant and efficient lung absorption.

Blood enters the lungs via pulmonary artery and proceeds through the arterioles into the alveolar capillaries. Oxygenated blood flows out of the alveolar capillaries, through venuoles, and travels to the heart.

The main means of oxygen transport in the body is via hemoglobin. Hemoglobin consists of iron-containing heme,[1] combined with the globin protein. Each iron atom in heme is attached to four pyrole groups by covalent bonds. A fifth covalent bond of the iron is attached to the globin part of the molecule and a sixth covalent bond is available to combine with oxygen. There are four iron atoms, i.e., four heme groups in each hemoglobin molecule. These four heme groups cooperate in the loading and unloading of oxygen. When oxygen binds to one of the heme groups, the others change shape slightly and their attraction to oxygen increases. Loading of the first oxygen molecule results in rapid loading of the next three in the process of oxyhemoglobin formation. A drop in pH lowers the attraction of hemoglobin to oxygen. In conditions of excess of carbon-dioxide, pH values decrease due to formation of carbonic acid, encouraging hemoglobin to release extra oxygen. Carbon dioxide forms in cell respiration during the production of ATP. Seven percent of carbon dioxide released from respiring cells dissolves into blood plasma, 23% of which binds to the multiple amino groups of hemoglobin and forms carboxyhemoglobin, and 70% of which is carried in blood in the form of bicarbonate ions. Carbon dioxide dissolved into blood plasma diffuses into the red blood cells, where most of it is converted to bicarbonate ions. It first reacts with

[1] Heme group is a prosthetic group consisting of a protoporphyrin ring and a central iron (Fe) atom.

water and forms carbonic acid, which dissociates to H^+ and HCO_3^-. Most of the hydrogen ions that are produced attach to hemoglobin or other proteins. In alveoli capillaries, bicarbonate ions combine with a hydrogen ion and form carbonic acid, which then dissociates into carbon dioxide and water. Carbon dioxide then diffuses into the alveoli and out of the body with the next exhalation (Figure 2.12.).

Adults inhale approximately 10,000 liters of air per day, introducing along with the air different chemicals from the environment, which may be various air contaminants, such as CO or NO_2, exhaust gases from car engines, pesticide aerosols, mineral particles and others.

The rate of gas exchange between the environment and the alveoli interspace depends on lung ventilation. While a person is exercising, the volume of air that enters the lungs during a single inhalation, which is for steady conditions usually 400-600 ml, increases significantly influencing the higher toxicant concentrations to be inhaled. Breathing frequency and also depth influence the retention time of the chemical in the lungs. The amount of inhaled air also depends on other parameters, such as the temperature and air humidity. While a person is physically active, the minute volume of inhaled air changes from 7 liters to approximately 30-60 liters in females, depending on the intensity of their activity. For men, the volume changes from 4.5 liters to approximately 16-24 liters.

In the lungs, gases and vapors are absorbed by simple diffusion and this depends on difference in concentration amount between the blood and alveoli interspace, as well as on the solubility of the gas in the blood. Gases with good solubility in blood and tissues are more readily absorbed. The absorption rate also changes in over time. At the beginning of the absorption process, the rate is the greater, due to significant differences in concentration. With the increased blood concentration, the absorption slows due to a decrease in the concentration gradient, i.e., the principal force of absorption by diffusion.

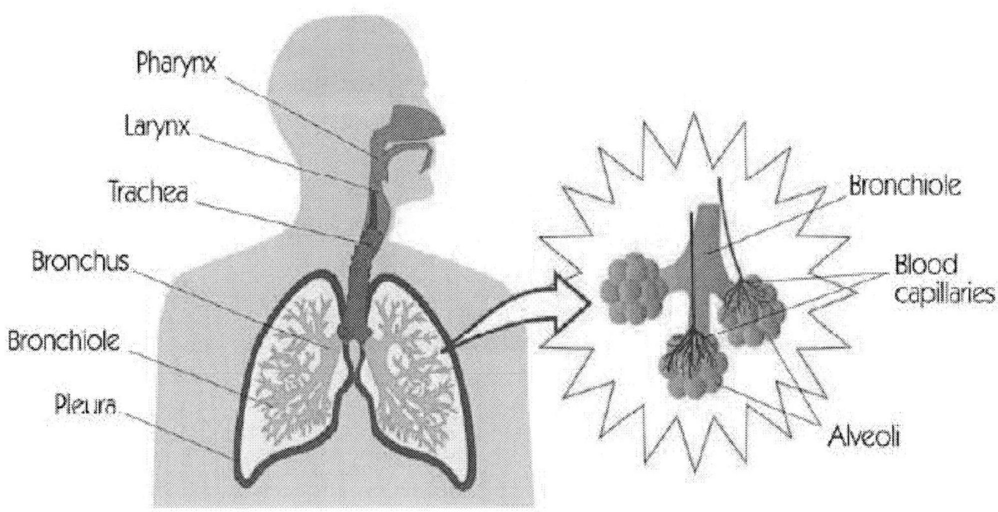

Figure 2.11. Human respiratory tract.

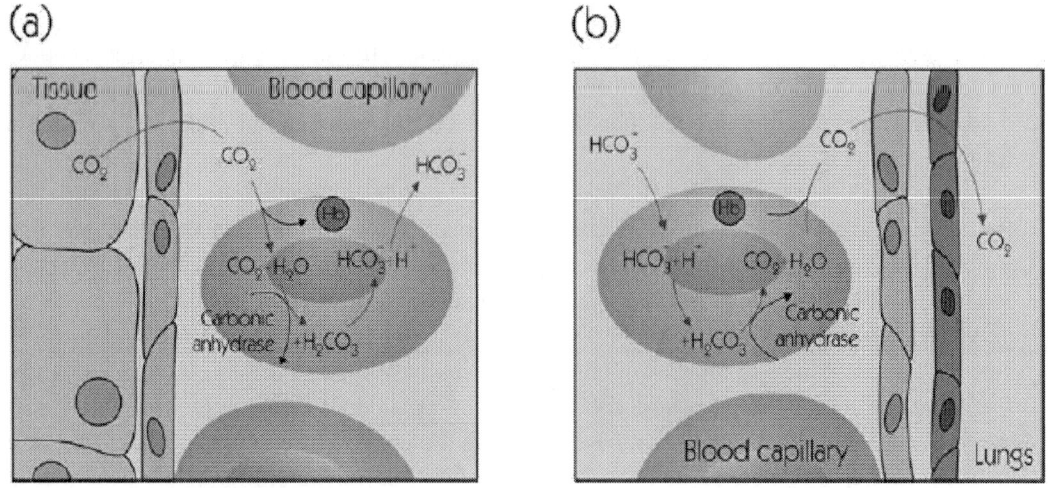

Figure 2.12. The mechanism of carbon-dioxide excretion. (a) Transport of carbon dioxide from tissue into the blood; (b) Transport of carbon dioxide from the blood into the lungs.

During absorption of aerosols, the size of toxicant particles is very important. Only solid particles with a diameter of 0.1 – 1 μm reach the alveoli, where they are absorbed. Smaller particles are expelled with a turbulent current of exhaled air, while larger particles are retained on the epithelium of the nose and the trachea. Soluble particles from these compartments of the respiratory tract may be absorbed after they dissolve. Insoluble particles are expelled by the movement of ciliated epithelium, And up to 80% of inhaled large particles also are removed by this mechanical mechanism. Expelled toxicant particles may enter the mouth and be swallowed, following the oral route of absorption. Another very important mechanism of the removal of fine toxicant particles arises from phagocitic activity of the leukocytes. Phagocite particles are partially removed by excreted mucilage, however some amount can enter the lymph nods, where it can rest for a long time. The deposition of solid particles can permanently damage the lung tissue and provoke pneumoconiosis. Inorganic materials, such as coal, silica, HgO, asbestos, and plant dust, damage the tissue in this way and are often encountered in occupational medical conditions, such as silicosis or asbestosis.

ABSORPTION THROUGH THE SKIN

If the skin is in direct contact with the surrounding environment, many chemicals in the environment may be transported into systemic circulation by dermal absorption. Dermal absorption occurs in occupational, environmental or accidental exposures to different chemicals. Many cases of pesticide poisoning occur through skin contact in agricultural workers during pesticide application without protection. Even handling of pesticide-treated fruits may pose a risk of intoxication, if the fruits are collected without protective gloves. Exposure to toxicants may also occur via bathing, showering or immersion in contaminated natural swimming water. Toxicologically significant absorption of contaminants in water usually occurs during prolonged exposure or during exposure to skin that is damaged, for example by sunburn.

Skin accounts about 15% of the body weight and is the largest organ of the human body. For the average adult human, the skin surface area is approximately 1.5-2.0 m^2. As the interface with the surrounding environment, skin plays the most important role in protecting an organism from external elements and factors such as pathogens, UV irradiation, mechanical injury, and other conditions. Other main functions of the skin are insulation and temperature regulation, sensation, excretion, and absorption and the synthesis of vitamins D and B. Vitamin synthesis is linked to pigmentation, with darker skin producing more vitamins. Skin contains a rich blood supply, which allows precise control of energy loss by radiation, convection and conduction. Dilated blood vessels in conditions of increased body temperature or exercise increase heat loss, while constricted vessels greatly reduce cutaneous blood flow and conserve the heat. The erector pili muscles (*arrectores pilorum*), which are directly bound to hair follicle, contract in the cold, lifting up the hair and augmenting the insulation that skin provides. Another important skin function is within the body's immune system. The skin`s immune system is a natural defense mechanism against microorganisms and is associated with skin-associated lymphoid tissue (SALT). Lymphoid epidermal cells also are involved in delayed hypersensitivity reactions in the skin. Skin exposure to a high dose of UV irradiation may suppress hypersensitivity reactions. However, lowered responsiveness of the skin's immune system may play a role in the development of skin cancers, real allergies and autoimmune diseases.

Mammalian skin is composed of an outer layer, the epidermis, and an inner layer the dermis (Figure 2.13.). The two layers are connected by finger-like structures of dermis embedded in the epidermal layer. Oxygen, nitrogen and carbon dioxide can diffuse into the epidermis in small amounts and some animals use the skin as their sole respiratory organ. The number of epidermal layers varies depending on the body part, with palms and the soles of the feet having the thickest epidermis. Epidermal cells are constantly renewed and a complete new layer of cells forms in about a month. Older cells are pushed to the surface by new cells, which are produced in a germinative layer at the border of epidermis and dermis and become filled with keratin. This keratinized layer (*stratum corneum*) absorbs water and represents a significant part of the skin's hydrophilic-lipophilic barrier. The *stratum corneum,* or outermost layer, of the epidermis is the most significant in providing physical protection from percutaneous absorption of chemicals. The rather complex composition of the cornified cells (those converted to hard tissue) in the *stratum corneum* provides a strong chemical resistance. The lipophilic part of skin's barrier arises from excreted lipids and waxes from various glands, embedded in dermis, which also play a role in lubricating, softening and nourishing the skin, as well as in antibacterial action. Organic solvents and surfactants remove this protective lipophilic layer and increase permeability to toxicants. The skin supports its own system of microorganisms, including yeasts and bacteria (*Staphylococcus epidermidis*), which cannot be removed by cleaning. The density of skin flora depends on the region of the skin. When the balance between microorganisms is disturbed, e.g., by antibiotics, or upon skin disinfection, an overgrowth of yeast may occur or a recolonization of the skin by other bacteria.

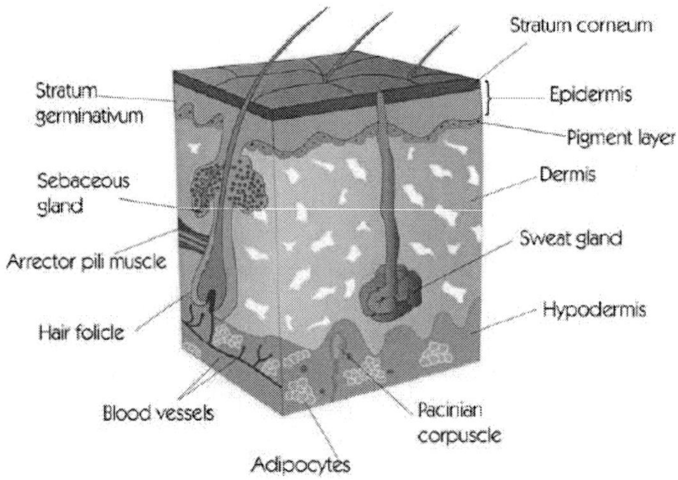

Figure 2.13. The structure of the skin.

At the base of epidermis, specialized cells called melanocytes, produce melanin, a pigment that protects the skin from sun-induced damage. Melanin is an oxygen scavenger and may reduce genetic damage to nuclear material of the skin cells by reacting with mutagenic oxygen radicals before genotoxic effects occur.

Blood vessels and nerve endings are present mainly in the dermal layer. Dermis also contains hair follicles, smooth muscle, glands and lymphatic tissue. The average square centimeter of skin contains about 250 sweat glands, 8 blood vessels, 24,000 melanocytes and several hundred nerve endings. Nerve fibers that are attached to skin receptors discharge during stimulus, or respond only when a stimulus starts and sometimes when stimulus ends. Some skin receptors respond to pressure or vibrations of different frequencies, while others respond to mechanical or heat stimuli.

In fish, exocrine glands in the dermis excrete lubricating substances aimed to decrease friction in water. Some amphibians' excretia contain very poisonous venoms. Their sweat glands open onto the skin via a pore through which the sweat is excreted. Humans contain about 2.5 million sweat glands which are most densely distributed on palms, soles of the feet, arm-pits and forehead. Sweating is also very important in temperature regulation. Besides water, minerals and salts, sweat contains urea in a concentration approximately 130 times lower than in urine.

The subcutaneous dermal layer, the hypodermis, lies below the dermis and is not part of the skin.. Its purpose is to attach the skin to underlying bones and muscles. It consists of connective tissue (elastin) and contains approximately 50% of total body fat, stored in fat cells or adipocytes. Besides adipocytes, the hypodermis is rich in fibroblasts and macrophages.

As described, the skin is an important physical barrier to the absorption of toxic substances, but it can also be a significant portal of entry for such compounds. The skin and its associated structures (the hair, the nails and the sweat glands) are also a route of excretion of toxicants and their by-products.

The developing skin of young children is thinner and less impervious to the passage of foreign substances and therefore more susceptible to damage from environmental agents. By the end of the first year of life, a child's skin permeability is closer to that of an adult. The skin's impermeability varies with the age and physical condition of the skin. Water absorbed by the keratinized layer can significantly enhance the entrance of toxic substances through opened pores. Toxicants may be absorbed by diffusion through many cell layers and, in this respect, lipophilic substances are more readily absorbed due to their more efficient passage through lipid bilayers. Highly hydrophobic substances, such as polychlorinated biphenyls (PCBs), polycyclic aromatic hydrocarbons (PAHs), dioxins and others, are dermally absorbed to a high degree. In general, chemicals that are highly absorbable through the skin are substances of low molecular weight, non-charged and hydrophobic. Toxicants also may follow the route of hair follicles or sweat tubules.

The rate of dermal absorption of a substance is proportional both to the concentration of the substance and to the surface area over which it is applied. Depending on skin thickness, absorption of a given substance by different regions of the body can vary. For example, hydrocortisone is absorbed over 50 times more efficiently by skin of the genital area versus the skin of the palms of the hands.

Any damage to the skin interrupts the physical envelope and significantly enchances toxicant absorption. Damage to the skin can occur through disease or direct environmental influence. With aging, the skin becomes thinner, less elastic and more easily damaged, and receives less blood flow and shows lower glandular activity.

Chemical substances may be activated or transformed in the skin by the energy of ultraviolet radiation and may provoke photosensitivity reaction. Photo-allergic responses are immune mediated and require previous sensitization to the photoactive substance, while phototoxic responses are not immune mediated. Applied to skin, a number of different substances and products, such as sun tanning products, lotions, soaps, can cause photosensitivity reactions. Citrus fruits, especially limes, and also common fig, celery, and some spice herbs, such as parsley, are sources of naturally occurring phototoxic agents known as psoralens or furocoumarins. Despite the photocarcinogenic properties of psoralens, they have been used as a tanning activators and tanning accelerators in sunscreens. Due to high UV absorbance by psoralens, they have also been used for the treatment of some dermal problems, such as vitiligo, eczema and psoriasis. After the application of the psoralens to the damaged skin area, the interaction with UV irradiation is allowed, followed by the removal of the damaged skin. This procedure removes the unhealthy skin and reveals the new skin layer. Exposure to substances such as mercury, silver, gold, arsenic, coal, tar, aromatic chlorinated hydrocarbons and polychlorinated biphenyls, is known to result in increased pigmentation. The effect on pigmentation is most likely systematic, caused by damage to melanocytes or by alteration of the pigment synthesis process. In some cases, hyper-pigmentation occurs as a result of exposure to a photosensitizing substance and subsequent exposure to UV radiation. In this case the effect is usually local. Decreased pigmentation also may occur as a result of exposure to industrial chemicals such as alkyl-phenols, but may also result from physical damage of the skin or chemical or thermal burns.

Blood- Brain Barrier

Blood-brain barrier (BBB) is a dynamic and selective interface within the capillary endothelium that limits entry of xenobiotics[1] in the blood into brain tissue.

This barrier also protects the brain from hormones and transmitters intended for other parts of the body and thus maintains a constant environment. The brain controls hormonal secretions through specialized sites where neurons receive information about the composition of the circulating blood. At these sites, the blood-brain barrier is interrupted to some extent. Because this barrier is not fully developed in fetuses and newborns, they are more susceptible to foreign substances. The existence of this barrier was first noticed in experiments by Paul Ehrlich in the late 19th century. Ehrlich noticed that if aniline dye is injected into the blood of experimental animals, all tissues except brain tissues become stained [3]. In 1913, the existence of the BBB was confirmed by Edwin Goldmann, who performed animal experiments where dye was injected into spinal fluid and only brain tissue and no other tissues were stained.

The primary function of the blood-brain barrier is to regulate the transport of nutrients, metabolites and drugs to and from the brain. Blood vessels in the brain are not in direct contact with the tissue, but are covered with a layer of specific endothelial cells. In order to be transported to the brain cells, a compound must cross this endothelial layer. Endothelial cells comprise only 0.1% of the brain weight, but together they extend about 644 km in length, with a surface area of 20 m^2. Tight junctions between endothelial cells (Figure 2.14.) prevent the free entrance of foreign substances into the brain. In the rest of the body, proportionally larger spaces between individual cells adjacent to blood vessels facilitate this type of transfer in other tissues. Some research results speculate that two proteins, zonulin and zot, can provoke the opening of the junction between endothelial cells after their binding to receptors, allowing a substance to be absorbed [3]. Although these findings are based on observation of absorption processes in the gastrointestinal tract, it is assumed that a similar transport mechanism might occur in the brain. In addition, endothelial cells metabolize certain molecules to prevent their entry into the central nervous system [4]. For example, L-DOPA, the precursor to dopamine, can cross the blood-brain barrier, whereas dopamine itself cannot. As a result, L-DOPA is administered for the treatment of dopamine deficiencies.

In addition to tight junctions between endothelial cells acting to prevent transport into the brain, glial cells surrounding capillaries in the brain, as well as low concentration of interstitial proteins in the brain, also prevent hydrophilic molecules from entering the brain [5].

---

[1] A xenobiotic is any chemical that would not normally be found in a living organism or be expected to be produced by it (Definition from Online edition, About.com: Chemistry,

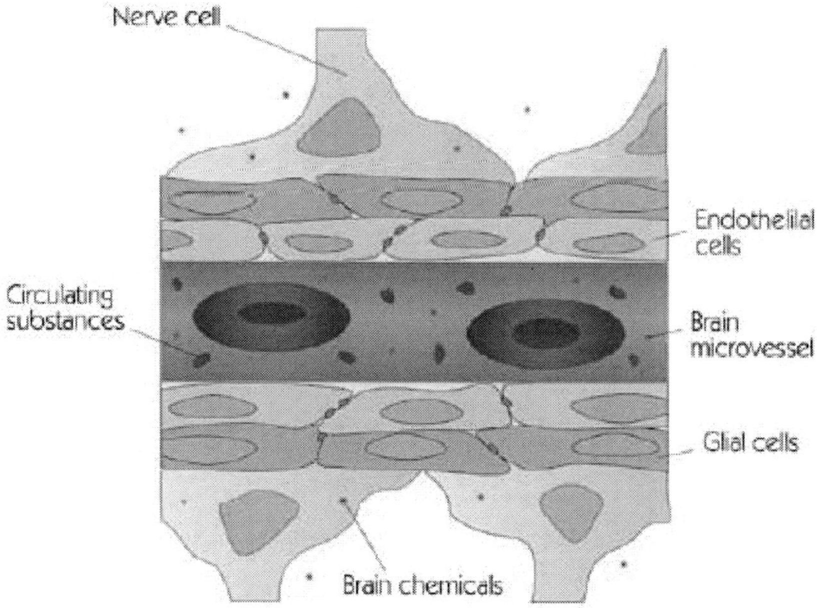

Figure 2.14. Blood-brain barrier.

The blood-brain barrier is impermeable to large molecules, but some smaller molecules, such as alcohol, caffeine and nicotine, are easily transferred into the brain. Nutritional molecules such as oxygen and glucose, which are necessary for brain functioning, also cross blood-brain barrier easily by means of lipid solubility. Specific transport mechanisms are responsible for the transport of sugars and amino acids to the brain. The specificity of these transport mechanisms also restricts therapeutic drugs from entering the brain. For example, P-glycoprotein transporter protects the brain from toxic compounds, but at the same time represents the primary obstacle to successful therapy for brain disorders. New generation drugs for the treatment of neural diseases are designed to enable transport into the brain via endogenous transporters. A "Trojan-Horse" concept applies genetically-engineered proteins that are attached to the drug molecules, enabling transport into the brain by endogenous transporters. Another idea is to pack the drug into liposomes that have an antibody attached to their surface. The antibody serves to activate brain capillary receptors, enabling the liposomes to pass and deliver their active component to the brain cells.

The Placental Barrier

The placenta is a temporary organ required for the development of the embryo and fetus that ensures the maintenance of pregnancy, as well as fetal growth and development. It allows the exchange of metabolic products between the fetus and the mother. Besides its involvement in the transport of the nutrients and fetal waste products, the placenta also has a significant role in endocrine secretion, immunological functions and metabolism. The placenta has a very large surface area, which facilitates the transport of substances both into it and out of it. The placental surface area at term is approximately 11 m^2. About 5 to 10% of the surface is only few micrometers in thickness. The exchange of gases like carbon-dioxyde and oxygen occurs via diffusion through cell layers of the placental barrier. During early pregnancy, the placental hormone somatomammotropin induces the synthesis of glycogen, cholesterol and fatty acids, which serve as sources of nutrients and energy for the embryo.

The placental barrier is composed of structures that keep maternal blood and fetal blood separate, not allowing their mixing. However, exchanges between maternal and fetal circulating substances take place in the chorionic villus, which is the functional unit of human placenta. Chorionic villi are hair-like projections that line the outside edge of the placenta. There are about 150 ml of maternal blood in the intervillous space, and it is exchanged 3 or 4 times in a minute. The villus consists of a central fetal capillary and an outer trophoblast[1] layer [6]. Trophoblastic cells are present as multinucleated cells called syncytiotrophoblasts[2] and mononuclear cells called cytotrophoblasts[3] [7] (Figure 2.15).

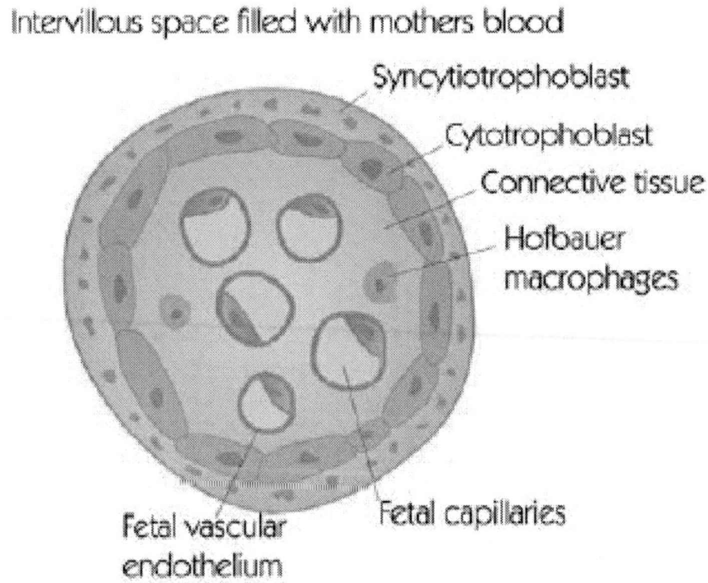

Figure 2.15. Placental barrier. The structure of the villus, the functional unit of the human placenta.

The syncytiotrophoblast has a huge number of microvilli - over 1 billion per cm^2 at term. The structure of the placental barrier changes over the course of pregnancy. In the first trimester it consists of the syncytiotrophoblast, cytotrophoblast, the villus mesenchyma[4] and fetal capillary walls. In the mesenchyma, many cells that exhibit macrophage properties are found. In addition to the trophoblast layer, fetal and maternal circulation are separated by the

[1] Trophoblasts are cells forming the outer layer of placenta, which provide nutrients to the embryo.

[2] Syncytiotrophoblasts are multinucleated cells in the outer layer of trophoblasts.

[3] Cytotrophoblast is the inner layer of the trophoblast, interior to the syncytiotrophoblast in an embryo.

[4] Mesenchyme, is a reticular connective tissue, a type of loose connective tissue, which is located within the embryo

trophoblastic basement membrane, a connective tissue space, the endothelial basement membrane and the fetal capillary endothelium.

The placental barrier may be overcome via water-filled channels that allow the transport of hydrophilic molecules, or via a transcellular route, where transport of lipophilic compounds occurs due to their diffusion through many lipid bilayers. When water diffuses into the placenta along an osmolar gradient, electrolytes follow the water, and iron and calcium are transferred in only one direction, from the mother to the child.

The supply of nutrients to the fetus, essential for its growth, is performed via different mechanisms. Chemical compounds cross the placenta mainly through simple passive diffusion. The properties of the substance that determine placental transfer by passive diffusion include molecular weight, pKa, lipid solubility and protein binding. Chemical compounds with a molecular weight over 500 Da are transferred poorly across the placenta. Passive diffusion alone is not adequate to fulfill fetal requirements for nutrients, and several nutrients, such as amino acids, fatty acids and glucose, as well as electrolytes and vitamins, are transported by specialized transport proteins [8]. Transport proteins are located in the microvillous and basal membranes of the trophoblast. Asymmetry in the kinetics of substrate binding results in differences in influx and efflux at the interface with maternal and fetal blood, allowing directional flux of the nutrients across the placenta. Transferrin, a maternal plasma protein, carries iron to the placental surface, where it is actively transported into the fetal tissues. Proteins too large to cross the placental barrier can be transferred via pinocytosis. Immunoglobulins, mainly IgE, are believed to cross the placental barrier by receptor-mediated endocytosis. Peptides and amino acids delivered to the fetus via active transport ensure protein synthesis in the fetus itself. Many transporters without known physiological substrates have been identified in the placenta. Some of these transporters prevent the entry of xenobiotics into the placental unit. The best known of these is P-glycoprotein, which acts as a pump that transports substrates from the intracellular to the extracellular compartment. In animal models, inhibition of placental P-glycoprotein results in greatly induced transplacental passage of drugs into the fetus [9]. The current hypothesis is that placental drug-transporting P-glycoprotein protects the developing embryo and fetus from toxic substances and suppresses teratogenesis. Several transporters facilitate the transfer of xenobiotics to the fetal compartment. Some amino acid transporters may be involved in the transport of pharmacologically active drugs that structurally resemble amino acids. Some exogenous compounds, such as cocaine, nicotine and cannabinoid compounds interfere with the placental transfer of endogenous compounds, such as amino acids [10].

The placenta represents an incomplete barrier against certain drugs and foreign substances. Toxic substances are transported across placental barriers by one of the endogeneous mechanisms. Research results suggest that methyl-mercury, for example, is actively transported as its cysteine conjugate via the neutral amino acid carrier system. Antibiotics, diazepams and corticosteroids similarly cross the placental barrier. Depending on their size, certain steroid hormones cross the placenta as well [11]. The placenta is also very permeable to alcohol and to some viruses (e.g., Cytomegalovirs, Rubella virus). These agents can cause birth defects.

It is mainly free, unmetabolized exogenous compounds that cross the placenta. It has been suggested that metabolites also cross the placenta to some extent, but more slowly than parent compounds; and that placental transfer of conjugate metabolites is negligible [12].

Ionization significantly affects placental transfer. Un-ionized substances cross the placenta more easily than ionized ones [13]. Fetal blood is more acidic than maternal blood, and compounds that are weak bases are therefore more ionized in the fetal circulation. This creates a concentration gradient of free substances towards the fetus.

In addition to the placenta representing a physical barrier to entrance of foreign substances, the placenta also expresses a wide variety of metabolizing enzymes, many of which are involved in conjugative and other biotransformational reactions of xenobiotics. Glutathione S-transferase, epoxide hydrolase, N-acetyltransferase, sulfotransferases and UDP-glucuronosyl transferase activities have been detected in human placentas [14, 15]. The activity of xenobiotic-metabolizing enzymes is highly affected by external factors, such as cigarette smoke or glucocorticoid therapy [16].

REFERENCES

[1] Kotyk, A; Janáček, K. *Cell Membrane Transport: Principles and Technique*. New York, Plenum Press; 1970.

[2] Abbas, AK; Lichtman, AH. *Cellular and Molecular Immunology*, 5th ed., Philadelphia: Saunders; 2003.

[3] Kobiler, DK; Lustig, S; Shapira, S. *Blood Brain Barrier: Drug Delivery and Brain Pathology*. Springer; 2002.

[4] Begley, DJ; Bradbury, MW; Kreuter, J. *The blood-brain barrier and drug delivery to the CNS.* New York, Marcel Dekker; 2000.

[5] Hamilton, RD; Foss, AJ; Leach, L. Establishment of a human in vitro model of the outer blood-retinal barrier. *J. Anatomy* 211(6), 2007, 707-716.

[6] Kaufmann, P. Basic morphology of the fetal and maternal circuits in the human placenta. *Contrib. Gynecol. Obstet.* 13, 1985, 5 - 17.

[7] Enders, AC; Blankenship, TN. Comparative placental structure. *Adv. Drug Deliv. Rev.* 38, 1999, 3–15.

[8] Knipp, GT; Audus, KL; Soares, MJ. Nutrient transport across the placenta. *Adv. Drug Deliv. Rev*. 38, 1999, 41 - 58.

[9] Smit, JW; Huisman, MT; van Tellingen, O; Wiltshire, HR; Schinkel, AH. Absence or pharmacological blocking of placental P-glycoprotein profoundly increases fetal drug exposure. *J. Clin. Invest.* 104, 1999, 1441 -1447.

[10] Ganapathy, V; Prasad, PD; Ganapathy; ME; Leibach FH. Placental transporters relevant to drug distribution across the maternal-fetal interface. *J. Pharm. Exp. Ther.* 294, 2000, 413 - 420.

[11] Pacifici, GM; Nottoli, R. Placental transfer of drugs administered to the mother. *Clin. Pharmacokin.* 28, 1995, 235 - 269.

[12] Reynolds, F; Knott, C. Pharmacokinetics in pregnancy and placental drug transfer. *Oxf. Rev. Reprod. Biol.* 11, 1989, 389 - 449.

[13] Audus, KL. Controlling drug delivery across the placenta. *Eur. J. Pharm. Sci.* 8, 1999, 161 - 165.

[14] Collier, AC; Tingle, MD; Keelan, JA; Paxton, JW; Mitchell, MD. A highly sensitive fluorescent microplate method for the determination of UDP-glucuronosyl transferase activity in tissues and placental cell lines. *Drug Metab. Dispos.* 28, 2000, 1184 - 1186.

[15] Collier, AC; Tingle, MD; Paxton, JW; Mitchell, MD; Keelan, JA. Metabolizing enzyme localization and activities in the first trimester human placenta: the effect of maternal and gestational age, smoking and alcohol consumption. *Hum. Reprod.* 17, 2002, 2564 - 2572.

[16] Paakki, P; Kirkinen, P; Helin, H; Pelkonen, O; Raunio, H; Pasanen, M. Antepartum glucocorticoid therapy suppresses human placental xenobiotic and steroid metabolizing enzymes. *Placenta* 21, 2000, 241 - 246.

Chapter III

The Metabolism of Toxicants

The toxicity of certain xenobiotics is highly influenced by the body's ability to metabolize and excrete the toxic compound. The rate of biotransformation also plays an important role in the toxicity of the specific substance and the intensity of the toxic effect. It is speculated that enzymes involved in biotransformation processes of foreign compounds were developed as a result of evolutionary adaptation intended to create protective mechanisms against new chemicals arising in the environment. Even though these metabolic reactions are part of a protective mechanism, biotransformation reactions also can result in the formation of more toxic metabolites. While processes of both activation and inactivation of the chemicals occur, the major metabolites of most xenobiotics are detoxication (detoxification) products. The metabolism of exogenous substances is based mainly on the conversion of lipophilic chemicals into more readily excreted polar products. This accelerates the elimination of the substance from the body and decreases the interaction time. At physiological pH, metabolites are usually more ionized and, thus, are less bound to plasma and tissue proteins. Furthermore, the passage across cell membranes is more difficult for ionized substances and their retention in fatty tissue is poor. These factors contribute to faster elimination of the substance from the body and to its inactivation in a toxicological sense.

In the body, two stages of metabolic biotransformation occur. Phase I reactions are reactions of oxidation, reduction and hydrolysis, and these usually prepare the compound for reactions of conjugation with some endogenous molecules that occur in the second phase. In phase I reactions, polar functional groups are introduced into the molecule, or they are unmasked. Metabolites that have gained a sufficiently hydrophilic character after the first transformation can be readily excreted after phase I. Phase II reactions are reactions of synthesis, requiring energy, which is provided by adenosine triphosphate (ATP).

The most important site of biotransformation is the liver due to the high concentration of metabolic enzymes in hepatocytes. High blood flow through the liver delivers substances readily after gastrointestinal absorption. Other organs, such as the kidneys, lungs and gastrointestinal mucosa, are also involved in metabolic transformation, but to a lesser extent. Most of the enzymes involved in phase I reactions are bound to the membrane of the endoplasmic reticulum. The endoplasmic reticulum membrane is hydrophobic and this "attracts" lipophilic compounds and makes their phase I biotransformation rapid. The endoplasmic reticulum, together with phase I enzymes, may be separated from other parts of the cell by differential centrifugation. After centrifugation, enzymes occur in the microsomal fraction together with the parts of the endoplasmic reticulum, and for this reason are also called microsomal enzymes. The enzymes catalyzing phase II reactions are located mainly in the cytosolic fraction, i.e., in the supernatant fluids obtained after centrifugation.

Anaerobic intestinal microbes also have an important role in biotransformation processes by performing reductive and hydrolytic reactions. It has been confirmed that the toxicity of some xenobiotics, such as cyclamate and nitroaromatic compounds, is directly dependent on the metabolic reactions performed by intestinal microflora. Intestinal bacteria also hydrolyze natural cyanogenic glycosides, releasing the toxic cyanides.

PHASE I REACTIONS

Many exogenous substances that enter the organism are lipophilic and for that reason they easily cross lipid membranes and are easily transported to body fluids by lipoprotein carriers. Biotransformation reactions of phase I are aimed to reduce the lipid character of foreign substances in order to avoid their retention in the body. During these reactions, polar groups are either introduced into the molecule or unmasked, resulting in more polar metabolites of the original chemicals. Phase I reactions involve reactions of oxidation, reduction or hydrolysis. If the metabolites of phase I reactions are sufficiently polar, they can be excreted at this point. Important enzyme systems involved in oxidation reactions include the cytochrome P450 monooxigenase system, alcohol and aldehyde dehydrogenases, the flavin-containing monooxigenase system, monoamine oxidases and various peroxidases. Reduction reactions mostly involve NADPH-cytochrome P450 reductases. Hydrolytic reactions are catalyzed by esterases, amidases and epoxide hydrolases. Other phase I enzymes include prostaglandin synthetase, molybdenum hydroxilases and glutathione reductases.[1]

Oxidation Reactions

Cytochrome P450 oxidase is a common term for a large number of evolutionary-related oxidative enzymes, important in animal, plant and bacterial physiology. Most cytochrome P450 oxidases are composed of about 525 amino acids and contain a heme group at the active site [1]. When heme iron is reduced with sodium dithionyl and complexed with carbon monoxide, the resulting colored product shows an absorbance of 450 nm, for which the enzyme group was named: P450, meaning pigment of 450 nm.

Most cytochrome P450 oxidases metabolize multiple substrates, and many enzymes can catalyze multiple reactions, which accounts for their central importance in metabolizing the variety of both endogenous and exogenous molecules. This class of enzymes plays an important role in hormone synthesis and breakdown, cholesterol synthesis and vitamin D metabolism.

The most important reactions catalyzed by this class of enzymes include the following:

Aliphatic and aromatic hydroxylation:

$R\text{-}CH_2\text{-}CH_2\text{-}CH_3 \rightarrow R\text{-}CH_2\text{-}CHOH\text{-}CH_3$

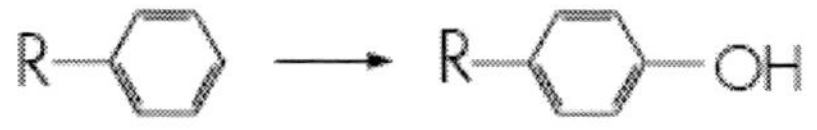

Alkyl side chains of aromatic compounds are easily oxidized to alcohols or hydroxylation can be performed at the carbon atom belonging to aromatic ring. For example, the first stage of the biotransformation of phenytoin, an antiepileptic drug, results from this type of reaction:

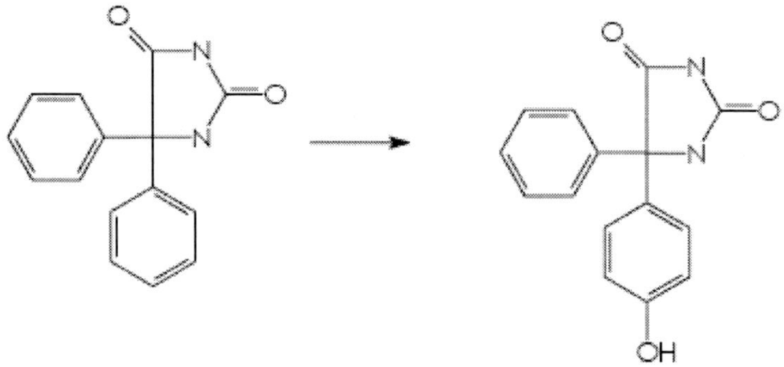

Dealkilation. The reactions may involve O-, N- and S-dealkilation. Many drugs and insecticides undergo N-dealkilation:

$$R\text{-}NH\text{-}CH_3 \xrightarrow{O_2} R\text{-}NH_2 + CH_2O$$

N-oxidation. The reactions may result in hydroxylamine, oxime or N-oxide formation. Imines and primary amines usually undergo N-hydroxylation:

$$R\text{-}NH\text{-}CO\text{-}CH_3 \xrightarrow{O_2} R\text{-}NOH\text{-}CO\text{-}CH_3$$

Desulfuration. These reactions convert the P=S bond to a P=O bond. In such a reaction, parathion is converted to paraoxon, a strong cholinesterase inhibitor.

S- and P-oxidation. Tioethers are oxidized to sulfoxides and trisubstituted phosphines are oxidized to phosphine oxides. Sulfoxides are further metabolized to sulfones and this is a common biotransformation reaction involved in the metabolism of insecticides, drugs and chlorinated hydrocarbons.

Many of the reactions described above can also be catalyzed by other enzymes, such as the *flavin-containing monooxigenase system (FMO)*. The flavin-containing monooxigenase system is an enzymatic group involved in oxidation reactions that contain flavin adenine

[1] NADPH oxidase (nicotinamide adenine dinucleotide phosphate-oxidase) is a membrane-bound enzyme complex

dinucleotid (FAD) as a cofactor (Figure 3.1.) The enzymes in this group belong to a microsomal fraction of cellular proteins and many isoenzymes of this family are known.

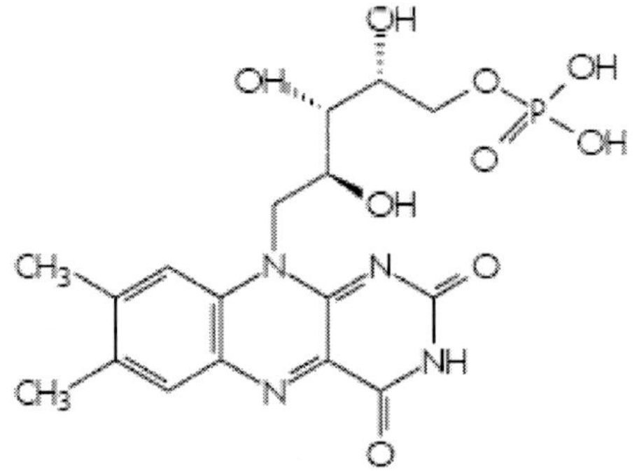

Figure 3.1. Flavin adenine dinucleotid (FAD), a cofactor of the FMO enzyme system.

FMO enzymes are involved in oxidation of inorganic and organic compounds containing nitrogen, sulphur and phosphorus:

X-NHR → X – NOH-R

2 R-SH → R-S-S-R ; R-S-S-R → R-SO-S-R

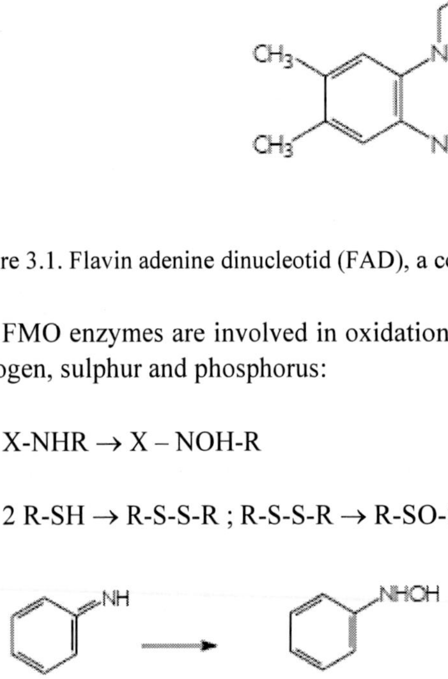

A major difference between this system and the cytochrome P450 monooxigenase system is that the FMO system does not oxidize carbon atoms; however, many reactions catalyzed by FMO can also be catalyzed by cytochrome P450. In contrast to the P450 monooxigenase system, the FMO system cannot be induced by substances other than substrates. To perform its activity, flavin-containing monooxigenases require the presence of oxygen and the assistance of NADPH. Substrates oxidized by the FMO system include inorganic compounds (HS^-, I^-, IO^-, I_2, CNS^-), organic nitrogen compounds (acyclic and cyclic amines, N-alkyl- and N, N–dialkylarylamines, hydrazine and primary amines), organic sulphur compounds (thiols, disulfides, cyclic and acyclic sulfides, mercaptopurines, thioamides), organic phosphorous compounds (phosphines, phosphonates), selenides and selenocarbamides [1].

Alcohol dehydrogenases are a group of enzymes involved in alcohol metabolism. They catalyze the destruction of alcohols, which are toxic to humans. The evolutionary purpose of alcohol dehydrogenases was probably the inactivation of alcohols naturally occurring in foods or produced by bacteria in the gastrointestinal tract.

The metabolism of primary alcohols is faster than that of secondary and tertiary alcohols. In yeast and many bacteria, *the alcohol dehydrogenases* class of enzymes catalyzes the opposite reaction, which is alcohol synthesis, in fermentation process.

In humans, these enzymes are produced in the stomach lining and in the liver. By alcohol dehydrogenase activity, ethanol is oxidized to acetaldehyde:

$CH_3\text{-}CH_2\text{-}OH + NAD^+ \rightarrow CH_3\text{-}CHO + NADH + H^+$

Alcohol dehydrogenases are also involved in the metabolism of other alcohols, such as methanol or ethylene glycol. In alcohol metabolism, acetaldehyde is the substance, responsible for "hangover" symptoms. The aldehyde requires further metabolism because this compound is very toxic, and because of its good lipid solubility, it is not easily excreted.

$CH_3\text{-}CHO + NAD(P)^+ + H_2O \rightarrow CH_3\text{-}COOH + NAD(P)H + H^+$

Formed acid can either be excreted or undergo phase II reactions to become even more polar. In a similar manner, aldehyde dehydrogenase detoxifies acrolein, the hepatotoxic metabolite formed from allyl alcohol. However the activity of dehydrogenases does not always lead to compound inactivation. In the case of 2-butoxyethanol, biotransformation leads to the formation of 2-butoxyacetic acid, which is potentially hematotoxic.

Humans have six slightly different alcohol dehydrogenases. All of them are dimmers, consisting of two polypeptides. A zinc ion in their structure plays an important role in fixing the hydroxyl group of alcohol in place at the catalytic site [2].

In yeast and bacteria, alcohol dehydrogenases play an important part in fermentation processes and glycolysis. As a result of glycolysis the glucose is converted into pyruvat, which is further transformed into acetaldehyde and carbon dioxide. Acetaldehyde is then reduced to ethanol by alcohol dehydrogenase. The purpose of this later step is to regenerate NAD^+ in order to continue the energy-producing glycolysis process. The main alcohol dehydrogenases in yeast are larger than those in humans and consist of four subunits, rather than two in humans. They also contain zinc at the catalytic site. Short-chain alcohol dehydrogenases, which do not contain zinc ions and whose subunits are shorter, are usually involved in oxidation of secondary alcohols. Some of these enzymes contain metals other than zinc, such as calcium or magnesium, but only for structural purposes, never in the catalytic center. Iron containing alcohol dehydrogenases, which are oxygen-sensitive, have been discovered in bacteria.

Monoamine oxidases (MAO) are enzymes that catalyze oxidation of monoamines and contain covalently-bound cofactor flavin-adenine-dinucleotids (FADs). They play vital role in inactivating biogenic amines, such as neurotransmitters. These enzymes are bound to the outer membrane of mitochondria in most cell types in the body.

Two types of monoamine oxidases (MAOs) occur in humans: MAO-A and MAO-B. Both are found in neurons. MAO-A is also found in the liver, gastrointestinal tract and placenta, and is particularly important in the catabolism of monoamines ingested in food. Outside the central nervous system (CNS), MAO-B is found mainly in blood platelets. Both MAO-A and MAO-B are essential for the inactivation of neurotransmitters, for which they display different specificities. The destruction of serotonin, adrenaline (epinephrine) and noradrenalin (norepinephrine) is mainly catalyzed by MAO-A, while phenethylamine is mainly the substrate of MAO-B. Both forms catalyze the biotransformation of dopamine.

Monoamine oxidases remove an amine group from monoamines in the process of oxidative deamination:

$R\text{-}CH_2\text{-}NH_2 + H_2O + NAD^+ \rightarrow R\text{-}C{=}O + NH_3 + NADH$

MAO dysfunction is related to a number of neurological disorders. High or low levels of MAO in the body are often associated with depression, aggression, attention deficit disorder and social phobias. Monoamine oxidase inhibitors are the major class of drugs prescribed for the treatment of depression and other neurological disorders.

Molybdenum hydroxilases are the class of molybdenum-containing enzymes particularly important in the oxidation of aldehydes and N-heterocyclic compounds. The most significant representatives include aldehyde oxidase and xantine oxidase.

Superoxidismutase catalyzes the conversion of very dangerous and reactive superoxide radicals ($O_2\cdot$) into molecular oxygen and hydrogen peroxide. Apart from their generation through external factors, superoxide radicals are also formed by endogenous processes occurring in the respiratory system and as a result of immunological activity of phagocytes.

The primary function of *prostaglandin H synthetase* is prostaglandin synthesis, but this enzyme also can actively take part in the metabolism of some xenobiotics, such as paracetamol and benz(a)pyrene.

Reduction Reactions

Reductive reactions are performed in anaerobic conditions with the implication of NADH or NADPH cofactors. Most such reactions are catalyzed by NADPH-cytochrome P450 reductases. Several functional groups are susceptible to reduction: nitro-, diazo- and carbonyl-groups. Disulfides, sulfoxides and alkenes are also substrates common to reductive enzymes. Mammals have a poor ability to reduce azo compounds because their tissues and body fluids are oxygenated; however, intestinal bacteria may contribute to azo reduction and to inactivation of such compounds. The reduction processes proceed through nitroso- and hydroxylamine intermediates.

Peroxidases, which are heme-containing enzymes, catalyze the destruction of dangerous organic peroxides:

$$ROOR' + 2e^- + 2H^+ \rightarrow ROH + R'OH$$

These enzymes are active with both hydrogen peroxides and organic hydroperoxides, such as lipid peroxides. Glutathione peroxidase is a peroxidase found in humans that contains selenocysteine. It uses glutathione as an electron donor and is active with both hydrogen peroxide and organic hydroperoxides substrates. These enzymes play an important role in protecting the cells from oxidative damage, and in selenium deficiency, the level of these enzymes drops together with their protective activity.

Hydrolysis

The very large group of enzymes that catalyze the hydrolytic reactions occurs mostly in plasma and various tissues, where it constitutes both the microsomal and cytosolic fraction. Esters, amides, thioesters and phosphates are susceptible to hydrolysis and are common substrates to hydrolytic enzymes. Some products of hydrolytic reactions, such as alcohols and acids, may be directly excreted or may undergo conjugation by phase II reactions.

One of important reactions catalyzed by hydrolases is the hydroxylation of aren oxides, as well as aliphatic and aromatic epoxides (Figure 3.2.). Epoxides are very reactive and dangerous electrophilic molecules that can covalently bind to essential nucleophylic molecules in the body, such as DNA and proteins. Aggressive epoxides can damage tissue cells and lead to pathological changes, such as neoplasm (tumor) formation. The products of hydroxylation reactions are less electrophilic diols,[2] which are also less reactive and, thus, less toxic.

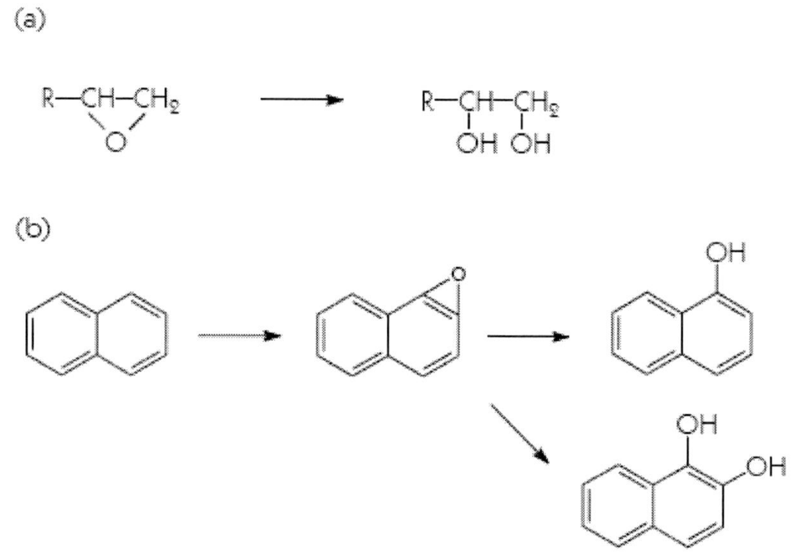

Figure 3.2. (a) Aliphatic and (b) aromatic hydroxylation, catalyzed by hydrolases.

PHASE II REACTIONS

Many phase I products are not eliminated rapidly and undergo subsequent reactions in which an endogenous substrate is combined with the newly incorporated functional group from the phase I reaction to form a highly polar conjugate. Phase II reactions are known as conjugation reactions in which the molecule is conjugated with glucuronic acid, sulfonates, amino acids, methyl compounds, glutathione and the like. Such biotransformation results in the further increase of xenobiotic hydrophilicity and greater ability for excretion. Phase II reactions are usually catalyzed by enzymes involving various cofactors which react with the polar functional groups belonging to the parent compound or introduced into the molecule by phase I reactions.

---

[2] Diols are organic compounds containing two hydroxy groups.

Some of the important enzymes involved in phase II reactions include glutathione-S-transferases, UDP-glucuronosyltransferases, N-acetyltransferases, amino acid N-acyl transferases and sulfotransferases.

Glucuronidation

Conjugation with glucuronic acid (glucuronidation)[3] is catalyzed by uridine-diphospho-glucuronosyltransferases (UDP- glucuronosyltransferases), and glucuronides are the most common conjugates of the toxicants that are excreted in bile. The level of UDP-glucuronosyltransferase in bodily tissue varies., In fetal tissues it is practically undetectable but increases with the age. Low level of conjugation systems in newborn are responsible for neonatal jaundice, caused by increased bilirubin levels. Bilirubin is normally excreted in bile in the form of the conjugate with glucuronic acid, but enzyme deficiency causes bilirubin to accumulate in the body of newborns. Most glucuronidation reactions occur in the liver, intestinal mucosa and kidneys, but they may also occur in other tissues and organs. Formed conjugates often form a substrate of β-glucuronidase (hydrolases), produced by intestinal flora. As a result of hydrolysis, the bond between the toxicant and glucuronic acid is broken down and the toxicant is released. The released toxicant may be reabsorbed in the intestines and is a subject of enterohepatic circulation. Enterohepatic circulation of some compounds is a cause of their slow excretion and the prolongation of the toxic effects. Also, some glucuronic acid conjugates are unstable under acidic conditions and can hydrolyze in urine.

Sulfation

The enzymes belonging to the sulfotransferase family may be both membrane bound or cytosolic. A sulfate group is transferred to a hydroxyl or amino group of the xenobiotic from 3`-phosphoadenosine-5`-phosphosulphate (PAPS). Formed sulfate esters are completely ionized and, thus, may be easily excreted mainly in urine and, to a lesser extent, in bile. Some sulfate conjugates are unstable and may be degraded by enzymes of intestinal microorganisms. Primary, secondary and tertiary alcohols, phenols and N-substituted hydroxylamines undergo sulfation to form sulfate esters. This type of enzymatic reaction is also important for endogenous processes such as biosynthesis of thyroid and steroid hormones, and of certain proteins and peptides.

Methylation

Methyl conjugates are generally less hydrophilic than the parent compound;, nevertheless, these reactions are considered to be detoxification processes. Methyl groups may be transferred from a donor compound, S-adenosyl-methionine or 5-

[3] Glucuronidation is the process of adding *glucuronide* to a toxicant or Phase I metabolite during Phase II biotransformation.

methyltetrahydrofolic acid (Figure 3.3.), to substrates that can accept a methyl group, such as proteins, nucleic acids or phospholipids. A methyl group may also be transferred to N, O or S atoms of xenobiotics. By the process of methylation, hydroxyl group of phenols and alcohols, amino groups of aliphatic and aromatic amines and groups of thiols (mercaptans) are masked. N-methylation involves primary, secondary and tertiary amines, followed by N-methylamines formation. Catechol-O-methyltranferase is involved in transferring a methyl group to a phenolic group of catechol. Catechol-like substrates, such as pyrogallol, flavonoids, pyrones and pyridenes, can irreversibly inhibit catechol-O-methyltranferase. S-methylation occurs in substrates containing thiol groups, and aromatic thiols are more readily methylated than aliphatic substrates.

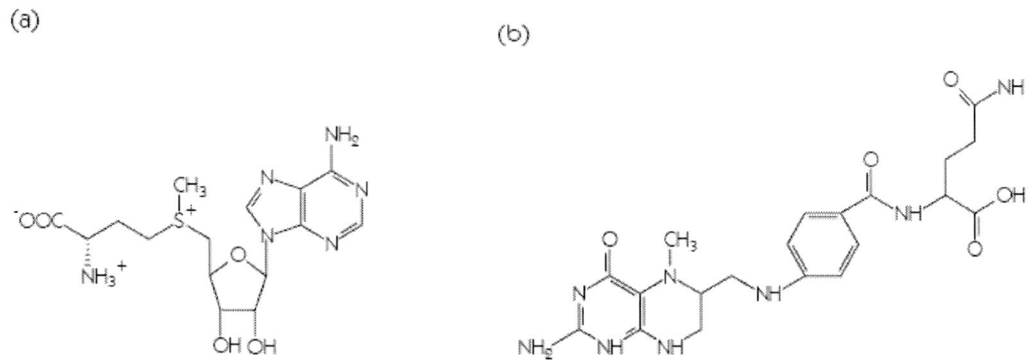

Figure 3.3. Donors of methyl groups: (a) S-adenosyl-methionine (SAMe) and (b) 5-methyltetrahydrofolic acid (5-Me-THFA).

Acylation

In acylation reactions, a carboxylic group of xenobiotics is conjugated with amino groups of endogenous amino acids, such as glycine, glutamine, arginine, taurine or ornithine (birds, reptiles). In order to be conjugated, the compound requires previous activation with acetyl-CoA to CoA derivative. Formed acyl-CoA derivative further reacts with aminoacid to form aminoacid conjugate and CoA:

$$\underset{\text{Acyl-CoA}}{\text{R-CO-SCoA}} + \underset{\text{Glycine}}{\text{NH}_2\text{-CH}_2\text{-COOH}} \rightarrow \underset{\text{Aminoacid conjugate}}{\text{R-CO-NH-CH}_2\text{-COOH}} + \underset{\text{Coenzyme A}}{\text{CoASH}}$$

Glutathione S-transferase

The glutathione S-transferase (GST) family of enzymes comprises a multitude of cytosolic and microsomal enzymes capable of multiple reactions with a multitude of endogenous and exogenous substrates. In some tissues these enzymes constitute up to 10% of cytosolic proteins.

Glutathione S-transferase contributes to metabolism of xenobiotics by conjugating them with reduced glutathione via the sulfhydryl group. Glutathione is an endogenous tripeptide (Figure 3.4.). Conjugation of foreign molecules by their electrophilic centers facilitates their dissolution in aqueous cellular and extracellular media. The reactions are useful in the detoxification processes of both endogenous and exogenous compounds such as peroxidased lipids, epoxides, haloalkanes, nitroalkanes, alkenes, organophosphates and methyl sulfoxide compounds.

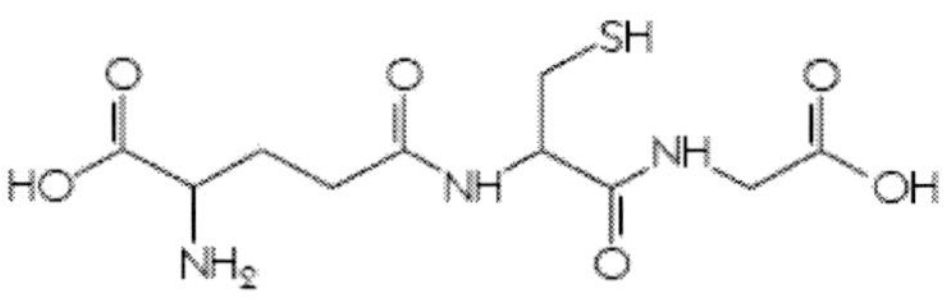

Figure 3.4. Endogenous tripeptide glutathione.

Factors Affecting Biotransformation

Various physiological and pathological factors influence biotransformation and the toxicity of xenobiotics. Passage through body membranes is highly dependent on physico-chemical properties of the toxicant, and this reflects the absorption rate and toxicant potential. Physiological factors that influence the metabolism include age, gender, individual variations, diet and others. The activity of certain metabolic pathways is influenced by toxicant concentrations, because the enzymes involved in biotransformation processes have a limited capacity. High concentrations of a toxicant quickly saturate the enzymes with high affinity to substrate, and the rest of the unmetabolized compound is transformed by activating low-affinity enzymes with a higher biotransformation capacity.

Dietary factors express a marked influence on the toxicity of substances. Fasting for 8h has been shown to reduce blood concentrations of glucose and produce changes in the activity of several metabolizing enzymes, including hepatic and renal cytochrome P450 activity. Fasting also results in glutathione depletion, reduced glucuronide conjugation, and overall decreased detoxification. Besides deficiency in conjugation molecules involved in phase II reactions, low energy status of the body contributes to inefficient detoxification. Low-protein diets have been shown to increase the toxicity of several pesticides, but may also be able to have a reverse influence in some situations. Reduced activity of primarily NADPH cytochrome P450 reductase and other P450 isoenzymes has been observed in protein-deficient diet, but the activity of glucuronyltransferase and glutathione S-transferase may also be compromised [3]. If toxicant biotransformation results in more toxic metabolites, as for example in heptachlor metabolism, a protein-deficient diet will depress the enzymes, producing more toxic metabolites and contributing to reduced toxicity. Lack of proteins in diet may also lead to changes in amino acid composition of the synthetized enzymes, which, in turn, may affect substrate binding. Dietary deficiencies of micro- and macronutrients that are required for the functioning of metabolizing systems, can lead to decreased biotransformation activities. Lack of sulphur-containing aminoacids directly reflects the glutathione content, which is very important in phase II reactions.

Fats provide building units for biological membranes, and since many enzymes are bound to the endoplasmic reticulum membrane, their activity may be influenced by fat intake. The influence of fat content in the diet has a less profound effect on the activity of biotransformation reactions in comparison to protein influence. The content of phosphatidylcholine in the diet is important with respect to impairment of the activity of phospholipases. Lipid substances such as cholesterol and fatty acids may occupy cytochrome P450-binding sites, displacing exogenous substances and prolonging their retention in the body.

Carbohydrate intake reflects biotransformation in the sense that some receptor sites contain carbohydrate components and in that glucose is a precursor of glucuronic acid-related phase II reactions.

Micronutrient deficiency does not impact microsomal enzymes as profound as proteins, but some vitamins have been shown to significantly affect biotransformation. Vitamin C facilitates the elimination of phase I products, decreases covalent binding of reactive intermediates, eliminates free-radical metabolites and chemically inhibits the formation of carcinogenic nitrosamines in gastrointestinal tract. Riboflavin (vitamin B2) is an essential component of NADPH-cytochrome P450-reductases with flavin adenine dinucleotide (FAD) and flavin mononucleotide (FMN) cofactors [3]. Deficiency in riboflavin thus adversely affects toxicant metabolism and can result in the uncoupling of electron transport. Folate also is important for toxicant biotransformation and chemical detoxification processes. Increased cytochrome P450 activity may be seen in dietary deficiencies of thiamine (vitamin B1) [4]. The reverse effect, i.e., the depression of the enzymes, is observed in tocopherol deficiencies, because lack of this vitamin results in loss of cytochrome P450 function. Heme, as an important cofactor of many metabolizing enzymes, is dependant on iron intake. Several enzymes, including glutathione-peroxidase, require selenium as an essential element. Another group of important enzymes contain manganese. Selenium deficiencies may also result in lipid peroxidation and destabilization of microsomal membranes, decrease in heme synthesis and loss of cytochrome P450 activity.

Many endogenous and exogenous substances can have a profound impact on the rate and course of biotransformation reactions. A great number of xenobiotics induce biosynthesis of microsomal enzymes and increase metabolic activity of the liver. Some substances can induce their own metabolism or can have impact on the biotransformation of other substances. Enzyme induction is independent of the chemical structure of the compound, but the structure determines which reaction (hydroxylation, deamination, and dealkilation) will be stimulated. Inductors of microsomal enzymes are generally substances that may represent substrates for more than a single enzyme. Inductors have a well-defined dose-response relationship, and below specific concentrations their stimulating properties are not expressed. Many pesticides (aldrin, chlordane, lindane, DDT), industrial chemicals (benzene, polychlorinated biphenyls), polycyclic aromatic hydrocarbons, drugs (phenobarbitone, barbiton, chlordiaepoxide), organosulphur compounds, and some naturally occurring compounds, such as diterpens from coffee and vegetable indols, have been shown to induce microsomal enzymes. Other agents, such as components from cigarette smoke or ethanol, exert similar effects. Most microsomal enzymes, specifically cytochrome P450 monooxygenase systems, are prone to activation. Enzymes catalyzing phase II reaction, with the exception of glutathione-S-transferase, cannot be activated by external factors. Histologically, activation of enzymatic systems can be observed as a proliferation of the endoplasmic reticulum membranes in the hepatocytes.

Competition between two or more substrates for the active site on biotransformation enzymes inhibits metabolic biotransformation. The metabolic product of the toxicant itself may express higher affinity to the active site of the enzyme. Compounds with higher affinity to certain enzymes can displace already bound substances and may inhibit their further biotransformation. As a consequence, longer half-lives and prolonged toxic impacts are observed for the displaced compound. Other possible mechanisms of enzyme inhibition may involve covalent binding of the substance to the protein part of the enzyme and disturbance of cofactor synthesis. Cobalt, for example, disturbs heme synthesis, which is crucial part of cytochrome P450 enzyme system, while vinyl-chloride covalently binds to nitrogen atoms in the heme. Enzyme inhibition may also occur due to the disturbance of electron transfer processes, such as in the case of carbon-monoxide which binds to heme and blocs its function. 3-aminotriazole, o-ethyl-o-p-nitrophenyl-thiophosphate, pentachlorophenol and CCl_4 have been shown to inhibit metabolic processes via different mechanisms. Some natural components of grapefruit juice, such as bergamottin, dihydroxybergamottin, and paradisin-A, express similar effects and make the consumption of grapefruit contraindicated during drug therapies [5].

Individual variations in susceptibility to certain toxic compound within one species arise due to differences in any of the toxicokinetic phases, including absorption, protein and receptor binding, enzyme induction/inhibition, biotransformation and excretory mechanisms. Slowly metabolising persons are exposed to a higher extent to toxic substance. Individual variations in the rate of biotransformation usually have a genetic background and are a direct consequence of differences in the amount and type of biotransformation enzymes. People with low activity of N-acetyltransferases, involved in Phase II reactions, are known as slow acetylators. Chemicals that undergo biologic acetylation are usually aromatic amines or hydrazines, including caffeine. Drugs undergoing this type of reaction include isoniazid, hydralazine, procainamide, sulfamethazine, sulfapyridine, dapsone and nitrazepam. Hereditary differences between slow and rapid acetylators are not due to differences in quantitative levels, but due to qualitative differences associated with isozymic variants of the enzyme. Rapid and slow acetylators differ in a single autosomal gene.

This variation may have dramatic consequences, as slow acetylators are more prone to dose-dependent toxicity. The rate of biotransformation is specifically low in Japanese people, so the dose of certain drugs should be carefully adjusted according to their metabolism. Glucose-6-phosphate dehydrogenase is a metabolic enzyme involved in the pentose phosphate pathway, and especially important in red blood cell metabolism. The pentose phosphate pathway is the principal source of NADPH in normal red blood cells. NADPH in return maintains the level of glutathione that helps to protect the red blood cells against oxidative damage. Glutathione plays a central role in many conjugation reactions, as well as for maintaining protein sulfhydryl groups in the reduced state, thereby preventing denaturation of enzymes and hemoglobin and preserving the integrity of the erythrocyte membrane. It has been shown that hereditary deficiency in glucose-6-phosphate dehydrogenase leads to severe hemolysis,[4] which arises in response to certain antibiotics (sulfanilamide, sulfapyridine, nitrofurantoin and others), analgetics (aspirin, bufferin, anacin

[4] Hemolysis is the lysis (break up) of the red blood cells resulting in the release of hemoglobin and intracellular potassium. Erythrocytes contain higher concentrations of potassium in comparison to plasma and the increase in potassium blood level can be an indicator of undergoing hemolysis.

etc.), antimalarials, anthelmintics, vitamin K, fava beans and other. Two types of people are especially susceptible to hemolysis due to glucose-6-phosphate dehydrogenase deficiencies: Afro-Americans and Mediterranean people. About half of Asians, including Chinese, Japanese, and Korean people, are deficient in mitochondrial aldehyde dehydrogenase isoenzymes, which are responsible for metabolizing acetaldehyde. The accumulation of acetaldehyde in the blood occurs after consuming alcoholic beverages and is accompanied with intense facial flushing after small amounts of alcohol ingested, tachycardia, and sensation of heat, headache and hypotension. Enzymes belonging to cytochrome P450 monooxigenase systems may also vary between individuals, with deficiencies occurring in 1 - 30% of people, depending on their ethnic background.

Mechanisms responsible for the diversity in toxicological responses between species include qualitative and quantitative differences in metabolism, i.e., differences in metabolic pathways and enzyme levels, differences in biliary and renal excretion, plasma protein binding, differences in tissue distribution and in response of target receptor sites.

Xenobiotics are poorly and much more slowly metabolized in fetuses, neonatants (newborns) and elderly people. Enzymatic systems in newborns are not fully developed and are unable to inactivate many compounds. With aging, levels of metabolic enzymes decrease, making elderly people more susceptible to many xenobiotics. Many age-related differences, particularly between the young and elderly, may be explained primarily by quantitative differences in detoxification processes. There is also clear evidence of the differences between newborn and adults in renal clearance of toxicants.

Liver pathology has an impact on elimination rate of toxins, because the metabolism is significantly impaired in affected hepatocytes, which are most important sites of toxin transformation. Hormone misbalances (thyroxin, insulin) also may lead to activation or inactivation of the metabolic enzymes. Hormone levels mainly affect enzymes involved in phase I reactions; however, conjugation with glucuronic acid, a phase II reaction, can also be influenced by deficiency of glucuronic acid, which commonly occurs in diabetes.

It has been demonstrated in animal studies that females have a much less powerful metabolizing capacity for most xenobiotics. This difference may be due to the influence of hormones, or it may be attributed to other physiological parameters influencing absorption, distribution and elimination of toxicants. Females also are more susceptible to the influence of alcohol, due to lower body content of water; thus, blood alcohol concentrations may become higher in females. In addition, testosterone and estrogens can directly stimulate metabolising enzymes involved in the biotransformation of xenobiotics. Pregnancy markedly increases the susceptibility of animals to pesticides, and lactating animals are more susceptible to heavy metals [6]. During pregnancy, the activity of biotransformation enzymes, such as the activity of o-methyltranferase and monoamineoxidase, is suppressed, as is conjugation with glucuronic acid.

TOXICOKINETICS AND TOXICODYNAMICS

Toxicokinetic phase comprises the physiological processes involved in absorption, distribution, biotransformation and excretion of toxicants. Biologically available amounts of the toxicant are the amount of the toxin that reaches the receptor provoking a specific response. The processes involved in the interaction between the toxicant and its sites of action constitute the toxicodynamic phase. In repeated exposure to certain toxicants, the body's response may be modified in both qualitative and quantitative ways.

Adverse effects resulting from repeated individual exposures to toxins may be intensified. Intensification may occur when the effect itself remains active in the body even after the toxicant has been fully excreted. Then the effect of the subsequent amounts of the introduced toxicant may overlap with the previous effect, resulting in new, accumulative and intensified outcome. This effect occurs in chronic exposure to organophosphorus pesticides, where inhibited activity of acetylcholinesterase, is the accumulation expression of all previous exposures to individual doses. Consequently, otherwise non-toxic doses may lead to severe toxic effects in repeated exposures. The accumulation of individual effects that are "remembered" in the body is specifically important in the case of carcinogenic and mutagenic substances. As an additive effect, it is claimed that a few sunburns during the childhood significantly increase the probability of carcinogenesis in adulthood. Namely, genotoxic agents provoke the damage to DNA, and this damage is transferred to future cell generations in replication processes.

Adverse toxicological manifestation often arises due to toxicant accumulation in the body. For high toxicant concentrations, inactivation processes as well as elimination rates are much slower. The capacity of the body to bind the toxicant to plasma proteins, and to biotransform and excrete the compound, may be exceeded where concentrations of the toxicant are high and where large amounts of the toxicant remain available to attack specific tissues.

Activation of certain metabolic pathways after exposure to a substance may result in weaker, less pronounced or qualitatively altered responses in subsequent exposures as a result of an increased level of tolerance. The production of certain biotransformation enzymes is initiated during first exposure, preparing the organism to readily metabolize the next amount of the substrate that enters the body. In the case of alcohol or nicotine tolerance, the effect is clearly demonstrated. Only the first few cigarettes provoke dizziness, and frequent alcohol consumption leads to organism "being trained" to tolerate higher alcohol concentrations without expressing the usual effects of alcohol. The tolerance also may be built up with respect to a group of chemically similar compounds. Ozone inhalation will influence the tolerance toward other aggressive gases, such as NO_2. However, achieved tolerance is reversible process, and an organism's sensitivity to a substance may be reconstituted after prolonged period without agent introduction.

Humans are simultaneously exposed to many different chemicals, and it is important to observe the effects of their simultaneous action. Overall effects may vary to a significant degree in comparison with the individual effects of the compounds. It is impossible to predict all possible interactions among the substances that may be simultaneously introduced. Assumptions and predictions of combined effects of different substances on the body is one of the most difficult tasks for toxicologists. Two basic types of interactions are common—synergism and antagonism.

In *synergistic interactions*, the final effect of two compounds may be represented as a mathematical sum of their individual effects or it may even exceed the sum of their individual effects. The recognition of altered effects of the substances that are introduced simultaneously is of highest toxicological interest, since the toxicity in those cases may be more pronounced. For example, various esters of phorbol, natural products of croton (*Croton tiglium*) oil, have important biological properties, the most notable of which is the capacity to promote the growth of tumors through activation of protein kinase C. It is important to note that even substances that seem harmless at first may promote toxicological responses of other substances to significant extent. In pesticides, synergists added to a formulation may significantly enhance pesticide activity, and this effect is used to reduce the necessary amounts of the pesticide. This factor is important for both environmental protection and for financial considerations. For example, some natural components of sesame oil, such as sesamin and sesamolin, as well as methylenedioxybenzene derivatives, express strong synergistic effects with pesticides. The compounds act by inhibiting the microsomal enzymatic system in insects, making the compound more efficient. Synergists may also act by increasing the absorption of a toxicant. Pesticide formulations also may be prepared in different solvents. The nature of the solvent is a very important determinant in dermal toxicity of certain formulations due to differences in absorption rates.

Antagonists have affinity but no efficacy for their cognate receptors, and their binding disrupts the interaction of the toxicant with the receptor and inhibits its function. Antagonistic effects between different substances are important for the treatment of poisoning, and this is the principal of interaction for antidotes.

Competitive antagonists reversibly bind to receptors at the active site of toxicant binding, but without activating the receptor. These antagonists often, but not necessarily, possess a very similar chemical structure to that of the toxicant or agonist. Agonists and antagonists compete for the same binding site on the receptor, and this can be observed as a rightward shift of the dose-response curve with no alteration of the maximal response. Once bound, an antagonist will block agonist binding. The final effect will be determined by the relative affinity of each molecule for the site and by their relative concentrations. While the competitive antagonist alone does not induce any biological response, inverse antagonists act like competitive antagonists, in the sense that they bind to the active site, but produce a distinct set of biological responses.

Non-competitive antagonists bind to a distinctly separate binding site from the agonist's, exerting the activity of the receptor. They do not compete with agonists for binding, but the bound antagonists may result in decreased affinity of an agonist for that receptor, or alternatively may prevent conformational changes in the receptor required for receptor activation. In dose-response curve, this is reflected as a depression of the maximal response and in some cases, rightward shifts also may be observed. For some substances, as for so-called *uncompetitive antagonists*, in order to act antagonistically, a previous activation by an agonist is required. Uncompetitive antagonists bind to a separate allosteric binding site and block the effects of the agonist, and are particularly efficient if agonist concentrations are high. *Partial agonists* compete with the full agonist for the active site on receptor and act as competitive antagonists if they are co-administrated simultaneously with the full agonist. In drug treatment, the significance of partial agonists is to enable a stable agonist-receptor interaction enhancing the effect of the agonist in agonist-deficient systems, while simultaneously blocking excessive activity of the agonist.

Many antagonists are reversible and, like most agonists, bind and unbind to receptor at rates determined by receptor-ligand kinetics. Irreversible antagonists covalently bind to a receptor target and, generally, cannot be removed. The receptor in such situations is inactivated until new receptors are synthesized.

Generally, the compounds that are used to treat poisoning can act on different levels and may affect absorption, stimulate excretion, or induce biotransformation enzymes. Apart from suppressing the activity of the toxicant on the receptor level, other physico-chemical procedures are frequently exploited in poisoning treatment. For example, chelating therapy for metal intoxication exploits physicochemical ways of decreasing the availability of a free toxicant. In oral poisoning, toxicants may be adsorbed on activated carbon in order to decrease oral absorption. Neutralization reactions with alkali are a way of overcoming toxic effects of acids in both oral and dermal exposures.

In respect to possible interactions between the substances in the body, overall effects may be expressed in simplified manner as 1+1=2, if the individual effects are added to each other in designing the overall response. Alternately, the response may be stronger than the sum of the individual ones (1+1=4). The final effect of an antagonist may be described as 1+1=0, independent of whether the interaction has occurred by competent binding to receptor site, physico-chemical interactions, such as adsorption or neutralization, or due to improved excretion or enzyme induction. Some compounds may potentiate the toxic effects of other substances. Indirectly, non-toxic or moderately-toxic compounds may exert adverse effects if administrated together with another substance (1+1=3). For example, isopropanol is a moderately toxic alcohol, but exposure to isopropanol significantly increases the hepatotoxicity of carbon tetrachloride.

REFERENCES

[1] Lee, JS; Obach, RS; Michael B; Fisher; MB. Drug metabolizing enzymes: cytochrome P450 and other enzymes in drug discovery and development. Informa Healthcare; 2003.

[2] M. Sardesai, VM. Biochemical and Clinical Aspects of Alcohol Metabolism. Thomas; 1969.

[3] Finley, JW; Daniel E;. Schwass, DE. Xenobiotic Metabolism: Nutritional Effects. *Am. Chem. Soc.;* 1985.

[4] Murray, M. Altered CYP expression and function in response to dietary factors: potential roles in disease pathogenesis. *Curr. Drug Met.* 7(1), 2006, 67-81.

[5] Bailey, DG; Dresser, GK. Interactions between grapefruit juice and cardiovascular drugs. *Am. J. Cardiovasc. Drug* 4(5), 2004, 281-297.

[6] Hood, RD. Handbook of Developmental Toxicology. CRS Press; 1996.

Chapter IV

TOXICANT EXCRETION

The body`s defensive mechanism functions by its ability to readily metabolize and excrete the toxic compounds. The toxicity of the substance is to certain extent determined by the rates of these reactions. Rapid excretion will reduce the time of a toxicant`s residence in the body and will contribute to less possibility for adverse toxicant/tissue interactions. In many poisoning treatments enhancement of toxicant excretion is the main principle of the therapy. As a result of biotransformation processes mainly occurring in the liver but also in other tissues, more hydrophilic metabolites are formed which are mainly excreted via kidneys. Toxicants with greater molecular mass are usually excreted into the feces via biliary emptying, both in the form of unmetabolized parent compound and in the form of its metabolites. Glutathione and glucuronide conjugates have a high probability of being excreted in bile. Volatile compounds and metabolites that are in equilibrium with blood concentration, as well as gases, are mostly pulmonarily excreted. Other routes of excretion, such is excretion in saliva, sweat, milk and dermal elements, like nails and hair, occur, but are less significant in quantitative respect. Nevertheless, they may be a reliable indicator of high concentration exposure.

URINARY EXCRETION

The main purpose of urinary tract is to maintain the chemical composition of the blood and to balance the volume of fluids and minerals in the body. Waste products of cellular activity, such as carbon dioxide, excess water and salts, nitrogenous compounds from protein decomposition, and xenobiotics, are excreted via the kidneys. The kidneys perform a central role in excretion by filtering the blood as many as 400 times a day and producing an average 1.5 l of urine. Good vascularization of the kidneys enable rapid and efficient transfer of the toxicants from the blood to the site of elimination. Twenty percent of total blood input is filtered in the kidneys.

While they are also blood filtering units, the kidneys are part of the urinary tract, which also contains the ureter and the bladder. The inner part of the kidney is called the renal medulla. The outer part is composed of the renal cortex and the renal pelvis. The renal cortex contains a multitude of nephrons, the basic functional units of the kidneys. Each kidney contains about one million nephrons. Each nephron is a small independent filtering unit and has its own blood supply and its own collecting tubule leading to ureter. Blood enters the kidney through the renal artery and purified blood leaves the nephron through a venule and the exits through the renal vein. The kidneys remove excess water, urea, and other waste products from the blood and they are then collected in the ureter. Nitrogenous wastes are initially brought to liver in the form of very toxic ammonia, which converted in the liver to less harmful urea and then transferred to the kidneys. Excretion of at least 500 ml of urine per day ensures the elimination of potentially toxic materials from the body and the maintenance of the blood's homeostasis.

Three main processes occur during renal excretion: filtration, reabsorption and secretion (Figure 4.1.). When blood enters a nephron through an arteriole, it flows into a network of 50 capillaries known as a glomerulus. The glomerulus is encased in the upper end by a Bowman's capsule, a cup-shaped structure with a thin double membrane at the beginning of the tubular component of a nephron. The walls of the capillaries and the Bowman's capsule are permeable and allow pressurized blood to be *filtered*. Impurities are filtered out through the pores and emptied into the collecting tubule.

Glomerular pores are larger than any other membrane pores in and other bodily cells. Their diameter is about 70 nm, which determines which molecules can pass through them (M < 60 000 Da) by simple filtration [1]. Plasma proteins, cells and platelets are too large to pass through these membrane pores and thus remain in the blood. Any binding of the toxicants to plasma proteins limits their ability to be filtered. The obtained filtrate contains water, urea, glucose, salts, amino acids and vitamins. Filtered material from the blood flows through the renal tubule, a long tube with permeable walls, consisting of the proximal convoluted tubule, the loop of henle, and the distal convoluted tubule. Approximately 180 liters of filtrate pass from the blood into the collecting tubules each day. However, a large part of this is *reabsorbed* in the renal tubules.

Most reabsorption occurs in the proximal tubule. There, of the 99% of water that is reabsorbed, about 75% is returned to the capillaries by osmosis. Glucose and minerals are returned to the blood by active transport. Small plasma proteins that are sometimes filtered through the glomerulus are also reabsorbed in the renal proximal tubule. Toxicants bound to small plasma proteins that have reached the proximal tubule may exert local toxicity. The nephrotoxic effects of cadmium arise due to its binding to plasma protein metallothionein. The size of the complex limits its filtration. Bound cadmium resides in the proximal tubules, where it exerts adverse effects. Part of the cadmium is reabsorbed and returned to blood circulation. In newborns, accumulation of nephrotoxic substances in renal tubules is reduced due to their undeveloped kidney function, making them less susceptible to nefrotoxic impact. Active transport, the principal mechanism of reabsorption for most substances, is inefficient in newborns.

Some additional reabsorption occurs when the filtrate reaches the distal convoluted tubule. Here, some substances pass into the filtrate through a *secretion* process. Tubular cells have systems for active transport of organic acids and bases, which are necessary for adjusting the pH of the blood. These transporters are of low specificity and may be used by some xenobiotic metabolites. However, unless a transport carrier system is available, polar compounds and ions are unable to diffuse back across the cell membranes and are secreted into the tubular urine. During World War II, a penicillin deficit led to the introduction of a therapeutic acid probenecid which competed with penicillin in the renal secretion process and prolonged the retention of penicillin and its antibacterial effects in the body.

Secretion processes may be active or passive. The passive secretion of weak acids and bases occurs as a result of differences in acidity. Un-ionized substances can reach tubules and can ionize there as a result of differences in acidity. Ionic substances can not be reabsorbed, but are secreted in urine. Such substances include wastes and toxic materials.

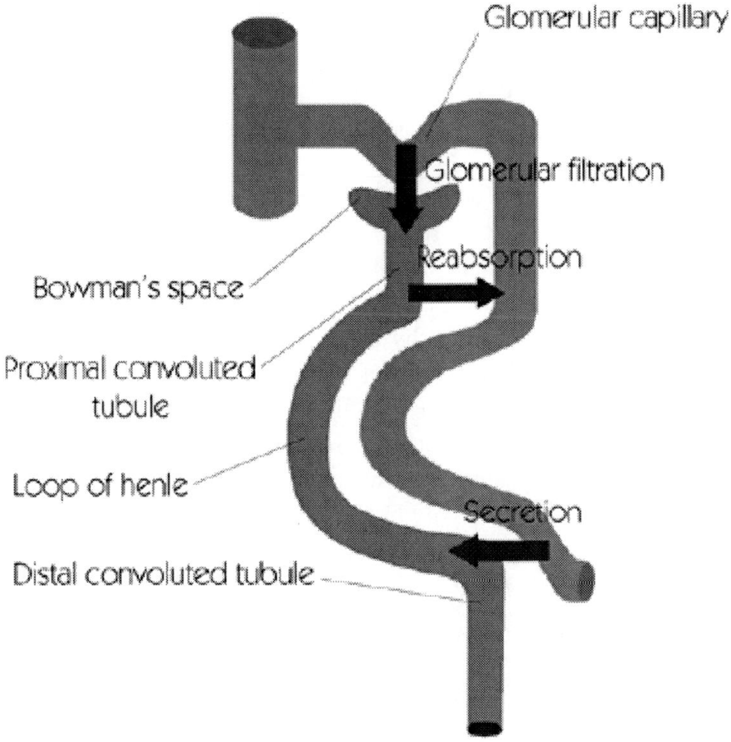

Figure 4.1. The mechanisms of renal excretion.

BILLIARY EXCRETION

After ingestion, digestion, and transport to the intestine, intestinal blood with absorbed substances enters the liver before entering systematic circulation. Xenobiotics and their metabolites formed in the hepatocytes are transferred back into the blood and eliminated via the kidneys, or are excreted to billiary tubules by diffusion or passive transport. In newborns, the billiary elimination function is not completely developed and the toxicity of xenobiotics is generally much higher. The concentration of substances excreted in the bile for some compounds reaches high levels and may induce hepatotoxic effects. Some metabolites may be eliminated from the bile in fecal matter along with unabsorbed matter from the intestines. In liver pathology. the liver's excretory potential may be significantly impaired and may give rise to risk of toxicant accumulation in the liver. Substances like biliary salts, bilirubin, lead, arsenic and manganese, concentrate in the bile reaching 10-1000 fold higher levels, than their plasma concentration [2]. These compounds are actively transported from plasma to hepatocytes and then, after transformation, to the bile. Substances that are excreted in bile may be classified in three groups, depending on their bile/plasma concentration ratio. The biliary excretion of albumines and ions like zinc, iron and chromium, is practically insignificant because their concentration in the bile is lower in comparison to plasma concentration. Glucose, sodium, potassium, cesium and cobalt are transferred to the bile to some extent. The ratio of bile and plasma concentrations for these substances is approximately one. Substances that concentrate in the bile, i.e., the substances that are significantly billiary excreted, like billirubin or lead, use protein transporters and consume energy during their transfer from plasma. Some drugs, like phenobarbital, induce biliary excretion and toxicant elimination.

Toxicant conjugates of high molecular weight, usually glucuronic acid and glutathione conjugates, are mostly biliarily excreted. Intestinal flora play an important role in toxicant metabolism, by catalyzing reductive and hydrolytic reactions. Thus, deactivation of some toxicants in the body occurs as a result of bacterial activity. However, other biliary excreted conjugates may be hydrolized in intestines by bacterial enzymes, releasing the toxicant for reabsorbtion. For such substances it is said that they are the subject of the *enterohepatic circulation,* which significantly increases the retention of toxicants in the body. In poisoning with substances that are biliary excreted but reabsorbed in intestines due to enterohepatic circulation, an important part of treatment is the interruption of the reabsorption cycle. In methyl-mercury poisoning for example, polythyol resins can absorb the released metabolite, interrupting the enterohepatic circulation and contributing to faster elimination.

Excretion Via the Lungs

Volatile compounds and metabolites, as well as gases, are excreted via lungs after they diffuse from the blood across alveolar membranes. Substances like organic solvents, alcohol, dimethylselenide, carbon-monoxide and the like, are eliminated from the body by exhalation. Since the diffusion process is the principal route of toxicant transfer from the blood to alveolar interspace, that lipid solubility clearly influences the elimination rate. However, the elimination of highly lipophilic volatiles may be retarded by their accumulation in fat tissue. The anesthetic halothane, for example, may be detected in the breath several weeks after its application due to its high hydrophobicity. The portion of the volatilate excreted by lungs depends directly on vapor pressure, lung ventilation, body temperature and atmospheric pressure. During physical exercise and labor, lung ventilation is intensified, giving a rise to more efficient elimination of toxicants from the body. Weather conditions such as temperature and atmospheric pressure also influence the amount of excreted toxicant; although their affect is less profound. A practical example of the connection between blood concentrations and vapor pressure is the in measure of alcohol blood levels in drivers. The more soluble a substance is in the blood, the more slowly it is eliminated via exhaled air. Chloroform, for example, is retained in the body for prolonged periods due to its high blood solubility.

EXCRETION IN MILK

Substances circulating in the blood may diffuse across the cell membranes and may reach the milk. This occurs mainly for lipophilic molecules, which are not readily excreted and which remain in the blood long enough for diffusion to occur. Hydrophilic substances are readily excreted and are not able to diffuse across the cell membranes. Fat content in human milk is usually relatively high (3-5%), contributing to concentration of excreted lipophilic substances in milk. This potential poses serious risks for nursing infants and for the contamination of dairy products. Nursing infants are much more susceptible to toxicant influence due to undeveloped metabolizing and excretory systems, and are also exposed to concentrated toxicants in mother's milk. Compounds such as DDT, polychlorinated biphenyls, dioxin and the like, are known to occur in human milk. Humans are at risk when consuming milk or dairy products contaminated with high levels of excreted metabolites, such as aflatoxin metabolites. Milk also may concentrate toxic elements that resemble calcium, such as lead.

Excretion via milk occurs by simple diffusion and is therefore also highly dependant on a substance's acidity. The pH of human milk is 6.5, promoting the retention of substances with weak alkaline character to be retained in the milk. Acidic compounds, on the other hand, will be in un-ionized form and will diffuse out into the plasma.

OTHER ROUTES OF EXCRETION

In addition to the principal routes of toxicant excretion, any bodily secretion, such as sweat, saliva and tears, as well as dermal components, like hair and nails, may be a site of toxicant elimination from the body. This is particularly true for metal and metalloid excretion in hair and nails. Hair grows from the cells belonging to dermal layer and therefore may indicate previous exposure to toxicants like arsenic, cadmium, lead, mercury, selenium and the like. The skin is supplied with the blood via capillaries embedded in the dermal layer. During the growth, the hair may incorporate the elements from surrounding cells, and thus also may be indicative of blood composition to a certain degree. The poisoning of Napoleon Bonaparte during his exile on St. Helena Island, was proven by analyzing and detecting elevated arsenic levels in his hair. Foreign substances and their metabolites also may be excreted via practically all dermal glands, including the mammary glands. Metals are also partially excreted via intestinal mucose and eliminated by feces because of the sloughing of the intestinal epithelial cells, which occurs approximately every 3 days. Toxicants excreted in sweat by simple diffusion also may provoke dermatitis. Excretion though the these routes is favorable for un-ionized, lipohilic substances because the main mechanism of elimination is simple diffusion governed by concentration differences and the substance`s fat solubility. Toxicants that are excreted in saliva may eventually be swallowed and be reabsorbed through the gastrointestinal tract.

ELIMINATION KINETICS

The rate of toxicant elimination depends on the activity of biotransformation enzymes and on the excretion rate. Both can be influenced by various external and physiological factors. A pathological condition existing in the excretory system may prolong the elimination rate, while enzyme inductors, such as diterpens from coffee or industrial chemicals like polycyclic aromatic hydrocarbons, may enhance the kinetics of elimination. The rate of elimination is best expressed in terms of half-life, t_{50}. Half-life is the time period in which 50% of a compound present in plasma is eliminated. Hydrophilic substances, which are rapidly metabolized and excreted, have low t_{50} values, while lipophilic substances, and especially substances that accumulate in bones, may have very high t_{50} values. The half-life of radioactive strontium, ^{90}Sr, which is accumulated in bones, is 4,000 days. Elimination rates of the same substance in different individuals may vary, depending on hereditary physiological factors, pathology of eliminatory organs and competition with other substances for binding sites on plasma proteins and tissues. Enzyme induction and inhibition, pH of urine and other factors may influence the rate of elimination. Slow elimination rates occur due to different physiological phenomena. Highly lipophilic substances, such as pesticides, dioxins and polycyclic aromatic hydrocarbons, tend to accumulate in fat tissue due to their good lipid solubility. Other substances, such as mercury, lead, fluoride, tend to accumulate because of strong chemical bonds with specific tissues. Methyl-mercury binds covalently and practically irreversible to brain tissue, causing severe neurological symptoms. Lead, strontium and fluoride displace similar ions from the bone structure and may reside in the bones for years.

The kinetics of the toxicant elimination may be defined by monitoring the excretion of the decayed substance in plasma levels. A one-compartment open model, establishes that the toxicant enters the body and estimates the approximate amount being excreted. However, this model neglects the distribution of the compound to other tissues and organs. To estimate the substance's exponential decay, its plasma concentration may be observed. Monitoring of plasma concentration establish the amount of the substance excreted in equal time intervals. This observation of first-order kinetics may be presented as a linear dependence on a semi-logarithmic scale (log C vs. time). Establishing this dependence enables the calculation of the toxicant concentration in tissues on the basis of plasma concentration.

In a two-compartment open model, toxicant entrance, toxicant distribution between plasma and tissues, and toxicant excretion are observed. Elimination kinetics in this type of system also may be described as an exponential curve on a semi-logarithmic scale (Figure 4.2.).

Elimination rates change over time and in this process, rapid (A) and slow (B) phases are observed. The values of k_{12}, k_{21} k_{el} constants express the relative influence of distribution and elimination on a concentration profile:

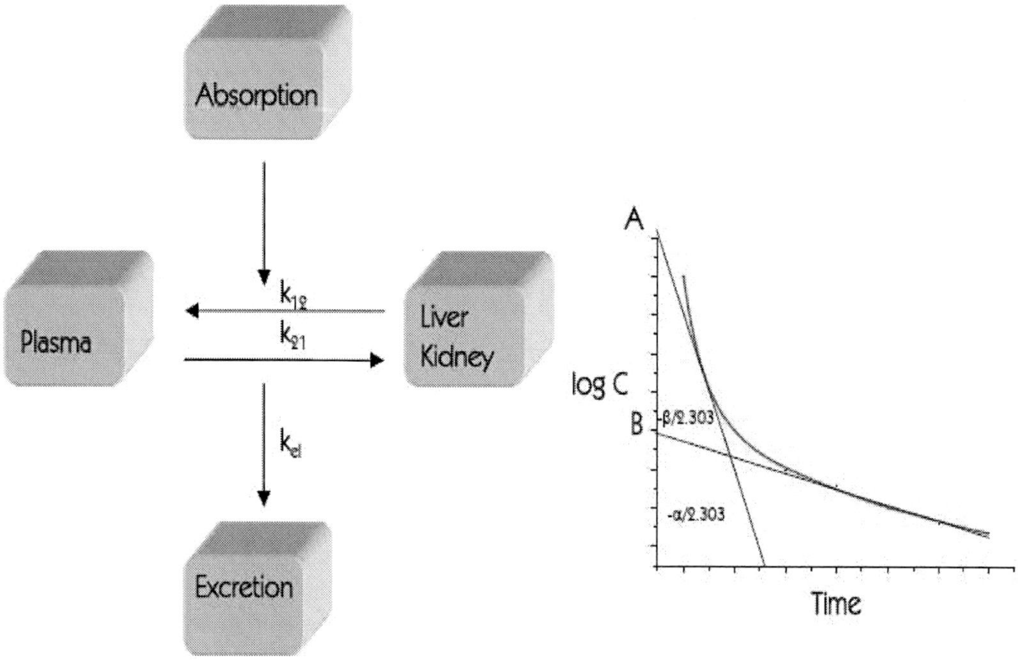

Figure 4.2. Elimination kinetics in two-compartment open model.

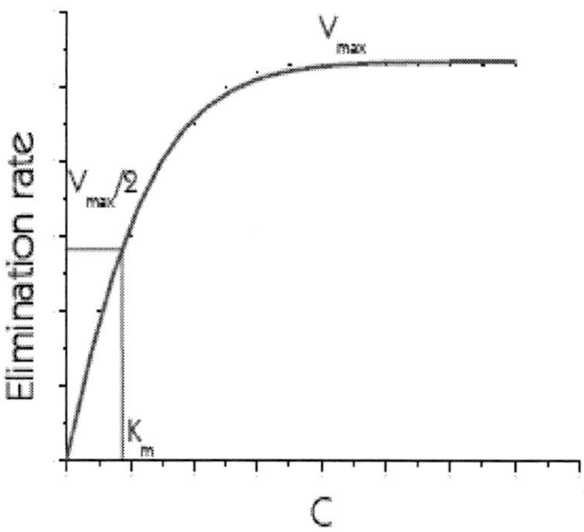

$$V = V_{\max} \cdot \frac{C}{C + K_m}$$

Figure 4.3. The saturation of elimination kinetics.

$$k_{21} = \frac{\alpha \cdot B + \beta \cdot A}{A + B}$$

$$k_{el} = \frac{\alpha \cdot \beta}{k_{21}}$$

$$k_{12} = \alpha + \beta - k_{21} - k_{el}$$

Most toxicants are eliminated via first-order kinetics. Initial excretion rate can decrease due to saturation of the excretory system. Membrane transporters have limited capacity and can be saturated for higher concentration levels (Figure 4.3.), leaving the free toxicant in the blood. For toxicant levels below k_m, elimination occurs via first-order kinetics. For higher concentrations, the kinetics resemble the Michaelis-Menten saturation curve. Determination of the k_m value is important because toxicants above this value in the body behave differently in comparison to those where linear elimination occurs. For toxic concentrations above the k_m level, the risk of severe toxic effects is most profound. For some toxicants saturation occurs at quite low concentration levels. Immediate saturation of the system occurs in zero-order kinetics, which is characterized by equal rate of elimination independent of plasma concentration. In zero-order kinetics real half-life and elimination constants cannot be defined. For ethanol, which follows zero-order kinetics, the saturation of the enzymatic system occurs quickly and the constant amount of alcohol is biotransformed at the approximate rate of 100 mg/kg/h [3]. This is one reason for rapid proliferation of the symptoms of alcohol intoxication, from euphoria to coma.

In chronic exposure to toxicants with high t_{50} values, plasma and tissue concentrations gradually increase. Repeated exposure to toxicants with high t_{50} values can cause their accumulation in the body if their elimination is slower than their rate of consequent exposure. Since equal amounts of the substance are eliminated during equal intervals, mechanisms of elimination are not harmonized with the dynamics of toxicant entrance during long-term toxicant exposures and any unexcreted amount remaining in the body and will added to the newly introduced amount. Finally a steady-state may be reached, in which a constant plasma concentration is observed. In a steady-state, the absorption and elimination of the substance are in dynamic equilibrium. It is important to note that toxic manifestations of certain compounds are more closely related to equilibrium concentrations of the free, active toxicant than on the total absorbed amount. If a toxin is being excreted as fast as it is taken in, and concentration of the free toxicant in the plasma does not reach the level of acute toxicity, the manifestation of intoxication is not necessarily observed, even though the total absorbed concentration may exceed acute toxicity levels.

References

[1] Costanzo, LS. *Physiology*. Lippincott Williams and Wilkins; 2006.

[2] Siegers, CP; Watkins, JB. *Biliary Excretion of Drugs and Other Chemicals*. VCH Publishers; 1991.

[3] Golan, DE; Tashjian, AH; Armstrong, AW. *Principles of Pharmacology: The Pathophysiologic Basis of Drug Therapy*. Lippincott Williams and Wilkins; 2007.

Chapter V

GENOTOXICITY

Genotoxic agents are agents which may damage the genetic material of the cells provoking their mutation. Severe DNA damage is usually lethal for the cell, while minor damage can be carried through many generations. Minor mutations are not necessarily obvious and always expressed on a phenotype scale, but may become evident in some situations, for example, if the two cells with the mutation at same point are recombined. Only genetic alterations that have occurred in germ cells may be transmitted to descendants, whereas mutations that have arised in somatic cells are often implicated in carcinogenesis and teratogenesis. Even though the incidence of altered genetic material to initiate carcinogesis or teratogenesis is high, these processes may also arise by completely different mechanism. Various chemical substances, some pathogens, as well as electromagnetic radiation, are potent genotoxic agents.

MUTAGENESIS

Human genome contains about 3,000,000,000 of nucleotides organized in 23 chromosome pairs. Within chromosome pairs, twenty two are autosomes and one pair is sex chromosome that determines the gender. Uncoiled, the DNA contained by each of those chromosomes measures between 1.7 cm and 8.5 cm in length. For spatial rationalization chromosomal DNA is packed into a compact structures implying specialized proteins called histones. The fundamental packing unit is known as a nucleosome, which is approximately 11 nm in diameter. The DNA double helix is wraped around an octamer, composed of eight histone protein molecules to form a single nucleosome. Another histone fastens the DNA to the nucleosome core. Nucleosomes are usually packed together, with the aid of histones, to form about 30 nm large fiber. Each chromosome has „p" (petit) or short arm, and „q" (queue) or long arm (Figure 5.1.). Chromosomes are classified as metacentric ($p = q$), submetacentric ($p<q$) and acrocentric ($p <<q$) according to the length of their short and long arms. Specific locatable region of genomic sequence, corresponding to specific heritable property, is called gene. Physical development and phenotype of the organisms depend on mutual interaction between genes, as well as interaction with the environment. Genes often contain regions that do not encode products, but regulate gene expression.

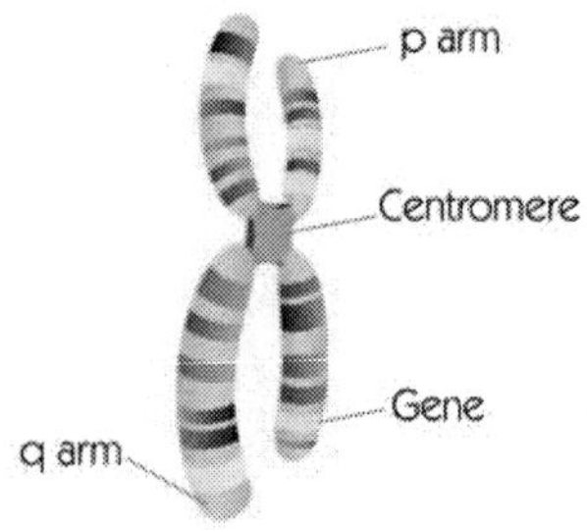

Figure 5.1. Metacentric chromosome (p = q).

Mutagenic agents are agents who alter the genetic material either at the gene level or at the chromosomal levels. In this respect small scale and large scale mutations are recognized. These changes can result in partially or completely non-functional proteins due to altered gene expression. Some mutations alter DNA base sequence but do not change the function of the protein coded by the gene.

Chromosomes (*chromo* – colored) derive their name from the fact that they can be easily stained and observed under the microscope. Large scale mutations that have occurred at one or more chromosomes can easily be spotted under the microscope after appropriate staining. If the test substance causes large scale mutations, i.e. chromosomal aberrations, the mutagenity tests performed on a cell cultures might reveal this. The test is performed by incubating the human chromosomes obtained from lymphocytes, fibroblasts or dermal cells with suspected mutagen. Cell division is stopped in the metaphase stage with the addition of colchicine. After staining chromosomal pieces can be seen under the microscope for mutagenic substances (Figure 5.2.). Small-scale mutations that occur at the gene level can not be spotted under the microscope.

Heavily mutated cells usually die during the division. Mutated gametes are usually not able to be fertilized, and if in rare cases are, formed zygote dies in the early stage of the development. Mutations are hereditary only if they have stricken the germ cells, whereas other that occur in somatic cells can not be transferred to offspring.

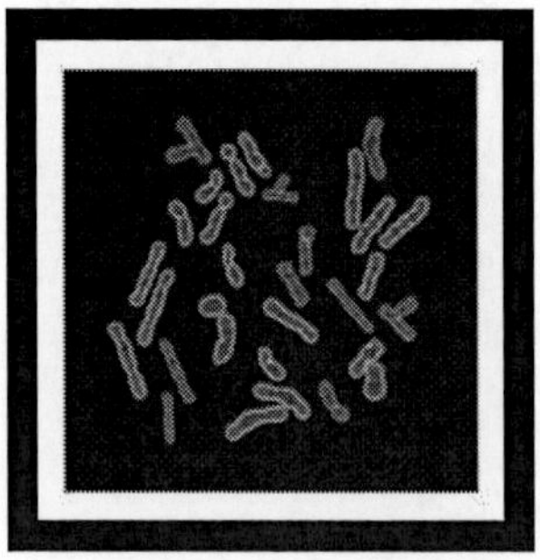

Figure 5.2. Chromosome fragmentation observed for mutagenic substances.

High level of DNA damage can be lethal for the organism. Nonlethal mutations that have occurred in germ cells can be passed to the descendants. Approximately 90 human diseases, i.e. 1/10 of all known diseases, are thought to occur as a result of one or more mutations (sickle cell leukemia, cystic fibrosis etc.). Also great number of minor mental and physical dysfunctions have the genetic background. In gene therapy that has developed in past decades, the treatment aims to fix the dysfunctional gene by introducing the corrected one. Although the technology is still improving, it has been used with some success to treat certain diseases. Corrected gene, produced by genetic engineering technology, is inserted into the genome of the specific tissue in order to replace an abnormal, disease-causing gene. To deliver the therapeutic gene to the patient's target cells a specific carrier called vector must be used. Currently, the most common types of vectors are viruses (retroviruses, adenoviruses) that have been genetically altered to carry normal human DNA. Target cells such as the liver or lung cells are infected with the vector. The vector unloads its genetic material containing the therapeutic human gene into the target cells. Functional proteins which are produced from the therapeutic gene restore the target cells to a normal state. Similar processes occur in virally-caused pathological conditions, in which encapsulated viruses deliver their content into the human cells, provoking various persistent diseases.

In the nature mutations occur frequently and spontaneously. These natural mutations are usually minor and are occurring at the gene level. As a result of a natural selection they contribute to evolution of the species. Some of the natural mutations are denoted as neutral drift, because it can not be defined whether they contribute to the development and improvement of the species and whether they are beneficial or not. When genetic material of the cell is damaged, the cell initiates natural repair system. Even though aimed to fix the damage, the result of reparation system in which cell enzymes are activated, may be more severe mutation. Apart from natural mutations, induced mutations can be provoked by various pathogens, chemical and electromagnetic agents.

Strongest mutagenic substances include purine and a pyrimidine structural analog, antibiotics who's primarily function is to disturb the DNA synthezis in bacteria, and alkilating substances that covalently bind to nucleic acids. The probability of bond formation between alkilating agent and DNA molecule is highest at the nitrogen atom in seventh position of the guanine molecule. Among chemical mutagens methylating and alkylating agents (N-ethyl-N-nitrosourea, ethyl methanosulfonate) can mutate both replicating and non-replicating DNA. These agents cause transition, transversion or deletion. On contrary, base analogs mutate the DNA as a result of the substances incorporation during the replicating process. Certain bromine compounds (ethidium-bromide) act as intercalating agents and can cause chromosomal breakage, while other (platinum) act as a crosslinker between two DNA strands. Crosslinking between guanidine pairs leads to chromosome fragmentation. Oxidative damage of the DNA molecule may also provoke the mutation and is usually caused by oxygen radicals or other strong oxidants.

Electromagnetic radiation of high energy alternates the nucleic acids. Nonionizing UV radiation excites the electrons from the DNA molecules, changing the base-paring properties. Cytosine and thymine are most vulnerable to excitation. UV light can also induce adjacent thymine bases in DNA strand to pair with each other. α-, β-, γ-radiation as well as X-rays have much higher energy that UV light and are considered more dangerous and potent in mutagenesis.

Small scale mutations are denoted as point mutations. Point mutations can result from the exchange of purine base with another purine base or pyrimidine base with another pyrimidine base (*transversion*) or the purine base may be exchanged with the pyrimidine one, and *vice versa* (*transduction*). The result of point mutations can be coding for the incorporation of the same or different aminoacid in the polypeptide, or signalization of the cessation of the protein synthesis. Both insertation and deletion of one or more nucleotides in the DNA sequence can cause the shift of mRNA reading frame or force mRNA to bind on a different site. Very common spontaneous mutations include *tautomerism*, *depurinisation* (loss of a purine base) and *deamination*. Loss of purine base intercepts the DNA chain. In *tautomerism* a base is altered due to the repositioning of the hydrogen atom. In a *deamination*, atypical base is formed, initiating the DNA repair mechanisms, however not all deamination mutations can be repaired by natural repair system. Nitric acid provokes deamination by producing hypoxanthine from adenine.

In large scale mutations, i.e. chromosomal aberrations, the whole gene region can be replaced, amplified or deleted (Figure 5.3.).

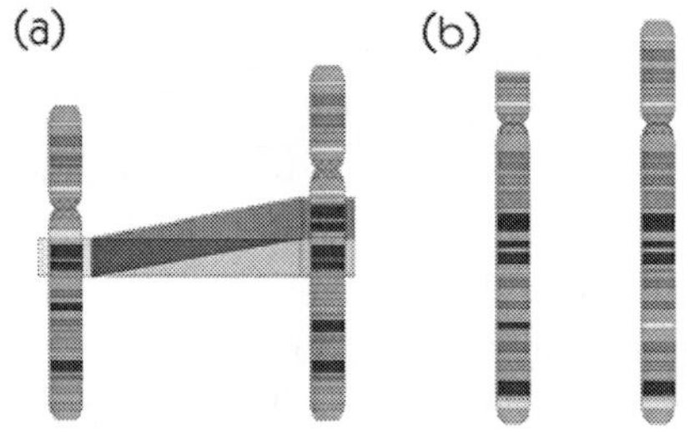

Figure 5.3. Large scale mutations. Chromosomal amplification (a) and deletion (b).

Chromosome translocation is caused by the rearrangement of the parts belonging to nonhomologous chromosomes and this mutation may be *reciprocal (non-Robertsonian)* and *Robertsonian*. *Reciprocal translocation* occurs when both “p” and “q” parts of two different chromosomes are recombined. *Robertsonian* rearrangement involves the fusion of two acrocentric chromosomes that fuse near the centromere region with loss of the short arms. The resulting karyotype in humans leaves only 45 chromosomes since two chromosomes have fused together. Most common Robertsonian translocation in humans involves chromosomes 13 and 14 and is seen in about 1 in 1300 persons. Carriers of Robertsonian translocations are phenotypically normal, but there is a higher risk of miscarriages or abnormal offspring. *Isochromosome* contains only “p” or “q” parts of the same chromosome [1]. If the orientation of the chromosomal region within a single chromosome is changed, such alteration is denoted as a *chromosomal inversion*. Large-scale mutations can also result from the loss of heterozygocity if one allele is lost either by deletion or recombination. Some chromosome abnormalities such as translocation, or chromosomal inversion, do not cause disease in carriers, although they may lead to higher incidence of having a child with a chromosome disorder.

By functional effects, large-scale mutations can be classified to *amorphic, neomorphic* and *antimorphic*. In *amorphic* mutations, which occur mostly in recessive phenotypes, new gene product has less or no function at all due to loss of the function of one allele. In dominant phenotypes mostly *neomorphic* mutations occur, in which the gene product has completely new and abnormal function. *Antimorphic* mutations have an altered gene product that acts antagonistically to the product of the normal phenotype and are specific for dominant and semi-dominant phenotypes. One of such mutations is *Marfan syndrome*, in which the defective connective tissue is produced. Instead of producing protein fibrillin, which is the structural unit of connective tissue and a support to other tissues, abnormal allele produces a glycoprotein which antagonizes with the product of the normal allele.

According to manifestations, mutations can be classified as *morphological* and *biochemical*, even though these two types are related. *Morphological* mutations are often a direct consequence of the altered biochemical pathway. They affect the outward appearance of an organism (different leaf shape, different height of the plant, smooth or rough seed etc.). *Biochemical* mutations result in stopping or changing the biochemical pathways.

Abnormal numbers of chromosomes or chromosome sets, *aneuplody*, may be lethal or give rise to serious genetic disorders. *Klinefelter`s syndrome* (XXY) carries one additional X chromosome in sex pair chromosomes [2]. Men with Klinefelter`s syndrome are usually sterile, and tend to have longer arms and legs. Boys with the syndrome are often shy and quiet, and have a higher incidence of speech delay and dyslexia. During puberty, without testosterone treatment, some of them may develop gynecomastia, an abnormally developed mammary glands. In such trisomy both boys (XYY) and girls (XXX) have learning difficulties and higher incidence of dyslexia [1]. XXX girls tend to be tall and thin. *Turner syndrome* is relatively frequent mutation and occurs in about 1 in 2500 females. Instead of normal XX sex chromosomes for a female, only one X chromosome is present and fully functional [3]. In rarer cases a second X chromosome is present but is abnormal, or both the cells containing a single and paired chromosome may be present. Sexual characteristics of females with this syndrome are usually underdeveloped. Females with Turner syndrome often have a short stature, low hairline, abnormal eye features, poor breast development, small fingernails, and are usually sterile. *Martin-Bell syndrome* (fragile X syndrome) is a genetic disorder caused by the mutation on the X chromosome [4]. The methylation of specific locus in X chromosome band results in constriction of the X chromosome which appears 'fragile' under the microscope at that point. In fragile X patients, during translation step excessive amount of protein is produced. Besides intellectual disability, prominent characteristics of the person with the syndrome include an elongated face, large ears, flat feet, larger testicles in men, and low muscle tone. Behavioral characteristics may include frequent hand-flapping, shyness and limited eye contact. *Edwards syndrome*, which is the second most common trisomy after Down`s syndrome, is a trisomy of chromosome 18. Symptoms include mental and motor retardation and numerous congenital anomalies that cause serious health problems. 90% of individuals with this mutation die in infancy. Survived persons are relativelly healthy thereafter. They have a characteristic hand appearance with clenched hands and overlapping fingers. Symptoms of trisomy on chromosome 13 are somewhat similar to those of trisomy 18, but the individuals do not have the characteristic hand shape. *Down`s syndrom* is caused by an extra copy of chromosome 21 (trisomy 21). Characteristics include decreased muscle tone, asymmetrical skull, slanting eyes and mild to moderate mental retardation. *Pallister-Killian* syndrome occurs due to tetrasomy of the twelfth chromosome. This leads to development of isochromosome 12p, made up of the two short arms of the chromosome. Symptoms include varying degrees of mental retardation, epilepsy, hypotonia and both hypopigmentation and hyperpigmentation. The persons with the syndrome exhibit a distinctive facial structure, characterized by high foreheads, sparse hair on the temple, a wide space between the eyes, and a flat nose. *Wolf-Hirschhorn syndrome*, which is caused by partial deletion of the short arm of the chromosome 4 is characterized by severe growth retardation and severe to profound mental retardation [4]. *Cri du chat,* which is caused by partial deletion of the short arm of the chromosome 5 means "cry of the cat" in French, and the condition was so-named because affected babies make high-pitched cries that sound like a cat. Affected individuals have wide-set eyes, a small head and jaw. The persons with the syndrom are moderately to severely mentally retarded and very short. *Williams's syndrome* is caused by the deletion of the genetic material from the q region at the seventh chromosome. The deleted region includes more than 20 genes. Some characteristic features, such as certain facial features and possible learning disabilities, are very close to fetal alcohol syndrome. The persons with Williams's syndrome also demonstrate some characteristics of autism. *Jacobsen syndrom* is a very rare disorder. The mutation involved is the 11q deletion. Affected individuals have normal intelligence or mild mental retardation, with poor expressive language skills. Most of them also have a bleeding disorder called Paris-Trousseau syndrome. *Prader-Willi syndrome* is another very rare genetic disorder in which seven genes on chromosome 15 are deleted. Specific features, like prominent nasal bridge, small hands and feet, high, narrow forehead and lack of complete sexual development are distinctive for affected individuals. Another mutation occurring at 15th chromosome is a neuro-genetic disorder, called *Angelman syndrome*. It is caused by the deletion or inactivation of genes on the maternally inherited chromosome 15 [5]. The condition is characterized by intellectual and developmental delay, sleep disturbance, seizures, frequent laughter or smiling, and constant demonstration of a good mood. Persons with this syndrome show delay in sitting and walking and absence or little speech. Their head size is usually below average, often with flattening at the back, and facial features include wide mouth, widely spaced teeth and prominent chin. *Rubinstein-Taybi* syndrome occurs in individuals with mutation in the gene responsible for coding the production of a specific protein that plays an important role in regulating cell growth and division in fetus, as well as for the control of other genes. Affected persons are characterized by short stature, moderate to severe mental retardation, distinctive facial features, and broad thumbs and first toes. People with Rubinstein-Taybi syndrome are at increased risk of developing tumor and leukemia.

Carcinogenesis

Cancer can be described as a pathological condition accompanied with the formation of tissue lesions and abnormal tissue development. The term *cancer* was coined by Hippocrates 300 BC due to similarity between the appearence of the neoplazmatic tissue and the crabs claws (*carcinos* - crab in Greek). The scientists agree that such tissue lesions existed even in prehistory, when the environment was free from industrial chemicals and toxic substances of antropogenic origin. Fact supporting this theory is the discovery of bone cancer in dinosaurus dating 80 millions years ago. The illnes was also identified in Egyptian mumies (3000 BC). In the past, abnormal tissue development was treated by cauterization. Later on microscope development offred more knowledge about the nature of this ilness.

Increased incidence of tumor formation during past 50 years is linked to certain external agents. Percivall Pott has pioneered this approach by recognizing high incidence of scrotal cancer in chimney workers. The condition was caused by exposure to polycyclic aromatic hydrocarbons from tar and this assumption was confirmed many years later.

Neoplasmatic lesions can be either malignant (carcinom) or benign (adenome). Tumor is a neoplazmatic lesion with well defined perimeter. Malignant cancer is characterised with its ability to metastize, i.e. to transffer the canceronous cells into another organs and tissues. The development of a benign tumor can be thretening due to mechanicall impact on the surrounding tissues. Tumors formed in blood vessels may cause the restriction of the blood flow, specifically dangerous in brain where they can provoke paralysis. In gastrointestinal tract formed tumors can disturb the digestion due to mechanical obstruction. The growth of malignant tumors is not localised. In metastasis malignant cells can infect other tissues and establish malignant growth in a different types of tissues. Metastized breast cancer, for example, can spread to bone tissue.

Carcinogenesis occurs due to impairement of the normal cell metabolism and this may be caused by different factors. One of the possible mechanisms involved in carcinogenesis is a mutation in somatic cells. Sole interaction that produces mutation is not sufficient to initiate carcinogenic process in a cell. Only permanent changes are effective in this respect. Fixation of mutation occurs through cell proliferation before the cell menages to repair the damaged DNA. Most cancers are thought to occur by one or more mutations of the somatic cells. Less than 10% of mutagenic substances, however, are shown to be carcinogenic and most potent carcinogens don`t express mutagenic activity in animal studies. On the other hand, many substances, such as ascorbic acid and caffeine, that demonstrate carcinogenic effect in test animals, do not express similar effects in humans. Other mechanisms of carcinogenesis include chemical, mechanical or electromagnetic damage of the cells integrity. Specific pathogens may also disturb cell metabolism and provoke the formation of neoplasms. Carcinogenic chemicals act by different mechanisms. In this process the reaction of the substance with specific cell receptors or cell proteins participating in cell metabolism, may be crutial. After binding of a chemical carcinogen to specific part of the cell, the cell metabolism becomes disturbed and it starts to devide in a chaotic way. Cells susceptability to certain carcinogenic agent is often genetically determined. Theoretically, a single mutated cell or a cell with altered metabolism, may initiate tumor formation, however the experts in this field feel that for tumor formation certain critical number of similar cells in the proximity must exist. Most of known carcinogens demonstrate organotropism, i.e. increased suspectability to tumor formation in specific organ or tissue.

Chemical carcinogens are usually very reactive and can react directly with the cell receptors. Such substances initiate carcinogenesis and are denoted as *primary carcinogens*. *Procarcinogens* are not directly involved in carcinogenesis, but are activated in the body by biotransformation reactions. *Promotors* of carcinogenesis are not able to initiate abnormal cell metabolism itself, however they promote and speeden the processes initiated by primar carcinogens. These substances can significantly increase the proliferation of the cancerous cells even if taken long time after the introduction of the primary carcinogen. Promotor effects are reflected in increased number of tumors and in shortening the period between the initiation and tumor expression. Promotors are efficient only when applied simultaneously or after the real carcinogen. If promotor is applied on the skin of experimental animals before the carcinogen, carcinogenesis may be ommited. However, if the promotor is applied even long time after primary carcinogen, tumor formation is much more profound in quantitative sense. It is important to notice that even substances, which seem safe, can activate the latent demage of the cells and may force ilness expression. Various esters of phorbol, isolated from croton plant, have important biological properties, the most notable of which is the capacity to act as tumor promoters through activation of protein kinase. Certain substances when introduced into the body simultaneously or just before the real carcinogen, significantly increase the tumour expression. These substances are known as *cocarcinogens* and the mechanism of their action can be different. They possibly increase the entrance of the primary carcinogen into the cell or they might inhibit natural cellular reparation processes. Cocarcinogenic action may also arise due to enhanced activation of procarcinogens. Another possible explanation of cocarcinogenic activity is the conversion of initiated cellular lesions to permanent changes. Cancer of respiratory system caused by asbestos is five fold more probable in smokers. Thus, the components of cigarette smoke and asbestos act as cocarcinogens. Similarly, smoking can cause severe carcinogenesis in the presence of uranium dust. Numerous examples of other joint activities and cocarcinogenic interaction exist. Silica induced tumor genesis is cocarcinogenically supported with benzpyrene and this is demonstrated in increased tumour quantity. In some cases the substances can act both as promoters and cocarcinogens. This hidden nature of many substances, which are not considered as primary carcinogens, shouldn't be neglected and should be included in basic toxicological studies.

The shape of the dose/response curve for carcinogenic substances is very specific, having an intercept at "effect" axis. Such form is related to inexistence of NOEL (Not Observable Effect Level). Even the lowest dose of carcinogenic substance has the impact on the body. Every dose of the substance leaves the irreversible damage (effect) in the body which is summarized with the next one in subsequent exposure, even if the period in-between two exposures is long. This fact distinguishes the toxicological consideration of carcinogenic substances from other toxic substances. In toxicants that accumulate in the body, the toxic effect depends on the concentration of its free, unbound form. For carcinogenic substances, remained free amount of the toxicant is less indicating and important, because the effect that is provoked earlier remains in the body "memorized" and is summarized with the next one. Carcinogenic effect is also characterized by long latent period from the initiation of the process to tumour expression.

Many naturall substances are well known carcinogens. Saffrol is a component of many plants (*Sassafras*) and in experimental animals it was demonstrated that it induces liver cancer. Saffrol was also detected in pepper, which becides this substance also containes another known carcinogen, the pyperidine analoque. Furanocoumarins are class of toxic and carcinogenic organic compounds produced by variety of plants, primarily species of *Apiaceae* and *Rutacea* among which are celery and parsley [6]. They are produced by plants as a defense mechanism against insects and mammals. Some of the furanocoumarins are photoactive, so their toxicity is enhanced in the presence of the ultraviolet radiation. Activated by UV radiation they provoke the crosslinking between two DNA straines, initiating carcinogenesis.

Some mycotoxins, like aflatoxin B1, exibit potent carcinogenic effect in experimental animals. In 1988, the IARC (International Agency for Research on Cancer) has placed aflatoxin B1 on the list of human carcinogens. This was supported by a numerous epidemiological studies that have confirmed a positive association between dietary aflatoxin intake and the incidence of liver cancer.

Carcinogens may be found among organic and inorganic chemical substances, among fibres and certain materials, as well as among pathogens. Polycyclic aromatic hydrocarbons, notrosoamines, alkylating substances (epoxides, β-lactones, ethylenamine derivatives), heterocyclic amines etc. are well known potent organic carcinogens.

Polycyclic aromatic hydrocarbons are formed during incomplete combustion of organic material such as fossil fuels and during pyrolysis of food components. Benzo(a)pyrene, one of the most potent carcinogens, has been detected in buiscuit and bread crust, broiled and barbecued meat and roasted coffee. Broiled hamburger containes about 43 ppb of PAH, charcoal-broiled meat 200 ppb, while in lean beaf lowest amount of benzo(a)pyrene (3 ppb) is produced during processing. Cigarete smoke containes about 1 ppt of PAH while about 4000 μg/1000 m^3 is emitted into the city air from car engine axhaust [7]. Major site for PAH induced carcinogenesis is the skin, bladder and gastrointestinal tract. Chronic PAH inhalation from tabacco smoke is shown to be implicated in cancer of respiratory system. Dietary exposure to PAH may induce cytochrome P450 and increase the sensitivity to subsequent exposures. During PAH metabolizm very reactive epoxides are formed and carcinogenic effects arise from the interaction between these reactive metabolites and DNA or RNA.

During food processing other carcinogenic substances may be formed. Whereas PAHs are produced as pyrolytic products of mostly fats, the formation of heterocyclic amines is implicated in processing of food which contain nitrogen compounds. Pyrolysis products of carbolines, quinolines and quinoxalines may also be a source of heterocyclic amines. Heterocyclic amines are mostly produced from aminoacids, such as tryptophan, glutamic acid, phenylalanine and lysine, but may also be formed from other nitrogen-containing food components. The products of aminoacid decomposition may be detected in broiled beef, fish, bread crust, fried potatoes and coffee. Major target organs of carcinogenesis trigered by heterocyclic amines are liver, small and large intestine, oral cavity, lungs, blood vessels, skin and mamary glands. The metabolizm of heterocyclic amines include cytochrome P450 activation by N-hydroxilation. Further biotransformation proceeds by O-acethylation or O-sulfonation. These intermediars form DNA adducts with guanosine in tissues of the target organs. Unreacted amount of heterocyclic amines participates further in phase II reactions and the products are excreted with urine and feces.

Nitrosoamines are another important group of food borne carcinogens. In food they are produced during food processing from nitrates and secondary and tertiery amines, whereas secondary and tertiery amines are indirectly formed from the food proteines. Nitrosoamines can be detected in various food and beverages, as well as in cigarette smoke. Nitrites can also contribute to the formation of nitrosoamines, because they are added to canned food together with nitrates, and can be easily oxidized during food processing to nitrates contributing to enchanced nitrosoamine formation. Ascorbic acid, as reductive agent, prevents the formation of nitrosoamines and reduces carcinogenesis incidence. Other antioxidants have similar effect. Due to high concentrations of nitrites and nitrates in some meat products, where they serve to preserve the red color, the adition of ascorbic acid is highly recommended, especially to bacon and other pickled meat products which contain very high concentrations. Nitrosoamine induced carcinogenesis is mostly related to tumor formation in liver, kidney, bladder, pancreas and respiratory tract. In the body, the activation of nitrosoamines is perfomed by cytochrome P450 and metabolic process involves oxidative N-dealkylation. The major metabolite, diethylnitrosamine, exibits powerfull carcinogenic activity and is the direct cause of the carcinogenic potential of nitrosoamines.

Alkilating substances, such as epoxides and β-lactones, are not naturally present, but are also common intermediat products of the chemical industry. People may also be exposed to this chemicals by using some chemotherapheutics. Alkilating substances react with nucleophylic compounds in the body, acting in a similar way as a ionising radiation.

Radon is a natural gas and it represens the source of ionizing radiation that can be implicated in carcinogenesis. This radioactive gas arises from the Earth crust and is present throughout the environment. General effects of radon to human body are caused by its radioactivity. As an inert gas, radon has a low solubility in body fluids and is distributed uniformly throughout the body. Radon absorbed at solid particles can be inhaled. In the lungs, radioactive decay takes place, followed by the emission of α-particles, which cause the formation of free radicals and DNA breakage. Inhaled radon is known strong pulmonary carcinogen.

Asbestos mineral fibers have been commonly used In the past in a variety of building constructions as an insulating material and fire-retardant. Today the use of asbestos is banned due to recognized adverse influence on health Chronic exposure to asbestos fibers has been linked to increased risk of chest, abdominal and lung cancers, as well as other lung diseases, such as mesothelioma and asbestosis. Smokers have higher risk of developing asbestos-induced lung cancer. After inhalation, asbestos fibers remain and accumulate in the lungs physically damaging the tissue. Similarly, some evidence of carcinogenesis with primarly mechanical mode of action was observed in *in vitro* studies for crystalline silica. *In vivo* studies were in accordance and have shown that crystalline silica is carcinogenic in rats and that tumor types appear to vary according to the route of administration.

Some elements, such as hehavalent chromium, nickel and arsenic are known carcinogens but may also act as progression agents. Progression is a phase when benign tumor begins to act as a malignant.

For some viruses, such as Human Papillomavirus, Hepatitis B and C, Epstein-Barr, HIV etc. is suggested that they express carcinogenic effects under certain conditions. 70-80% of all uterus cancers are considered to be caused by Human Papillomavirus. Hepatitis B i C are most likely involved in the formation of liver and kidney cancers. Epstein-Barr virus and viruses of a herpes family are linked to nasopharyngeal carcinoma and certain types of lymphoma.

Carcinogenesis may also be induced by physical interaction with tissue which may disturb cell metabolism. Implantation of an inert object under the skin of experimental animals provokes the formation of sarcom, independently on a chemical composition of the object, whereas same material applied in the form of powder doesn`t induce carcinogenic effect.

Carcinogenesis is a complex process that depends on many factors, both internal and external. Individual hormonal and immunological statuses, as well as hereditary factors are dominant endogenous factors. Dietary habits, stress, smoking, diseases and environmental pollution may significantly contribute to carcinogenesis. It should be noted that this complex process is a result of combined simultaneous impact of many different parameters, among which many of them remain unrecognized and undefined. In trying to define carcinogenesis, the process can be described only in respect to probability and it is immeasurable which influence is principal with the most profound effect.

TERATOGENESIS

Teratology (*teras* – monster, *logos* – science) is a science which deals with causes, frequency and development of congenital malformations. Through history deformed persons were faced with different social attitudes and reactions, significantly varying in different cultures. Such persons were frequently rejected and completely excluded from the society, whereas in many cultures deformed persons were considered to posses extraordinary skills and were highly respected. Birth defects occur in 3-5% of newborns and their causes remain mostly unknown. Only about 5% of congenital anomalies are recognized to occur due to maternal exposures to certain exogenous agents, i.e. teratogens [8]. A teratogen is an agent that produces a permanent alteration in the structure or function of the organism, which was exposed during embryonic or fetal life. Teratogenic effect should be distinguished from embryotoxic effect, which is manifested as fetal death. During organogenesis (3 – 7 week) growing fetus is most susceptable to morfological malformations of organs, tissues and sceleton. In subsequent period, during hystogenesis (7 – 12 week) and funtional maturation (8 - 38 week), teratogenic substances provoke mostly slow growth and functional disorders. Depending on the type and severity of the anomaly, the damage can be reversal or irreversal. Minor anomalies can be fully or partially repaired during child`s growth and development.

Detailed study on placental structure and transfer mechanisms across placenta, revealed that fetus is not completely protected from the toxic substances and viruses in the uterus, despite mother's ability to metabolize and readily excrete the foreign substance. Most xenobiotics and many viruses (cytomegalovirus) cross placental barrier by one of the natural mechanisms aimed to transfer the nutrients to fetus. The implied transport mechanism is dependent on the physico-chemical properties of the substance. Nevertheless, it is certain that existence of placenta limits the transfer of the foreign substances to fetal blood and reduces the chances of adverse effects on developing fetus. The protective properties are related to multilayed structure of placenta, in which syncytiotrophoblast layer plays an important role. Furthermore, the metabolic capacity of placental cells (trophoblast, macrophages) contribute to barrier function by biotransforming some substrates. The functionality of trophoblast is particullary active in respect to biotransformation of some hormones (corticosteroids), and nutrients (amino acids, iron) transffer.

Different mechanisms of teratogenesis can be recognized. Becides chemical substances, other agents, such as pathogens and ionizing radiation may provoke fetal anomalies. Some substances cause the gene mutations and chromosomal abberations in germ and somatic cells of the fetus, while others disturb the cell division processes or DNA synthesis and function. Abnormal cell growth or proliferation, as well as cell death or migration, may be in the core of teratogenesis.

Metabolic misbalance in mother, such as diabetes, folic acid defficiency and tumors, strongly influence fetal development and can be a cause of mental retardation and congenital malformations in newborns. Some studies have shown an increased risk for defects of fetal neural tube in babies of women who were exposed to hyperthermia in early pregnancy.

Hyperthermia most often occurs from a fever. Extremely heavy exercise or prolonged exposure to heat sources such as hot tubs, very hot baths, or saunas can also raise the body temperature. Studies have suggested the increased risk for miscarriage and possible associations between high fever and heart and abdominal wall defects in fetus. Phenylketonuria is an inherited disorder that influence fetal development and course of pregnancy. In this condition increased amount of phenylalanine is observed in mothers blood, influencing high concentrations of the amino acid to be transffered to fetus. Babies born from mothers with high levels of phenylalanine have a high risk of mental retardation. These infants may also grow more slowly and may have heart defects or other heart problems, microcephaly and behavioral problems. Thyroid hormones are critical for the development of the fetal and neonatal brain, as well as for many other aspects of the fetal growth. Endemic cretinism in pregnant women, caused by severe iodine deficiency, is a major cause of mental retardation in children in some parts of the world (Asia, Africa, Latin America). The condition can be prevented by iodine supplementation before conception.

Workplace exposure during pregnancy and the likelihood of fetal abnormalities resulting from such exposures is studied by occupational reproductive toxicology. Environmental pollutants (PCBs, diethylstilbestrol, methyl mercury) and occupational chemicals (organic solvents, arsenic, pesticides) may be implicated in the formation of congenital defects. Lead is known to cause spontaneous abortion and stillbirth and some evidence exist that exposure to very high levels may cause the damage to the fetal central nervous system. After the birth, toxic agents may be transferred from mother to child via breast milk. Workplace lipophylic contaminants are concentrated and excreted in the breast milk. In addition, special precaution should be taken in drug therapy during pregnancy, since most of the drugs (lithium, captopril, enalapril, aminopterin, 13-cis-retinoic acid, coumarin anticoagulans) cross the placenta and posses very potent teratogenic effect. Antibiotics, such as penicillamine and tetracyclines, should be avoided during pregnancy. In addition, the use of any drug should be carefully revised and balanced in respect to expected benefits and potential teratogenic risk. The use of any xenobiotic (cocain, trimethadione) or hormone supplements is contraindicated in pregnancy. The list of potential chemical teratogens is very long and is given in Appendix I [9]. The presence of the substance on the list doesn`t certainly mean that the compound produces teratogenic effect, however the use of listed chemicals should be taken with particular precaution. *Vice versa* is also truth, i.e. the absence of the compound from the list doesn`t claim this chemical as a safe to be used during pregnancy. Diet, dietary supplements and overall habits during pregnancy, should be carefully planned and rewieved, because even common substances, such as caffeine, piperidine (pepper), folic acid, vitamine A or oxygen, posses potential for provoking teratogenic effects and pose a risk in certain concentrations or in combination with other factors. For many substances (nicotine, aspirine, contraceptives, agent orange) suspected teratogenic effect still can not be claimed because it wasn`t prooved due to unsufficient unambigouisy of the animal test results. After FDA approval of specific substance, it is expected that within 6.0±4.1 years a potential teratogenic risk in humans will be revealed and 9.1±4.5 years are sufficient to confirm the safety in humans. Animal teratology studies are valuable but false positive (chlorpheniramine, hydroxyzine, propoxyphene) and false negative (captopril, enalapril, carbimazole, methimazole, misoprostol) results can be obtained.

In 1953 small pharmaceutical company, while searching for inexpensive method of manufacturing antibiotics, has synthesized thalidomide, a drug which didn't demonstrate any antibiotic activity. However, it was noticed that lethal dose for the compound is very high, so it was promoted as a safe drug. It was prescribed first as an anticonvulsant and soon it was sold in 50 countries under 40 different names as an antiemetic. The drug was prescribed to pregnant women to beat morning sickness and as an aid to sleep. As a consequence about 8000-12000 infants were born with deformities caused by thalidomide, and only 5000 survived beyond childhood. Babies were born with phocomelia – abnormally short limbs, with malformed internal organs, ears and eyes. The defects weren't hereditary, since thalidomide provokes teratogenesis by mechanisms other than mutagenesis. In repeated animal studies, teratogenic effect of thalidomide was demonstrated only in some species of monkeys and rabbits. After decades of research other physiological effects of the drug were revealed. The drug is now used in combination with dexamethasone for the treatment of multiple myeloma, Behcet`s and Crohn`s disease and other cancer diseases, where chemotherapy is least effective.

Adverse effects of chemicals on developing fetus and strong potential for provoking birth deformities were demonstrated in mass teratogenesis that has occurred during Vietnam War. Great number of malformed babies was born as a result of high quantities of Agent Orange that were used in Vietnam. Agent Orange was a 50:50 mixture of two phenoxy herbicides, defoliants, which were used to defoliate the trees in order to prevent hiding of the enemy. It is believed that strong teratogenic effect was caused by dioxin, TCDD (2, 3, 7, 8 – tetrachlorodibenzo-para-dioxine), the contaminant of the herbicide. Teratogenic effect of sole Agent Orange still hasn't been confirmed.

Fetal exposure to radiation may induce growth impairments, malignancy, and chromosome fragmentation. Preimplantation exposure of embryo most probably leads to fatal outcome, even for very small radiation dose, such as 10 rads and in the past irradiation with X-rays was used to provoke abortion [8]. Fetus may be exposed to X-rays during Röntgen diagnosis or in some incidental situations such as nuclear weapon. Impaired cell division, cell death, and various mutations may be observed, depending on the radiation dose. Newborns which were exposed during embyonic or fetal life usually demonstrate growth impairment, which can possibly improve during the life. Brain and eyes are mostly affected and microcephaly, hydrocephaly, optic atrophy and cataracts are common. Maternal exposure to ionizing radiation increases the risk of leukaemia in the child. Less common effects are skeletal and genital abnormalities, as well as malformed internal organs. Small doses of radiation may alter developing germ cells, thereby postponing obvious defects to next generation.

Many naturally occuring agents may potentially induce teratogenic effect. Severe birth defects, including neurological damage and fusion of two eyes into one, were observed in sheep and cattle. It was established that certain active compound from *Veratrum californicum* plant, naturally growing in meadows mountains, was the cause of teratogenesis that has occured in cattle. The mechanism underlying these defects still remaines unclear and should be elucidated, as in the case of many other teratogens.

Some naturally occuring parasites, viruses and bacteria, such as cytomegalovirus, herpes virus hominis I i II, rubella, siphylis, Venezuelan encephalitis virus etc., have strong teratogenic potential. High incidence of deformed frogs has been reported in North America, speculating the impact of some chemical environmental pollutant, ultraviolet radiation or parasites. Deformities were related to limbs and usually completely missing limb or three legs were observed in frogs. After years of reaserch it was prooven that the deformities were caused by the trematode parasite. It is assumed that this parasite and many others, such as toxoplasma parasite, provoke similar effects in humans.

Animal studies may offer some hints in defining the teratogenic potential of the substances. In such experiments extrapolation of the experimental results to humans is seriously limited. The results may vary depending on the experimental animal, and false results can be observed, such as in the case of thalidomide which wasn't demonstrating teratogenic effect in first animal testing. For these reasons toxicological studies directed to revealing the teratogenic effect of the test substance are performed on several animal species. The introduction of the test substance begins after the implementation of the fertilized ovum. In this respect, seventh day is the earliest that the teratogen can induce malformation. Usually four doses are administrated. The highest dose shouldn't cause severe toxic effects to mother. Usually one positive and one negative group are included into the study. Well known teratogen, such as vitamin A or trypan blue, is administrated to the positive control group. Many factors influence the teratogenicity course, such as dose, route of administration, frequency of exposure, duration of exposure, nature of the agent and gestational timing. Concurrent exposures, concurrent illness and genetic susceptibility of both mother and fetus strongly influence the final outcome.

Teratogenic testings precede with reproductive and developmental toxicity tests, in order to form an overall picture about the substances toxicity. Reproductive tests are of high significance for the modern population because of increased number of sterility, possibly arising from the damaged germ cells by ubiquitous chemicals. Heavy metals (cadmium, lead, mercury, arsenic), pesticides (DDT, parathion, dichlorvos, 2, 4, 5 – T), industrial chemicals (PAH, hydrazine), quinine, caffeine, thiazide have demonstrated strong adverse effects particularly on the male reproductive ability. Reproductive testings are three-generation tests in which the effects of the test substance on fertility, on the developing fetus and on the mother and her offspring are monitored. After exposure transplacentally and via milk, adverse effects may arise in offspring, influencing their growth, development and sexual maturity. Reproductive system starts developing from the embryonic stage and continues to sexual maturity. In the period of gonads formation the fetus is very susceptible to the influence of various substances. Substance that is considered toxic on the reproductive level effects gonadal function, mating behaviour, conception rates or implementation of the fertilized ovum. Substance may also be classified as reproductive and developmental toxicants on the basis of adverse influence on lactation and nurturing. The tests directed to revealing the reproductive toxic potential of the substances are complicated by the lack of adequate experimental models, long duration and by difficulties of data interpretation.

References

[1] Gardner, RJM; Sutherland, GR. *Chromosome Abnormalities and Genetic Counseling*. Oxford University Press; 2004.

[2] Obe, G; Natarajan, AT. *Chromosomal Aberrations: Basic and Applied Aspects*. Springler Verlag; 1990.

[3] Ford CE ; Jones KW ; Polani PE ; de Almeida JC ; Briggs JH. A sex-chromosome anomaly in a case of gonadal digenesis (Turner's syndrome). *Lancet* 273 (7075), 1959, 711–713.

[4] Dykens, EM; Hodapp, RM; Leckman, JF. *Behavior and Development in Fragile X Syndrome*. Sage Publication; 2003.

[5] Marga, H. *Living with Genetic Syndromes Associated with Intellectual Disability*. Jessica Kingsley Publisher; 2001.

[6] Baer-Dubowska, W; Bartoszek, A; Malejka-Ciganti, D. *Carcinogenic And Anticarcinogenic Food Components*. CRS Press; 2005.

[7] Gelboin, HV; Paul On Pong Ts'o, POPT; Andrews, LS. *Polycyclic Hydrocarbons and Cancer*. Academic Press; 1981.

[8] Schardein, JL. *Chemically Induced Birth Defects*. Informa Healthcare; 2000.

[9] Sax, NI; Lewis, RJ. *Dangerous Properties of Industrial Materials*. Van Nostrand Reinhold; 1989.

Chapter VI

ALLERGIC REACTIONS

The term *allergy* was coined in 1906 by a Viennese pediatrician, Clemens von Pirquet (1874-1929). The term derives from the Greek words *allos* meaning other and *ergon* meaning work. Allergy thus refers to an altered reaction, i.e., the exaggerated reaction of the immune system in response to contact with exogenous substance. Allergens, which may provoke abnormal response of the body, are substances that are foreign to the body. Pollen, dust mites, moulds, dander and specific foods are common allergens for many people. With the exception of specific chemical substances, such as drugs, the majority of allergens are proteins, but not all proteins are allergens. In general, only relatively large proteins may trigger the allergic response. The proteins that provoke allergic reaction are often food born or part of a microorganism's hull. Smaller proteins called haptens, which cannot individually trigger the allergic response, may become real allergens upon binding to an endogenous protein in the body.

The frequency of food allergy is highest in childhood and decreases with the age. During infancy, the gastrointestinal epithelial membrane is immature and allows more proteins to be absorbed from intestines into the blood, where they can provoke allergic response. With age, the mucosal barrier matures and becomes more efficient in digestion and less permeable.

In it most general sense, the term allergy refers to several kinds of immunological reaction, as well as some specific non-immunological responses. Immunological response may be either mediated by immunoglobulin E (IgE mediated) or non IgE mediated, but real immunological allergy, denoted as type I hypersensitivity, is always IgE mediated (Figure 6.1.). A non-immunological body response may have manifestations similar to an immunological allergy, but their mechanisms are different. Final manifestations of allergic responses can be instantaneous or may arise after a prolonged period .

An allergic response may be manifested at specific organs and tissue, or may strike the body systematically, provoking anaphylaxis. Localized response may be demonstrated in the gastrointestinal, respiratory or cardiovascular system, or on the skin. Respiratory reactions include throat tightness, wheezing, asthma, shortness of breath, repetitive coughing and runny nose. Gastrointestinal symptoms include tongue and throat swelling, lip and mouth swelling, nausea, vomiting, diarrhea, abdominal pain, cramps and colic. Skin reaction is accompanied with itching, flushing, rash, tissue swelling, dermatitis and eyes and ears itching. Hypotension, arrhythmia and chest pain related to the affected cardiovascular system may also occur.

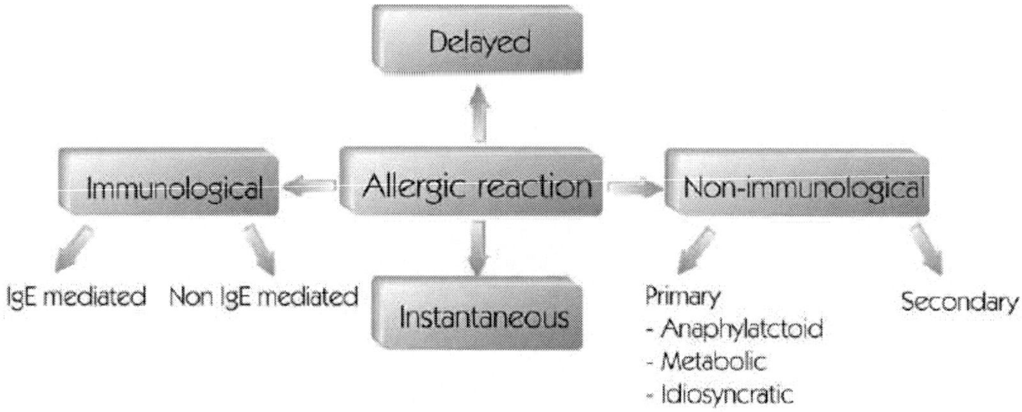

Figure 6.1. The types of allergic responses.

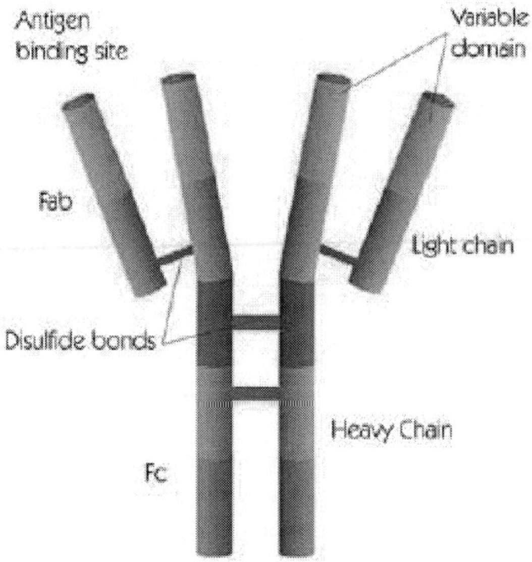

Figure 6.2. The structure of immunoglobulins.

IMMUNOGLOBULINS

There are numerous ways by which the body fights invading pathogens and each antigen triggers different immune mechanisms. Pathogens and viruses that succeed in entering a cell, such as HIV virus, may be protected from the immunological attack, thus, hindering treatment. The ability of some antigens to hide may also explain the chronic nature of many skin diseases.

Antibodies, or immunoglobulins (Ig), are glycoproteins found in blood, tissue fluids and body secretions, and are part of immune response, important in identification and neutralization of bacteria, viruses, cancerous cells and other foreign bodies. If separated from other plasma proteins by electrophoresis, immunoglobulins are located in the cell's γ-region (gamma region). Antibodies are involved in immunological response in many different ways. They may block viruses directly, preventing them from infecting the cells by binding to their receptors, or they may act in more complex way in a cascade of biochemical reactions. Immunoglobulins are also important for activating specific cells, such as mast cells or phagocytes, which precede immunological battle, and the binding of immunoglobulins to membrane receptors initiates the degranulation (release process) of mast cells or phagocytosis, the process the human body uses to destroy dead or foreign cells.

Antibodies (immunoglobulins) are synthesized and secreted by plasma B cells. Only an activated cell bound to specific antigen has the ability to produce and secrete immunoglobulins. The interaction of B cell with a T helper cell is also prerequisite to complete cell activation.

Immunoglobulins are "Y"-shaped molecules, consisting of two heavy and two light polypeptide chains, bonded by disulfide bridges (Figure 6.2.). Heavy polypeptide chains are composed of between 450 and 550 amino acids. Each heavy chain has a constant region which is the same for all immunoglobulins of the same class, but differs between classes of immunoglobulins, and a variable region, which is the same for all immunoglobulins produced by the same B cells. The approximate length of a light chain is between 211 and 217 amino acids, and each light chain has one constant and one variable domain. Two regions of the immunoglobulin are important for performing its function: an Fc region and a Fab region. The Fc region (fragment, crystallizable) enables the binding of the antibody to various cell receptors, complements proteins, and mediates different physiological effects. The Fab region (fragment, antigen binding), composed of one constant and one variable domain of both heavy and light chain, defines the paratope, which is the the antigen binding site. These two variable domains bind the epitope, or recognizable fraction, of their specific antigens.

The N-terminal of the antibody is glycolylated, but in some immunoglobulins (IgA1 and IgD), the sugar molecule may be bound to O-terminal as well. The variability of variable domains of the immunoglobulins is ensured by the ability of an activated B cell to mutate and adjust to a newly encountered antigen. This is one reason that the immune system is able to respond to an enormously large number of foreign objects, i.e., antigens. Each antibody recognizes a specific antigen, specific to its paratope, and acts like a lock to a specific epitope, or a key, located on the antigen.

A secreted antibody may be dimeric, tetrameric or pentameric, with a corresponding number of associated monomer "Y" unit. In mammals, five types of immunoglobulins exist. They are labeled IgA, IgD, IgE, IgG or IgM, according to differences in their heavy chain constant domains. In humans, four subtypes of IgG and two subtypes of IgA are present. Each immunoglobulin class differs in its biological properties and in the type of antigens that it deals with.

IgG is a major class of immunoglobulins in the human body, representing about 80% of all immunoglobulins. The four forms of IgG initiate the majority of immunological responses against pathogens by their binding to phagocytes and to components of the complement system. IgM is secreted from B cells in pentameric form and is bound to its surface. This class of immunoglobulins plays a role in the early stage of immunological response, and is most active until a sufficient amount of IgG is produced. Present in the blood and lymph system, IgM functions by binding to phagocytes. IgA represents about 6% of all bodily immunoglobulins and is found mainly in areas containing mucus, such as the gastrointestinal, respiratory and urogenital tracts, as well as in blood and lymph systems. It can be monomeric or dimeric and their main function is to prevent the colonization of the mucus membranes by pathogens. IgD, present in about 0.2%, is a monomeric immunoglobulin, located on the membrane of naive, immature B cells. IgD functions mainly as an antigen receptor. Real allergic reaction, or type I hypersensitivity, is triggered by IgE, but this class of immunoglobulins is also important for providing protection against helmints (parasitic worms). Immunoglobulin E is present in blood and the lymph system in very low concentrations (0.05%) and during sensitization they bind to the membranes of basophiles and mastocytes. High levels of blood IgE reveal that the body is sensitized to a certain antigen or may indicate a parasitic infection. Of all the immunoglobulins, IgG are present in the highest concentrations in blood, and they also circulate for the longest period, having a half-life of 7 to 21 days. The shortest half-life, observed for IgE, is about two days. Other immunoglobulins have half-lives of 6, 10 and 3 days, for IgA, IgM, IgD, respectively [1].

Immature B cells express on their surface only IgM, which remains bonded to the membrane. Upon the maturation of B cells, both IgM and IgD are produced. B cells become activated upon binding to antigen and begin to divide and differentiate into the cells that produce and secrete antibodies, rather than keeping them bonded to the surface. The daughter cells may produce different antibodies, IgD, IgA or IgG, depending on their heavy chains, giving rise to a variety of defense agents. Another mechanism for generating antibody diversity is related directly to mature, activated B cells. Activated B cells are prone to somatic mutations in the genes, thus determining the structure of the variable, both light and heavy chains of secreted antibodies. By binding their specific antigens (protein fragments of the virus hull or foreign molecule), formed antibody-antigen products may agglutinate and subsequently be precipitated. Agglutination products may block viral receptors or may be primed for phagocytosis by macrophages. After the agglutination process, other immune responses, such as complement pathways, may be initiated.

The complement system is another important part of humoral immune response (HIR), representing a biochemical cascade whose participants are different proteins. Recognition that certain components of blood serum possess antiseptic properties occurred in the late 19th century. The heat-labile component of blood serum demonstrated non-specific antimicrobial activity and loss of effectiveness with heating, while the heat-stabile component was found to confer immunity against specific microorganisms. The heat-labile component constitutes what is called the complement system, which is another important protective mechanism for the body.

The complement system consists of a number of small proteins found in the blood, which disrupt the target cell's membrane. Over 20 proteins and protein fragments constitute the complement system, including serum proteins and cell membrane receptors. These proteins are synthesized mainly in the liver and they account for about 5% of the serum globulins. The complement system is not adaptable and cannot be influenced by external factors. The system acts to protect the body in cooperation with specific antibodies or independently in a non-specific way. The activation of this immune response can be achieved in one of the three mechanisms: classical complement pathway, alternative complement pathway and lectin pathway.

The classical complement pathway typically requires antibodies for activation means that it is a specific immune response. The alternate complement pathway may be activated by hydrolysis of membrane complexes or by an antigen itself (non-specific immune response). The lectin complement pathway assists immunity also in a non-specific way. The complement system is regulated by complement control proteins, which are present in much higher concentrations in blood plasma than the active complement proteins, preventing host cells from being destroyed. It is speculated that a dysfunctional complement system might play a significant role in many diseases with an immunological component, such as asthma, arthritis and many autoimmune diseases.

The classical complement pathway is triggered by binding the proteins of the complement system to a formed antigen-antibody complex or directly to the surface of a pathogen. This binding provokes conformational changes in protein components, which activate other proteins (enzymes) of the complement cascade. The result of alternative pathway response (membrane attack pathway) is the formation of the protein complex directly on the cell membrane of the pathogen. The formed protein complex acts as a transmembrane channel and causes cells collapse (cell lyses). The alternative pathway is triggered by hydrolysis of one of the complement proteins (C3) directly on the surface of a pathogen. The lectin pathway acts like the classical pathway, but it is activated by binding mannan[1]-binding lectin to mannose residues on the pathogen surface.

Immunological Reactions Mediated by IgE

Real allergy, or type I hypersensitivity, is provoked by excessive activation of mast cells and basophils by IgE. The principal difference between type I hypersensitivity to allergen and normal humoral response against a foreign body, is the favorable secretion of IgE by plasma cells, as opposed to either IgM or IgG, which are secreted in normal immunological response. At first contact with an allergen, allergic manifestations are not observed and the organism is sensitized to the allergen with abundant production of IgE. The next contact with the allergen triggers the release of inflammatory agents from sensitized cells and initiates typical allergic manifestations.

The manifestations of allergic reaction are directly dependent on the substances that are released from the granules of previously activated basophils and mastocytes. With sensitization, excreted immunoglobulins bind to Fc receptors on the surface of mastocytes and basophiles, which are both directly involved in acute inflammatory reaction. IgE-coated cells are ready to trigger the allergic response at the next contact with the allergen. Subsequent exposure to the same allergen causes reactivation of the immunoglobulins bounded to the membranes of basophils and mastocytes, and signals their degranulation (Figure 6.3.).[2] During degranulation, the content of cell granules is released into surrounding cells provoking inflammatory symptoms. Basophils and mastocytes release histamine and other inflammatory chemical mediators, such as cytokines, interleukins, leukotriens and prostaglandins, which provoke vasodilatation, mucous secretion, nerve stimulation or smooth muscle constriction. After the chemical mediators of the acute response subside, late phase responses may occur. These may arise due to migration of other leukocytes from remote body sites. Late reaction is usually seen 4-6 hours after the original reaction and can last 1-2 days.

[1] Mannan is a polysaccharide composed of mannose sugars that is normally present in bacteria, yeast and plants.

Cytokines released from the mast cells may also play a role in the persistence of symptoms and long-term effects [1].

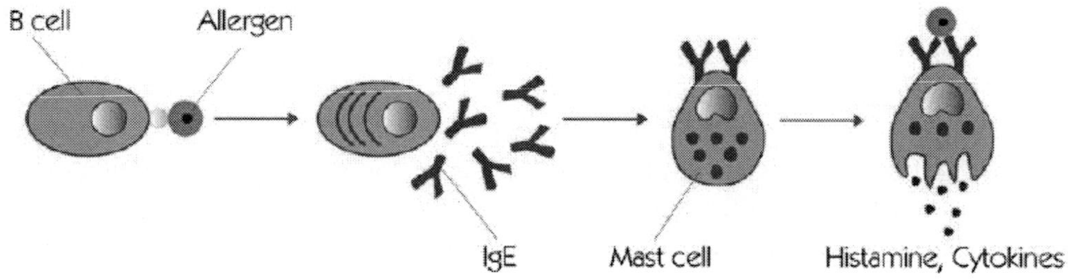

Figure 6.3. Type I hypersensitivity - IgE-mediated allergy.

Table 6.1. Comparison of toxic reaction and hypersensitivity of type I.

Toxic reaction	Allergic reaction
Manifestation specific to the substance	Manifestation not specific to the substance
Severity is dose dependent	Severity is not dose dependent
Manifestation with first and every subsequent contact	Never manifests at first contact, but with every subsequent contact
Affects normal organisms	Affects only sensitized organisms
Results from direct interaction	Results from the release of histamine and other mediators

Table 6.2. Comparison of different types of immunological hypersensitivity

	Type-I	Type-II	Type-III	Type-IV
Mechanism	Anaphylactic	Cytotoxic	Formation of immune complex	Delayed
Antibody involved	IgE	IgG, IgM	IgG, IgM	None
Response time	15-30 minutes	Minutes-Hours	3-8 hours	48-72 hours
Histology	Basophils, eosinophil	Antibodies, neutrophils	Antigen/antibody complex containing complement proteins, neutrophils	Monocytes, lymphocytes
Initiator	Antibody	Antibody	Antibody	T-cells

In comparing toxic reaction with hypersensitivity of type I, it should be pointed out that in allergic responses the severity of symptoms does not depend on the dose and that serious reactions may be initiated even with the traces of allergen, while in classical toxic reactions, the symptoms are directly dependent on the dose (Table 6.1.). Allergic response is never

[2] Deregulation is a cellular process of the rupture of vesicles (granules) found inside certain cells that releases antimicrobial cytotoxic molecules.

observed in first contact with the substance (sensitization), while a physiological response to a toxic agent follows the first and every subsequent contact or dose.

The symptoms of allergic reactions can be manifested locally, for example as swelling of mucosal membranes, eyes redness and itching or bronchoconstriction, or may systematically provoke anaphylaxis. Rashes, urticaria (severe rash) and contact dermatitis are also locally experienced in dermal contact with the allergen.

Real allergy is impossible to cure; however, many efforts have been made in the past decade to suppress its manifestations. Drug therapy, such as antihistamines, cortisone, theophylline or adrenaline, usually aims at alleviating the symptoms of the final manifestation and cannot restrain physiological reaction. These chemotherapeutics are used to reduce the activation of the mast cells and degranulation processes. Immunotherapy, on contrary, is directed toward suppressing the immunological reaction by aiming at one of the steps in this complex event. One of the procedures in the therapy involves gradual vaccination of the patient with progressive doses of the allergen in order to favor the production of IgG, as opposed to excessive IgE production. Another possibility of conducting immunotherapy includes intravenous injection of monoclonal anti-IgE antibodies which bind to free IgE. Injected anti-antibodies are not able to bind to IgE already bound to Fc receptors of basophils and mast cells, but only to free, circulating portions.

Currently, scientists are interested in designing a new generation of allergy drugs that would be cheaper than monoclonal antibodies. Drug design applies an approach that is similar to that of previous efforts, but instead of quite expensive monoclonal antibodies, examines the use of small drugs able to interact with free IgE.

Immunological Reactions not Mediated by IgE

The manifestations of immunological response not mediated by IgE generally arise later than in type I hypersensitivity, and these reactions involve many elements of the humoral immune system (Table 6.2.). After contact with allergen, inflammatory mediators are released from the cells by non-IgE mediated mechanisms. These mediators, such as histamine, tryptase and kininogenase, which already have been synthesized and stored in the granules of limphocytes in the body, are released into surrounding tissues as a response to an antigen. The mediators are responsible for bronchoconstriction, mucus secretion, vasodilatation, proteolysis, edema and attraction of eosinophils and neutrophils. Mediators such as leukotrienes B_4, C_4 and D_4, prostaglandins D_2 and PAF (platelet aggregating factor), which is responsible for platelet aggregation and heparin release, are produced after initial contact with an allergen. Leukotrienes C_4 and D_4 provoke same symptoms as histamine, but are more than 1,000-fold more potent. Prostaglandins are responsible for edema and pain.

In food hypersensitivity, after an offending food is ingested, inflammatory substances released locally damage intestinal tissue and provoke diarrhea, bloating, fatigue, anemia and muscle and bone pain. A typical example of non-IgE-mediated primary food sensitivity is celiac disease, in which afflicted persons are sensitive to gluten. Susceptible persons most probably lack the enzyme necessary to digest gliadin, a simple protein present in gluten and consisting of 43% of glutamine [2]. A non-IgE-mediated food allergy does not necessarily occur immediately after exposure, although it may; and the features that distinguish it from type I hypersensitivity are that the development of symptoms depends on the ingested amount and the symptoms are manifested only if a sufficient amount has been ingested. In some cases the reaction proceeds a few days after ingesting allergen-containing food.

In non-IgE-mediated hypersensitivity, reactions most commonly develop slowly and manifestations of symptoms usually occur many hours after contact with allergen. The symptoms may subside 72 h to 96 h later, depending on the mechanism involved. Immunological reactions involving non-IgE mediated mechanisms arise due to the release of inflammatory agents which may be initiated with both endogenous and exogenous substances. Non-IgE-mediated immunological hypersensitivity is speculated to be at the core of many unelucidated diseases.

Type II hypersensitivity is primarily mediated by antibodies of IgM or IgG classes and complement proteins, but phagocytes and natural killer cells (NKs) also have a crucial role in tissue damage and the development of symptoms. NK cells lyse (split apart) antibody-coated mastocytes and basophils by attaching themselves to antibody molecules and provoking the cells destruction by the mechanism of antibody-dependent cellular cytotoxicity. During cell destruction, the content of the mastocyte`s and basophil`s granules is emptied into the surrounding tissue, provoking an inflammatory reaction. Lesions formed in allergic reactions of this type contain mostly antibodies and neutrophils. The treatment of type II immunological hypersensitivity involves administration of anti-inflammatory and immunosuppressive agents. Type III hypersensitivity, also known as immune complex hypersensitivity, is mediated by soluble antigen-antibody complexes, which contain mostly IgG class immunoglobulins, although IgM immunoglobulins may also be encountered. Components of the complement system also participate in the formation of soluble immune complexes and are incorporated into them. Macrophages infiltrate in the process in the later stages and may contribute to the healing process and the suppression of symptoms. In type III hypersensitivity, the resulting tissue lesions contain primarily neutrophils and deposits of immune complexes with complement proteins. The reaction may be general or may involve individual organs such as skin, kidneys, lungs, blood vessels, joints (rheumatoid arthritis) and other organs. This type of hypersensitivity may also possibly explain the symptoms of many diseases caused by microorganisms. Type IV hypersensitivity is also known as cell mediated or delayed type hypersensitivity. This type of immune response results from the interaction between allergens and sensitized, tissue bond T-cells. Type IV hypersensitivity is involved in the pathogenesis of many autoimmune and infectious diseases, such as tuberculosis, leprosy, toxoplasmosis, as well as in contact dermatitis caused by various chemicals, dust or heavy metals. Tissue damage in type IV hypersensitivity arises due to activity of T lymphocytes, monocytes and macrophages. Cytotoxic T cells cause direct damage to tissue, whereas helper T cells have a role in secretion of cytokines, which activate cytotoxic T cells, monocytes and macrophages, which also participate in antigen attacks. Major cytokines involved in delayed hypersensitivity include a monocyte chemotactic factor, interleukin-2, interferon-gamma, tumor necrosis factors (TNF) alpha and beta, and others. [3]. Delayed hypersensitivity lesions contain mainly monocytes and T cells. For treatment of this type of immunological response, corticosteroids and other immunosuppressive agents are commonly used.

In non-IgE-mediated hypersensitivity, the identification of allergens is very difficult due to slow, cell-mediated, non-IgE mechanisms. The usual tests performed for detecting allergens in type I hypersensitivity, such as skin prick tests and the serum measurement of antigen-specific IgE levels are useless in detecting the allergens involved in other immunological mechanisms. Diagnostic tests for non-IgE hypersensitivity usually include detection of circulating antibodies as well as biopsy tests. The diagnosis of a particular immunological type of hypersensitivity and the identification of allergens are further complicated by the fact that in some cases a combination of IgE-mediated and non-IgE-mediated mechanisms is encountered.

Primary Nonimmunological Hypersensitivity

Hypersensitivity to certain agents with nonimmunological background is not considered a real allergy, and is usually denoted as intolerance. In food intolerance, the person is able to tolerate a certain amount of the offending food without adverse effects. In discriminating the immunological from the non-immunological response, it is important to notice that many nonimmunological food sensitivities involve different food components, often components other than proteins, and in many cases the offending substance remains unidentified. While true food allergies are attributed to food proteins, nonimmunological reactions may be associated with both naturally occurring non-protein substances as well as synthetic ones.. Clinical pictures characterizing allergy and intolerance may be identical, and distinctions between food allergy and food intolerance thus depend on whether the involvement of the immune system can be verified. Diagnostic tests such as the skin prick test and the measurement of specific IgE or histamine levels, offer the possibility of discriminating between the two possible mechanisms.

In *anaphylactoid* non-immunological food sensitivity,[3] food components provoke the spontaneous release of chemical inflammatory mediators, like histamine, from tissue mast cells. Even though anaphylactoid reaction is not IgE mediated, it is clinically indistinguishable from such. The manifestation of the body`s response does not require previous exposure to an antigen. It is assumed that the offending substance reacts with the destabilized membrane of tissue mastocites, provoking the release of mediators and mimicking a real allergy. This type of sensitivity is treated in the same way as real allergies, with the use of antihistamine drugs, including steroids and the like. Strawberries, cheese and chocolate intolerance, as well as sensitivity to many drugs, are anaphylactoid types of reactions occurring quickly after contact with the provoking substance.

A *metabolic food disorder* is another type of non-immunological food hypersensitivity arising as a response to certain food components. The manifestations of this disorder are a consequence of altered food metabolism or due to genetically linked enhanced sensitivity to some substances. Genetically determined enzymatic deficiencies affect the host's ability to metabolize specific foods, or enhance the sensitivity of the host to some food-borne chemicals due to alteration of the metabolic pathways. Intolerance of lactose, disaccharide, found in milk and other dairy products, is the most common food intolerance and occurs as a result of an inability to completely break down lactose. In sensitive individuals, lack or reduced levels of β-galactosidaze leads to impaired ability to digest lactose. Undigested lactose passes into the colon, where it encounters intestinal bacteria. Bacteria metabolize the lactose to CO_2, H_2

[3] Anaphylaxis is an acute systemic (multi-system) and severe type I hypersensitivity allergic reaction in humans and other mammals.

and H_2O and the resulting products are responsible for such symptoms as abdominal cramping, diarrhea, and flatulence. The symptoms vary in intensity depending on the level of intestinal β-galactosidase activity and the amount of ingested lactose. Lactose is not responsible for all cases of milk intolerance, and it is assumed that milk proteins may also cause idiosyncratic sensitivity. In some cases it is difficult to distinguish between hypersensitivity to milk proteins and a real allergy to milk, which is IgE mediated, but the difference may manifest itself in a delayed reaction indicating the intolerance type of reaction.

Hereditary fructose intolerance is a rare genetic disorder of defective fructose metabolism due to deficiency of aldolase B enzyme, which plays a role in the metabolic conversion of fructose into glucose. Lack of aldolase B results in the accumulation of fructose-1-phosphate in the liver, kidneys and intestine. The accumulation of this fructose metabolite in the liver inhibits glycogen breakdown and glucose release, thereby causing severe hypoglycemia following fructose ingestion. Clinical symptoms of fructose intolerance include severe abdominal pain, vomiting, convulsions and excessive sleepiness.

Favism is a hereditary medical condition which is caused by glucose-6-phosphate dehydrogenase deficiency. Glucose-6-phosphate dehydrogenase is an important red blood cell enzyme responsible for maintaining glutathione and NADPH, which help prevent oxidative damage. Fava beans contain the natural oxidants vicine and convicine, which may trigger hemolysis of erithrocytes in glucose-6-phosphate dehydrogenase deficiency. Glucose-6-phosphate dehydrogenase-deficient individuals may experience hemolytic anemia as a result of exposure to fava beans or inhalation of pollen from the *Vicia faba* plant. The condition is most common in people who live in the Mediterranean region, the Middle East, China and Bulgaria [4]. Foods other than fava beans and some pharmaceuticals can provoke similar reactions. Sensitive persons may feel tired or may experience a headache after contact with fava beans and related legumes. Other symptoms include abdominal pain, nausea and vomiting. If left untreated, favism can result in coma. Early warning signs of favism include jaundice, dark urine and fatigue.

Individual nonimmunological intolerance to a specific food or drug. without any defined mechanism, is known as an *idiosyncratic reaction*. Many reported adverse reactions to food additives are placed in this category because of a lack of understanding of their mode of action. Sulfites, various dyes, tartrazine, monosodium glutamate, as well as numerous drugs, such as quinidine, oxaliplatin, naltrexon and aspirin, can provoke very serious reactions. For most idiosyncratic reactions the mechanism is not yet clarified. The occurrence of migraine after the consumption of aspartame and chocolate; chronic urticaria as a response to BHA, BHT and benzoate; or aggressive behavior after consuming sugar, is yet to be explained. To aid in resolving the causative relationship between certain symptoms and specific foods, controlled double-blind study should be performed. The results of these controlled studies, however, are not always clear. For example, the results of a double-blind challenge directed at establishing a relationship between the consumption of synthetic food colorants and hyperkinesis in children brought questionable results. On the other hand, association of some food components and certain symptoms is firmly defined, as is the case with sulfite-induced asthma or the association of aspartame consumption and urticaria. Also, idiosyncratic reactions in response to food ingredients intended to provide color, enhance flavor or protect against the growth of bacteria have been reported. Some substances, such as sulfites, may occur in food products naturally, as in red wines, or may be added to prevent the growth of molds. Sulfite sensitivity may be possibly explained by a deficiency of sulfite oxidase in sensitive individuals. Some sulfite-sensitive persons exhibit a positive skin test to sulfites, indicating a possible IgE-mediated mechanism.

Vast numbers of processed foods and restaurant dishes contain monosodium glutamate, which boosts the taste of the food by stimulating nerves on the tongue and in the brain. Numerous reports indicate that some people are sensitive to glutamate. Some studies have found that people who suffer from real allergies or asthma may be prone to glutamate sensitivity. There have also been reports of people with asthma having more severe asthma attacks after consuming glutamate. Some individuals react immediately after consuming glutamate, while others may experience symptoms up to 48 hours later. Reactions vary from individual to individual, and it is unclear whether the expressed adverse reactions arise from underlying health conditions, or whether possible cumulative effects are created after ingesting the substance over a period of years. The reaction symptoms to monosodium glutamate vary in both qualitative and quantitative sense, with some people experiencing mild and temporary symptoms such as headache, while others suffer from long-term symptoms ranging from depression to insomnia.

Idiosyncratic drug reactions are unpredictable reactions that usually imply the existence of processes that cannot be attributed either to the substances' primary metabolite or to common immunologic mechanisms. In these cases, the body`s responses are a display of adverse events that are thought to be the consequence of an unusual or aberrant metabolic pathway of a chemical, which in some cases may even lead to significant morbidity and mortality. However, these reactions are rare, and it is speculated that some of them might be provoked by unexplained cytotoxic mechanisms. Despite progress in identifying reactive metabolites involved in idiosyncratic drug reactions, their basic mechanisms remain elusive. Drugs such as quinidine, oxaliplatin, nalteroxen and aspirin may provoke very serious reactions with possible fatal outcome. Individuals expressing idiosyncratic reaction to aspirin may also be sensitive to salicylates naturally occurring in many fruits, vegetables, nuts, coffee, juices, beer and wine.

Secondary Nonimmunological Hypersensitivity

In some cases, hypersensitivity to a substance or a certain food ingredient may arise as a consequence of other conditions, such as drug therapy or different pathological conditions. Secondary food sensitivity usually disappears within a few weeks after recovery from an illness or after a drug therapy is discontinued. Enhanced sensitivity to tyramine is sometimes observed among patients receiving a monoamine oxidase–inhibiting therapy. In such cases, the body`s abnormal responsiveness is temporary and hypersensitivity may be lost after cessation of the drug therapy. Lactose intolerance may occur secondary to another intestinal illness or infection, such as Crohn`s disease, ulcerative colitis or viral gastroenteritis. Secondary lactose-intolerance is usually a short-term illness, because the level of the operative digestive enzyme, β-galactosidaze, rapidly recovers after the illness subsides.

Exercise-induced food allergies are a subset of hypersensitivity manifestations that involve immediate reactions occurring only when a specific food is consumed just before or just after the exercise. Ingestion of shellfish, peaches, wheat and celery is frequently observed in food-dependent exercise-induced hypersensitivity. Many cases of exercise-induced allergies are not related to food and may be provoked by other allergens. While the exact mechanisms of increased sensibility after physical exercise are unknown, it is assumed that an increase in mast cell responsiveness may be involved [5]. The symptoms are individualistic and are similar to those involved in other food allergies.

Allergy Testing

To confirm the immunological background of allergic response and to identify allergens, both *in vivo* and *in vitro* immunological tests may be applied. *In vivo* testing is performed either by double-blind simulation of natural exposure or by skin tests. Natural exposure may be mimicked by introducing the suspected inhalant allergen into the nasal cavity or by performing ingestion testing for identifying food allergens. Skin testing is performed either intradermally or epicutaneously (with a prick, scratch or puncture test). In intradermal tests, the tested substance is injected into the deeper layers of the skin, providing better specificity and more sensitive reaction.

In vitro allergen-specific IgE tests are performed to identify allergens and to determine the level of antigen-specific IgE in type I hypersensitivity. The tests are convenient when skin testing cannot be performed or interpreted due to developed generalized dermatitis, or if antihistamine therapy is under way. Allergen-specific tests are generally denoted as radioallergosorbent tests (RASTs) and are used for qualitative, quantitative and semi-quantitative measurements of specific IgE, and include various immunological test methodologies, such as radioimmunoassay, enzyme immunoassay, chemiluminescent immunoassay and fluorescent immunoassay. All RAST immunological tests follow a common general principle, and the tests vary only in methods of detection. Usually a specific allergen is bound to a solid-phase support. Serum sample containing IgE, both specific and not specific for the tested allergen, is incubated, allowing an antigen/antibody reaction. After completion of the reaction, excess serum and non-allergen specific IgE are washed away. In the next step, labeled anti-IgE antibody, which is specific for bound IgE, is added and a new complex is formed. The second antibody is applied which is specific to the complex formed between the immobilized antigen and serum IgE, which is usually rabbit IgG, or mouse monoclonal anti-human IgE that is labeled with radioactive iodine (^{125}I), or an enzyme such as horseradish peroxidase, alkaline phosphatase or urease [6]. Elevated results of the level of specific IgE usually indicate the hypersensitivity of type I; however, the amount of specific IgE does not necessarily predict the potential severity of the allergic reaction.

Tests relying on IgE determination may lead to erroneous conclusions due to insufficient sensitivity of the tests and because increased levels of specific IgE may be demonstrated years after exposure. The advantage of *in vivo* tests is that they are more specific and indicative; however, in *in vitro* testing, the risk of possible systemic reactions is avoided, even for very low allergen concentrations. In some cases of food-borne allergens, besides measurement of the allergen-specific IgE, the determination of serum IgG1 and IgG4 levels in the specific food component may also be indicative. Other blood tests, such as cytotoxicity, total IgE level or white blood cell differential, especially in respect to eosinophils and basophils, may indirectly indicate an ongoing allergic process. Elevated levels of these substances in tests results may suggest an allergy, but they may also be misinterpreted as resulting from other immunological reactions taking place in the body.

The Enzyme-Linked Immunosorbent Assay (ELISA) is a biochemical technique used to detect the presence of an antibody or an antigen in a sample and is, thus, useful for identification of specific IgE in plasma. As in all RAST tests, the technique implies the existence of two antibodies. The first antibody is specific to an antigen, while the other is labeled with an enzyme such as horseradish peroxidase, alkaline phosphatase, or urease, and reacts specifically to formed antigen-antibody complex. The role of second antibody with the coupled enzyme is to provoke a chromogenic or fluorogenic reaction with subsequently added substrate. This technique may be successfully applied for the determination of serum antigens, such as HIV or other viruses, and for detecting food allergens. The advantage of the technique is that it is possible to detect the virus soon after the infection, since antibodies applied in the test are specific to the virus component. The assay may also be used to determine the level of circulating antibodies, specific for allergens, thus distinguishing the type of immunological allergic response and indicating the level of sensitization. For antibody detection in serum, it is necessary to use an ELISA plate that is covered with the antigen in question. Serum samples, in which the presence of a specific antibody is to be confirmed, are added to the plate wells, where reaction with the bound antigen takes place. The plate is then washed, so that the unbound components are removed and only antibody-antigen complexes remains attached to the well. The second antibody, complementary to the formed complex, carries the enzyme and will bind to antigen-antibody complex. Upon substrate addition, the substrate-modifying enzyme will provoke chromogenic or fluorescent signal. Depending on the signal's intensity, the serum antibody level can be measured.

A sandwich (layered) ELISA test plate is coated with a capture antibody. The sample is added to the plate and any antigen in the sample will bind to the capture antibody (Figure 6.4.). In the next step the detecting antibody is added, which will bind to formed antibody-antigen complex. Enzyme-linked secondary antibody (anti-antibody) and substrate are subsequently added provoking a fluorogenic reaction.

In competitive ELISA testing, the intensity of the fluorogenic signal decreases with the antigen concentration in the sample. The first stage of the assay is to incubate the sample in which the allergen is to be detected with a known amount of the antibody. Complex, formed between the sample's antigen and a known amount of the antibody, is added to an antigen-coated well. The excess amount of the antibody which is not bound in reaction with the sample will react with the antigen on the reaction plate and the unbound antibody is removed by washing. A chromogenic reaction is provoked by reaction with the substrate after the addition of second antibody, which carries the substrate-modifying enzyme. The more antigen is present in the sample, the weaker signal is obtained. The antigen from the sample consumes a certain amount of the antibody from the reaction mixture, leaving less of the available antibody for binding to antigens in the wells. The higher the concentration of the antigen in the sample the fewer antibodies will bind to antigens from the plate.

A modern approach to confirming allergies is by detecting cysteinyl leukotriens using a Leukotriene ELISA test. The advantage of leukotrien tests in comparison to IgE tests is that they reveal not only sensitization to an allergen, but also the intensity of the allergic reaction. The tests may also indicate immunological response other than type I hypersensitivity. Leukotrienes are involved in common allergic symptoms like bronchoconstriction, smooth muscle contraction, increased capillary permeability and anaphylaxis. These mediators, together with others, are actual and direct causes of allergic manifestations. In comparison to IgE tests, tests that measure the amount of these mediators better illustrate the actual allergic status of the body. The great advantage of the latter tests is that they are successful at detecting both IgE mediated and non-IgE mediated allergies.

Another type of test, the Enzyme-Linked Immunosorbent Spot (ELISPOT) immunological assay, is most often used to determine the number of activated antigen-specific cytotoxic T-cells. ELISPOT assays are much more sensitive than conventional ELISA tests and are useful for monitoring antigen-specific responses applicable to a wide range of areas of immunological research, including cancer, transplantation, infectious diseases and vaccine development. In the ELISPOT test, process wells are coated with cytokine-specific antibody and harvested mouse T cells. The technique relies on the principle that T cells secrete cytokines after activation with the antigen. T-cells specific to a certain antigen are activated in its presence. Because the plates are pre-coated with the antibodies specific to the cytokines of interest, cytokines secreted by activated T cells are locally bound in the wells. Following incubation, the T cells are washed away and secondary antibodies specific to the same cytokine are added. The secondary antibody is usually biotinylated and can be visualized by adding streptavidin-alkaline phosphatase reagent. This reagent catalyses the conversion of a substrate, bound to the secondary antibody to a colored stain. By counting colored spots, the fraction of T cells activated by a given antigen can be calculated.

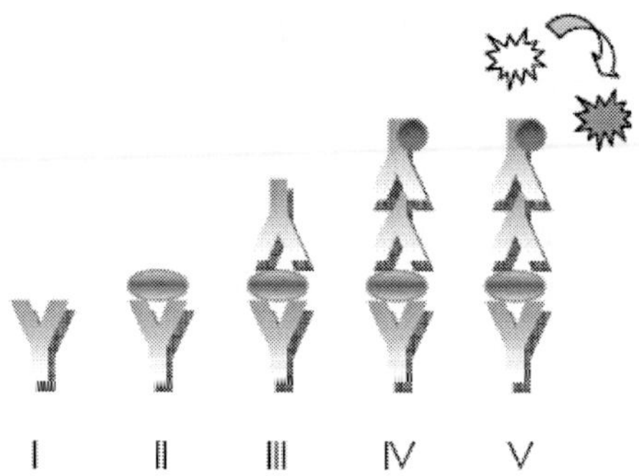

Figure 6.4. Sandwich ELISA test.

REFERENCES

[1] Deshpande, SS. *Handbook of Food Toxicology*. CRC Press; 2002.

[2] Omaye ST . Food and Nutritional Toxicology. CRJ Press; 2004.

[3] Pier, GB; Lyczak, JB; Wetzler. LM. *Immunology, Infection, and Immunity.* ASM Press; 2004.

[4] Sherman, WS. Hypersensitivity: mechanisms and management. Saudes; 1968.

[5] Smith T. Allergy testing in clinical practice. *Ann. Allergy* 68, 1992, 293-300.

[6] Winter, CK. *Food Toxicology.* CRC Press; 2000.

Chapter VII

Toxicological Effects of Inorganic Toxicants

Exposure to inorganic toxicants occurs through contaminated food, water and air, or as a consequence of inadequate occupational environments (i.e., contaminated) in specific work places (silanosis, azbestosis). Many factors influence the development of symptoms in both acute and chronic exposure. Inorganic toxicants with very poor elimination rates may accumulate in certain tissues and organs, either by physico-chemical or chemical affinity. In chronic exposure, an element may be deposited in the tissue for a long time, so that relatively small subsequent doses may provoke the symptoms of intoxication after the toxic level has been reached. Unexposed persons require higher doses to initiate the expression of symptoms in comparison to persons whose metabolizing and excretory abilities are limited due to saturation from previous exposures. Other factors influencing the toxicity of mineral elements include absorption and excretion toxicokinetics and mobilization of the stored toxicant under specific conditions, for example from bone or fat tissue. Variability in the detoxification mechanism caused by different factors and individual variations in susceptibility may significantly alter the manifestation of intoxication with inorganic toxicants.

Among inorganic contaminants, toxic metals and metalloids attract attention because of their cumulative effects and their inability to degrade in the environment, as well as the extreme levels of biomagnification that they may reach. Some heavy metals, such as mercury or cadmium, provoke adverse effects even in ppb levels and are considered strong and potent toxicants. Heavy metals, like lead, cadmium, tin and others, may migrate into the food from metal equipment during food processing or from the packing materials during product storage. In the past, many cases of accidental poisoning with heavy metals due to ignorance demonstrated that knowledge of chemistry is a prerequisite to avoiding the most ordinary risks that may occur in unexpected situations, for example when burning of treated or painted wood, which is likely to release potent contaminants.

In the past, chemical substances used for wood protection and wood paints usually contained arsenic, lead and other metals. Burning of older wooden objects may release toxic volatile compounds, which may contaminate surrounding air and food. A case of this type of "naive" poisoning was reported during the 1930s in Brest, France, and was caused, over a period of time, by consumption of bread that was typically baked over a fire produced by burning the old boat boards, painted with lead-containing paints. The health of more than 37 people was severely affected due to chronic lead poisoning.

The mechanism provoking toxic effects of metals generally comprises the formation of complexes with organic compounds in the body. Metals usually express affinity to binding with biological compounds containing oxygen, sulphur and nitrogen, thus disturbing metabolic processes and affecting the protein structure and enzyme activity. In acute poisoning, large excesses of metal ions provoke disruption of cell membranes and disturb mitochondrial function by generating free radicals [1]. Some light toxic metals may compete with other essential elements of similar mass and charge in performing their normal biological functions. Lithium toxicity, for example, results from the competition with sodium and disturbance in the transmission of nerve impulses, heart activity and certain metabolic functions. The body resists toxicants by binding them into reversible temporary complexes with particular plasma proteins and by metabolizing free metals. Metallothionein plays a significant role in the physiological detoxification of heavy metals, particularly cadmium. Metallothionein is a low molecular weight intracellular protein which can bind more than 18 different metals, decreasing free metal concentrations and reducing their toxicity. Metallothionein is also thought to be involved in the regulation of copper and zinc metabolism. High concentrations of metallothionein are found in the liver, kidneys, intestine and pancreas. The synthesis of metallothionein can be induced in response to many elements, including zinc, cadmium, copper, mercury, gold and bismuth. Due to its high affinity for binding many metals, this multifunctional protein is taken into consideration for use in the treatment of metal poisoning. Other proteins in the human body are also actively involved in processes of detoxification of the toxic elements. The protein hemosiderin binds excess iron in the body. Lead, bismuth and mercury form insoluble protein complexes in renal tubules in the form of eosinophilic bodies.

Common chemical reactions encountered in the metabolizing processes of inorganic toxicants are known as methylation processes. Excess selenium, for example, is excreted as trimethylselenium or in the form of volatile dimethylselenium, which is formed in high selenium concentrations, giving intoxicated persons a characteristic odor of garlic on the breath. On the other hand, methylation of some metals produces more toxic metabolites, as in the case of mercury, lead and tin [2].

ALUMINIUM

Aluminum is the most abundant (8.13%) metallic element in the Earth's crust and it can be found naturally in air, water and soil. Domestic application of aluminum is related to its use in the production of cooking pots and pans, utensils and foil. Some pharmaceutical products, such as analgetics and anti-inflammatory drugs, douche preparations, antiperspirants and toothpastes, may also contain aluminum. Aluminum is very efficient in water treatment and powdered aluminum is used in paint industry and pyrotechnics. The typical daily dietary intake of aluminum varies from 3 to 10 mg. Significant sources of aluminum include baked food, processed cheese, grains, vegetables, herbs and tea. Even though it is not as toxic as other metals, aluminum ingestion may cause serious health effects . Aluminum can accumulate in the brain causing seizures and reduced mental abilities, and certain compounds, such as aluminum fluoride (commonly used to treat municipal water supplies) as well as aluminum sulfate, readily cross the blood/brain barrier. The speculation regarding a link between aluminum and Alzheimer's disease is based on the increased aluminum content found in the brain of people affected by Alzheimer's disease. One additional fact supporting this theory is the observation of increased concentrations of aluminum in the soil and drinking water in areas whose population exhibit an unusually high incidence of Alzheimer's disease [1]. In acute aluminum poisoning, problems are first displayed in the gastrointestinal tract in the form of colics (abdominal pain which starts and stops abruptly), followed by extreme nervousness, headaches, memory loss and speech problems. In long-term exposure, aluminum interferes with the metabolism of calcium, provoking the softening of bones and the disturbance of liver and kidney functions. Also, a small percentage of people express allergy to aluminum in the form of contact dermatitis as a response to antiperspirant products or as digestive disorders, and also in the inability to absorb nutrients from food cooked in aluminum pans.

ANTIMONY

Antimony is widely used in glass, paint, ceramics and the printing industry. It also has an important role in the production of explosives and in semiconductor industries. Antimony compounds are used as fire retardants, in medical diagnosis, and in the treatment of some diseases. Antimony is released in environment by coal-fired power plants and smelters. The element is also present in tobacco smoke and is a common component of the sewage sludge.

The toxicity of antimony can be compared to that of arsenic due to their similar chemical properties, even though antimony is less toxic. It is assumed that the element is carcinogenic, but the speculation is not yet confirmed.

The most toxic form of antimony is gaseous stibine (SbH_3), which, when inhaled causes headache, nausea, vomiting, jaundice and anemia. Acute exposure to other forms of antimony cause skin problems and hair and weight loss. It has been demonstrated that in high concentration exposures damage to the cardiovascular system, liver and kidneys may be observed.

In chronic exposure, long-term inhalation of antimony compounds causes irritation of eyes and lungs and may also induce heart problems, stomach pain, diarrhea, vomiting and stomach ulcers. Exposure by other routes results mainly in skin irritation and dermatitis [2].

ARSENIC

Arsenic exists in four valence states: -III, 0, III and V, and it naturally occurs in water [3], soil, ores and the atmosphere. This metalloid is emitted into the atmosphere due to volcanic activity and the burning of fossil fuels primarily in the form of As_2O_3, which adsorbs onto solid particles and settles into the soil. Anthropogenic activities such as electronics manufacturing and the use of pesticides and wood preservatives influence arsenic distribution, as well. In the past, lead hydrogen arsenate was used as an insecticide, but its application in agriculture was replaced with monosodium methyl arsenate, a less toxic organic form of arsenic. In the past, most wood was treated with chromated copper arsenate to prevent rot or insect damage. Later, studies showed that arsenic leaches out of the wood and that a risk of arsenic release is especially high when the treated wood is burned.

Many cases of mass poisoning with arsenic also occurred due to its application as a component of wallpaper paints and were explained years later, giving the impetus to withdraw it from the further use. Also, glue used in the past to attach the wallpaper was at the time produced from natural products that were ideal substrate for microorganism growth. The paints were metabolized from the wallpaper, volatile arsenic compounds were released, provoking contamination of the surrounding air. A significant microbial transformation of arsenic compound into volatile compounds such as arsine, diethylarsine, ethylcacodyloxide, is specific for Penicillium species (Penicillium glaucum, Penicillium brevicaule). In Stockholm in 1911, more than 150 employees were poisoned by arsenic as a result of such a microbial conversion: their offices were in old building which had arsenic-based wallpapers. It is speculated that Napoleon`s poisoning with arsenic occurred in similar manner, as the room where he stayed during his illness was not ventilated properly and he spent lot of time in breathing arsine-contaminated air. This incidence was proven by analysis of his hair, because when intake levels of arsenic are high, partial excretion of arsenic biotransformation products occurs via hair.

Through history arsenic compounds were much used for various purposes and one in particular relied on the high toxicity of trivalent inorganic arsenic. During the Renaissance, virtually all professional poisoners used arsenic (III) oxide, which is convenient for the purpose due to its absence of taste and odor. At that time it was nearly impossible to detect the presence of the poison in food or beverage. Highlanders even used white arsenic powder to improve their strength and physical endurance. Although they consumed high concentrations of arsenic (III) oxide, they did not develop symptoms of poisoning because a gradual increase in the dose enabled the activation of biotransformation enzymes and a decrease in the absorption rate. The same compound also was used in the past by ladies as a power to produce beautiful, pale skin.

Many biochemical aspects of arsenic behavior in the human body are not completely defined, and by one unproven theory arsenic in low ppb concentrations is even considered essential for humans. Although its role in humans has not yet been clearly defined, it is speculated that arsenic may have a physiological role in the metabolism of sulfur-containing amino acids, methionine and cysteine. Studies indicate that arsenic plays a part in the conversion of methionine to its metabolite taurine, as well as in the methylation of various biomolecules, such as histones. It is assumed that approximately 1.2 μg of arsenic fulfills the daily human requirement [4]. It is also suggested that arsenic independently on omega-3 fatty acids, increases bleeding time and that low serum arsenic is correlated with central nervous system disorders and vascular diseases.

Many previous applications of arsenic are not in use today due to demonstrated adverse health effects. However, arsenic currently has an important industrial use as gallium arsenide, a semiconductor used in integrated circuits. Arsenic is also widely applied in bronzing and pyrotechnics. The average arsenic intake in humans through food and drink is 20-300 μg/day [5]. Smokers inhale approximately 10 μg/day of arsenic and non-smokers, about 1 μg/day.

There are great differences in the toxicity of different arsenic compounds. The toxicity of a particular arsenic compound depends on its valence state, on whether the compound is organic or inorganic, as well as on the physico-chemical properties that influence absorption and elimination rates. Generally, inorganic arsenic compounds are more toxic than organic compounds and trivalent compounds are considered the most toxic. Toxicity of organic arsenic compounds, including compounds such as monomethylarsonic acid, dimethylarsinic acid, arsenobetaine, and arsenocholine, is relatively low in comparison to inorganic compounds. Besides their toxicity, organic arsenic compounds have been shown to induce sensitization reactions in some individuals [1, 6].

Absorption of inorganic arsenic species, either orally, dermally or by inhalation, is followed by its accumulation in the liver, spleen, kidneys, lungs and gastrointestinal tract. Part of the absorbed element is excreted through keratin-rich tissues such as skin, hair and nails. A lethal dose of inorganic arsenic is estimated to be between 120 mg and 200 mg for a person of average weight. Pentavalent inorganic arsenic compounds are methylated in the liver, forming methyl arsenic and dimethylarsenic acids. This biotranformation system can easily be saturated, leading to an accumulation of inorganic arsenic in the soft tissue. The toxicity of pentavalent arsenic can be explained by the substitution of phosphate ions by As^{5+} in normal biochemical processes and ATP hydrolysis. Trivalent arsenic forms, like mercury compounds, bind to sulfhydryl groups of enzymes, such as pyruvate dehydrogenase system and enzymes of the Kreb's cycle, interfering oxidative phosphorylation.

Symptoms of arsenic poisoning can differ, depending on the compound, amount, route of exposure and individual susceptibility. The main clinical signs of acute arsenic poisoning are caused by capillary injury. Ingested arsenic species first provoke inflammation and necrosis of gastrointestinal mucosa, and this is accompanied with gastroenteritis, nausea, vomiting and abdominal pain. Great amounts of lost fluid disturb the electrolyte equilibrium, leading to muscle cramps. These symptoms occur between 1 and 12 hours after exposure. The cardiovascular system is also severely affected by acute intoxication with arsenic compounds and this is manifested as hypotension, tachycardia and hemolysis. Persons intoxicated with arsenic compounds are easily recognized by their blue fingers and toes and altered facial expression. Symptoms such as headache, irritability, muscular weakness, convulsions and coma may also arise due to adverse effects on central nervous system. Acute inhalation of arsine gas is accompanied by nose and throat irritation and severe hemolysis occurring within 3-4 hours of exposure, leading to acute tubular necrosis and renal failure [7]. Manifestations of sub-acute poisoning with arsenic compounds are similar to symptoms of acute poisoning, with a delay in the signs of development. In chronic exposure to arsenic compounds, the skin and nervous system are most affected, and this is accompanied by their damage and expressed through the gray coloring of the skin, hair loss and ulcers. Long-term exposure to arsenic is accompanied by numbness, tingling and quadriplegia (limb paralysis), which in many cases may be irreversible. Patients may also display "garlic breath," dermatitis and keratitis [1]. In chronic exposures, death may occur in several months or in as long as several years. Histopathological signs of gastrointestinal damage occurring in chronic arsenic exposure include redness of gut epithelial cells, fatty liver, and kidney damage. Arsenic is a proven lung and skin carcinogen in inhalation and ingestion exposures, but some toxicological tests suggest that it might also be carcinogenic in the bladder, kidneys and colon.

Chelating agents such as dimercaprol or less effective penicillamine are commonly used in the treatment of arsenic intoxication [8]. Fluid imbalances are usually corrected with proper electrolyte therapy.

Barium

Barium is a soft metallic element that chemically resembles calcium. Naturally occurring barium is a mixture of seven stable isotopes. Twenty-two other radioactive isotopes are known to exist. A mixture of barium sulfate and zinc sulfide is used as a pigment with good covering power. The sulfate is also used in paints, in X-ray diagnostics and in the glass industry. Barite is extensively used as a well-drilling fluid and in rubber production. The carbonate was used as a rat poison, while the nitrate and chlorate are used as pigments to provide colors in pyrotechnics. Impure sulfide phosphoresces after exposure to the light. All barium compounds that are water or acid soluble are very poisonous.

After absorption of barium compounds from gastrointestinal tract, about 54% of the element is bound to plasma proteins. Upon distribution, it accumulates in muscles, lungs and bones, due to similarity with calcium. Small amounts are stored in the kidneys, liver, spleen, brain, heart and hair. In the muscles the element remains about 30 hours, followed by gradual decrease in concentration [9]. The excretion of barium is primarily in feces. At low doses, barium acts as a muscle stimulant, because it replaces calcium in cell membranes, increasing their permeability and providing stimulation to muscles. At higher doses it affects the nervous system, eventually leading to paralysis. Acute and sub-chronic oral exposure to barium provokes vomiting and diarrhea, excess salivation, followed by decreased heart rate and elevated blood pressure [10]. Higher doses result in cardiac irregularities, weakness, tremors, anxiety, dyspnea, arrythmia and convulsions. Fatal outcome can occur due to cardiovascular or respiratory failure. A lethal dose of barium for an adult of average weight is around 0.8 g [11].

Beryllium

Beryllium is an element highly toxic to humans and it can be found in small quantities in air due to burning of coal and oil. Tobacco smoke also contains traces of beryllium. The element has strong corrosion-resistance properties and for this reason is used in many industries. Beryllium is used mainly in the form of its alloys for the production of electrical instruments, aircraft parts, electrical connectors and the like. Ceramics used as electrical insulators, microwave ovens and mirrors also contain incorporated beryllium. Occupational exposure to this element is particularly important and dangerous in specific industries.

The absorption of beryllium from the gastrointestinal tract depends on its form, but generally, only very small portion (less than 1%) is absorbed. After distribution and biotransformation, absorbed beryllium is excreted very slowly, mostly in feces. Its toxicity partially arises due to its tendency to replace calcium in bones, where it can accumulate. Generally, ingestion of beryllium compounds does not carry serious health risk, because only very small portion is absorbed via gut mucosa. Dermal exposure can result in contact dermatitis, rashes and ulcers.

In contrast to oral ingestion, which does not pose a high risk for moderate doses, inhalation of beryllium compounds is considered much more dangerous and toxicologically serious. After inhalation, soluble beryllium compounds are excreted in the urine, with the half-life of about 2-8 weeks [7]. Acute inhalation exposure is accompanied by symptoms ranging from inflammation of the nasal passages to severe pneumonia. Chronic beryllium disease develops during long-lasting inhalation of small element doses. A specific disease caused by chronic beryllium inhalation is a type of immune response accompanied by the development of lung fibrosis, which is observed only in sensitized people. The latency period can last up to 25 years before symptoms occur. The U.S. Environmental Protection Agency (USEPA) describes beryllium as a likely human carcinogen.

Bismuth

Metallic bismuth is widely used in the manufacture of low-melting solders and fusible alloys. Its industrial application is mostly related to production of ceramics and various pigments. In comparison to other metals widely applied in the industry, bismuth is considered one of the least toxic heavy metals. The element is also a constituent of many pharmaceutical preparations for the treatment of constipation.

The effects of acute bismuth poisoning include metallic taste, gingivitis, loss of appetite and weight loss. Central nervous system exposure leads to insomnia and depression. Acute poisoning with bismuth is characterized by rheumatic pain and less specific symptoms such as headache and fever. In chronic exposure to bismuth, the functions of the liver and kidneys are severely affected. Anemia and black line formation on gums, as well as ulcerative stomatitis also may develop [2]. Bismuth and its salts are known by their ability to cause kidney damage, although the degree of the injury is usually mild. Large doses of bismuth may cause the development of a reversible encephalopathy. Other toxic symptoms following bismuth absorption, such as vague feeling of bodily discomfort, diarrhea, skin reactions and dermatitis, also may develop. In cases of bismuth poisoning, toxicological results may indicate the presence of albumin and other proteins in the urine.

Cadmium

Cadmium is a common occupational hazard associated with industrial processes such as metal plating, production of nickel-cadmium batteries, pigments and plastics. The element is also a constituent of many alloys and fertilizers. Wide industrial application contributes to cadmium`s presence in sewage sludge and effluents and in polluted groundwater. Wide-spread use of cadmium in industry defines it as one of the most important environmental metal pollutants. Exposure to cadmium may occur occupationally in the industries mentioned, or through consumption of contaminated water and foodstuffs, especially grain and leafy vegetables, which readily absorb cadmium from the soil. Fish, which bioaccumulate cadmium from contaminated water, may be another source of serious cadmium poisoning. Human exposure to environmental cadmium is primarily the result of the burning of fossil fuels and municipal wastes. The element is also found in cigarette smoke in the amount of approximately 0.007-0.35 μg per cigarette, as well as in car engine exhausts. An important consideration of cadmium`s toxicity is that after absorbtion, a significant amount of it is retained by kidneys and liver, concentrating cadmium and provoking also organ-specific local toxic effects.

Average daily intake of cadmium by ingestion and inhalation is estimated to be around 20 to 40 μg; however, only 5 to 10% of the amount is absorbed [1]. After absorption, cadmium binds to the plasma protein albumin. Cadmium absorption triggers the synthesis of metallotheionein in the liver. Metallotheionein is a protein active primarily in zinc metabolism, but due to similarity of cadmium and zinc with respect to size and charge of the ions, the process of metallotheionein synthesis may be initiated. Cadmium-metallotheionein complex is formed in the liver and is further transported to kidneys. In the kidneys the size of the complex enables glomerular filtration, but in the proximal tubule, the cadmium is reabsorbed [1, 2]. In renal tubular cells, small fractions of cadmium-metallotheionein complex are degraded by digestive enzymes, releasing the free cadmium. As part of a protective mechanism, renal tubular cells also have the ability to produce metallotheionein; however, the synthetic capacity of tubular cell is not high. With high cadmium concentrations, the synthetic capacity of renal tubular cells is exceeded, enabling cadmium accumulation. Accumulated cadmium may cause irreversible damage to the renal cells [12] because of insufficient efficiency of renal cells to eliminate cadmium complex, as the half-life of cadmium in the kidneys is very long, between 15 and 30 years [1].

The toxic effects of cadmium arise mainly due to its inhibition of many bodily enzyme systems. Cadmium is able to inactivate enzymes containing sulfhydryl groups and it also disturbs oxidative phosphorylation in the mitochondria [13]. The element competes with other metals such as zinc and selenium for inclusion into metallo-enzymes, as well as with calcium for binding sites on regulatory proteins such as calmodulin [1]. Calmodulin is a calcium-binding protein that binds a number of different protein targets, thereby regulating many different cellular functions, such as inflammation, metabolism, apoptosis, muscle contraction, intracellular movement, short-term and long-term memory, nerve growth and immune response. It is speculated that cadmium may also interfere with biological processes involving magnesium.

Among the most prominent specific symptoms of acute cadmium poisoning is testicular damage, which is expressed as necrosis and degeneration of the testes with complete loss of spermatozoa within a few hours of exposure. Testicular injury is thought to occur due to reduction of blood flow to these organs [13]. Acute ingestion of cadmium produces severe gastrointestinal irritation, which is manifested as severe nausea and vomiting, abdominal cramps and diarrhea. A lethal dose of cadmium for ingestion is estimated to be between 350 and 8900 mg for a person of average weight [1]. Acute inhalation of cadmium-containing fumes can initially result in fever and may progress to chemical pneumonitis, pulmonary edema and death. When inhaled, cadmium induces severe lung irritation and damage, accompanied with chest pain, dyspnea, cyanosis and tachycardia.

The effects of chronic cadmium exposure depend on the dose and route of exposure. Long-term ingestion of cadmium compounds causes kidney damage, which is first displayed as proteinuria and disorders of calcium metabolism, which lead to frequent and painful bone fractions (Itai-itai disease) [7]. The effects of long-term exposure to cadmium were demonstrated during the World War II, when Japanese mining operations contaminated the Jinzu River with cadmium and other toxic metals. As a consequence, cadmium accumulated in rice crops growing along the riverbanks. Itai itai disease developed in local communities together with renal abnormalities, including proteinuria and glucosuria. Chronic inhalation of low cadmium concentrations causes obstruction of the airways, which is often manifested only after many years. Cadmium and several cadmium-containing compounds are also known to be carcinogenic and, in conducted toxicological studies, have been linked with lungs and prostate cancers [5]. At present, there is no effective treatment for cadmium intoxication, and patients are given supportive treatment according to their symptoms. However, improved efficiency in some of new chelating agents is predicted, but has yet to be clinically confirmed [13].

CHROMIUM

The industrial use of chromium is related mainly to the stainless steel and chromate production industries and welding. It is also used in the production of inks, paints, paper, floor coverings and magnetic tapes, and as a wood preservative. The element exhibits a wide range of possible oxidation states, the most common being II, III, and VI. Compounds with trivalent chromium are the most stable, and compounds with monovalent, tetravalent and pentavalent chromium are quite rare. Hexavalent chromium compounds are very powerful oxidants. Both chromium III and chromium VI occur in the environment and they are released by the burning of fossil fuels. Tobacco smoke contains between 0.0004 μg/cigarette and 0.069 μg/cigarette of chromium.

Before we consider chromium toxicity, it is important to mention that chromium III is essential element for humans. Trivalent chromium enhances the action of insulin, a hormone critical to metabolism of carbohydrates, fats and proteins in the body [14]. Besides potentiating insulin activity, trivalent chromium also stabilizes blood sugar levels, balances the good/bad (HDC/LDC) cholesterol levels and participates in the transport of amino acids. It is also thought that chromium III is involved in appetite control [15-21]. A diet low in trivalent chromium may influence the occurrence of osteoporosis and parathyroid disfunction; however, antagonistic action of other elements found in the diet, such as copper, potassium, selenium and vanadium could give the impression of a chromium deficiency, even with adequate chromium intake. It is assumed that with age, chromium depletes from the body, and so chromium should be supplemented in the diet. Dermally applied, high doses of chromium III may cause skin irritation.

In contrast to trivalent chromium, hexavalent form is very toxic and carcinogenic, possibly due to its strong oxidizing properties. After absorption, chromium VI is first reduced to pentavalent chromium. In this reduction, surrounding tissues are oxidized and damaged, provoking adverse effects. Pentavalent chromium is a known carcinogen and may lodge in any tissue and initiate cancerous growths. According to some findings, pentavalent chromium exposure may also be a factor leading to premature senility.

Ingestion of large amounts of hexavalent chromium can cause stomach ulcers, convulsions, kidney and liver damage, and in most cases is fatal. Inhalation of high levels of hexavalent chromium can cause nose irritation, with manifestations such as runny nose, nosebleeds, and ulcers in the nasal septum.

Skin contact with certain hexavalent chromium compounds may provoke the formation of skin ulcers. Some people are extremely sensitive to both chromium (VI) and chromium (III) and may develop allergic reaction.

COBALT

Cobalt is found naturally in food in trace quantities, and the burning of coal and oil releases it into environment. The element is also a constituent of phosphate fertilizers. Tobacco smoke contains around 0.0002 μg/cigarette of cobalt. Home- or work-related sources of cobalt are pottery, paints, some cosmetics, jewelry, antiperspirants, hair dyes, dental plates and other items.

Cobalt is an essential element, being an integral part of vitamin B12 (cobalamin), which is necessary for the formation of myelin, an insulating layer encasing the nerves. It also supports the production of red blood cells and the conversion of folate into its active form. This element is also essential for the metabolism of fats and carbohydrates, as well as for protein synthesis. The average adult retains approximately 2-5 mg of vitamin B12, most of which is stored in the liver. Long-term vitamin B12 deficiency may result in depletion of the myelin layer of the large nerves and the spinal cord. An excess of cobalt is rarely reached in the normal diet. After ingestion of large amounts of cobalt chloride, the function of the heart muscle may deteriorate and a cardio-myopathy may develop. Cases of epidemic cardio-myopathy and congestive heart failure were reported among drinkers of beer during a period when cobalt salts were added to beer as foam stabilizers, thus posing the risk of an overdose of cobalt for beer drinkers. When cobalt intake exceeds 5 mg/day, abnormal thyroid functions and overproduction of red blood cells may appear. High cobalt concentrations influence the kidneys in their production of erythropoietin, a hormone that regulates the production of red blood cells. This hormone is used in the treatment of certain forms of anemia and was frequently misused in sports doping. This type of doping could not be detected or confirmed, since its effect is a part of the normal biochemical processes that take place in the body. Supplementation of erythropoietin may increase the number of red blood cells by 25-35%, but the risk of blood clot formation, strokes and heart attacks also is enhanced. Inhalation of high levels of cobalt also affect the lungs, causing asthma, pneumonia and wheezing. While estimated to be rare, allergic reactions, dermatitis and asthma also may be triggered by dermal exposure to cobalt, especially in individuals with nickel hypersensitivity.

COPPER

Copper is a metal that occurs naturally throughout the environment in rocks, soil, water, and air, and it is released into the environment by mining, farming and manufacturing operations and through waste-water release into rivers and lakes. This metal is used for manufacturing many different kinds of products such as wire, plumbing pipes and sheet metal. In agriculture, it is a common component of pesticides, as well as being an ingredient in wood, leather and fabric preservatives. Copper released into environment from natural sources, like volcanoes, decaying vegetation and forest fires, usually attaches to organic particles, clay, soil or sand. Tobacco smoke contains about 0.19 μg/cigarette of copper.

Copper is an essential element for plants, animals and humans. Daily copper requirements for adults are 1.5-3 mg and for children 1-2 mg. In the human body, copper is present in virtually all tissues, and the average adult retains about 80 to 120 mg of copper, most of which is stored in the liver. Copper is considered an essential element and plays a catalytic role in hemoglobin synthesis, one of its most important roles. Several other physiologically important enzymes such as cytochrome oxidase, superoxide dismutase and histaminase also contain copper. Copper is also a constituent of lysil oxidase, which is necessary for the formation of the cross-links of collagen and elastin; tyrosinase, which is associated with normal pigmentation and keratinization of the hair; and dopamine-beta-hydroxylase, where copper serves as a cofactor in the synthesis of both norepinephrin and adrenal hormones, which directly affect fatigue, mood and depression.

After oral absorption, copper is carried by plasma protein albumin to the liver where it binds to ceruloplasmin, a copper transport protein. Serum copper levels may increase during acute infections, pregnancy, estrogen hormone therapy and other conditions. Chronically elevated copper levels may result from diet rich in copper-containing compounds, such as chocolate products, coffee, tea, seafood, nuts and liver, or from water and other beverages stored in copper containers. Due to the synergism of copper and aluminum, higher intake of aluminum results in retention of higher levels of copper in the body. High cellular copper levels are considered to be among most prominent causes of many physical and mental health problems, such as Alzheimer's disease. In genetically caused Wilson's disease, large amounts of copper are accumulated primarily in the liver, provoking chyrosis due to inability of the impaired liver to eliminate this element. The retention of copper accumulated in the brain may then also provoke severe mental impairment.

Ingestion of large amounts of copper can affect the cardiovascular, gastrointestinal and nervous systems. In cases of high acute copper intake, hemolysis may be observed. Vomiting and diarrhea due to gastrointestinal irritation and convulsions and coma due to affected central nervous system may occur. Excessive copper levels in the body have been associated with physical and mental fatigue, depression and other neurological problems, such as schizophrenia, learning disabilities, behavioral problems, memory and concentration. In copper intoxication, muscle and joint aches are commonly displayed and are accompanied by sleep disorders and headache. Copper plays an important role in formation of new blood vessels, and in cases of excess copper intake, vascular degeneration may occur, as well as increase in cancer formation through angiogenesis, i.e. the vascularization of tumor tissue. On the other hand, low copper intake can increase the risk of high blood cholesterol, coronary heart disease and arithmia. Copper deficit also has been linked to thyroid problems, arthritis and fragile bones. Menke's disease is the rare problem of copper mal-absorbtion in male infants, accompanied by chronic copper deficit. It is speculated that some relationship also exists between copper levels and fatigue, anemia, vitiligo and premature graying of the hair. Impaired immunity, i.e., low neutrophil production, and allergies as well as nervous disorders, such as frequent mood change, may also develop in cases of inadequate copper intake.

FLUORIDE

Fluorine is an essential element for humans and is present in bones in about 4% as a calcium fluoride. This element has a profound influence on the mineralization and remineralization of tooth enamel. By these effects, fluoride greatly reduces occurrence of dental caries. Fluoride acts by changing tooth crystal into a less acid-soluble crystal fluoropatite.

As a common procedure in many developed countries, fluorination of drinking water raised many health concerns by drastically increasing the number of arthritis-like symptoms. Many controversial public-health issues relate to the use of fluoride in water treatment and to its benefits in the reduction of dental caries in children. Many studies are cited to demonstrate that high fluoride intake is responsible for lowered IQ and learning abilities in children, as well as for increased incidence of bone fractures due to fluoride-enhanced osteoporosis. According to some research, adverse effects of fluoride include the conversion of the bones into a more brittle structure and the formation of more porous dental enamel. It is also believed that fluorides are carcinogenic, inducing not only bone cancer, but also the cancers involving other tissues and organs. The latency period in fluoride-induced cancers is considered to be shorter compared to that of other known carcinogens. However, observations about fluoride-induced carcinogenesis have not yet been proven in humans. Controversy regarding fluoride toxicity has forced many countries to abandon the fluorination of public water supplies, while the research on this topic continues.

In animal studies, fluoride was shown to cause infertility in experimental animals, and the U.S. Food and Drug Administration (FDA) suggests that there may be a correlation between an increase in infertility in women and water fluorination. In research studies, fluorides are shown to lower the intelligence capacity of humans, with children being most susceptible [22]. It is thought that the effects of fluoride on brain function depend on the age of the subject and fluoride levels. The fetus may be exposed to fluoride through the placenta, since fluoride ions readily pass through it through natural transport mechanisms [23]. Prenatal fluoride exposure has demonstrated post-natal as well as life-long hyperactivity effects in children . When young children are exposed to fluoride, the opposite effects may be observed: depressed activity in exposed children is demonstrated. Neurotoxic effects of fluorides are thought to result from the depletion of constituents of brain cell membranes. Several studies demonstrate a link between aluminum and fluoride intake, showing that bioavailability of aluminum increases in the presence of fluorides, causing the accumulation of aluminum in brain tissue and possibly some pathological conditions, such as Alzheimer's disease. It has been proposed that fluoride impairs thyroid function and that virtually all 150 recognized symptoms of hypothyroidism may be linked to fluoride intake. The mechanism of the disruption of thyroid function may be explained by iodide replacement due to the resemblance between iodide and fluoride. Due to similarity of fluoride ions to OH^- ions, in both size and charge, excess fluoride accumulates in the bones, replacing OH^- ions in calcium hydoxyapatite bone structure. Fluoride accumulation is accompanied by severe bone pain and bone deformity, known as fluorosis.

Many observations concerning fluoride toxicity are based on toxicological studies conducted on experimental animals and are not yet definitively confirmed in humans. Many research observations await verification. As with many other inorganic toxicants, the dose-effect relationship is very important, especially with elements that are essential. For such elements the reliable borderline between essential and toxic dose should be accuratelly established. Even though many adverse effects of fluoride are evident, water is still fluorinated in many countries and fluoride toothpaste continues to be recommended by stomatologists.

Gold

Gold has been known and highly valued since ancient times. In nature, gold most often occurs in metallic state or in the form of alloys, usually with silver. Oxidation states of gold in its compound ranges from -I to V, but the most common are compounds of Au(I) and Au(III). Although most gold is used in jewelry, it also has a variety of medical uses. Isotope ^{198}Au is used to treat cancer, and some gold-thio compounds are used to treat arthritis. Modern industrial uses of gold include dentistry and electronics due to gold's superior resistance to oxidative corrosion. Pure gold is sometimes used as a food decoration or as a component of some alcoholic drinks because it is considered non-toxic and non-irritating when ingested.

Elemental gold is considered virtually non-toxic, because the element is very inert and does not react with biogenic molecules after ingestion. In oral ingestion studies conducted on experimental animals, soluble gold compounds, such as potassium gold cyanide, used in gold electroplating, have demonstrated hepatotoxic and nephrotoxic effects.

Gold compounds, such as disodium aurothiomalate, aurothioglucose and auranofin, are used in the treatment of inflammatory diseases such as rheumatoid arthritis or psoriatic arthritis. These compounds are efficiently transferred to the synovium where they suppress inflammation. However, the mechanism by which gold compounds efficiently reach inflammation sites is not completely clear. It is known that in plasma, gold compound first bind to albumin. Cysteinyl groups of albumin are replaced with gold compounds via sulfhydril group of the drug. After distribution to the synovium with the blood, the same process of thiol exchange occurs with the cell membrane transport proteins, enabling the introduction of the drug into the intracellular space. The, in the cells, gold impairs mitochondrial activity and induces cell apoptosis. All immune response cells are affected by gold, so the release of histamine and other inflammatory agents is reduced. First the components of the complement system are inactivated by the gold compounds. Gold compounds are taken by macrophages and then stored in lysosomes, where they reduce the activity of lysosomal enzymes. However, due to nonspecific absorption and strong pharmacological action, gold compounds also may provoke a myriad of side effects. In gold therapy, the occurrence of gold deposits in the lens may provoke visual impairment which subsides 3-6 month after cessation of the drug use. Pneumonitis, dyspnea and other pulmonary side effects also may be observed after a total dose of 300-100 mg [24].

Iron

Iron is an essential element that constitutes hemoglobin, the oxygen-carrying protein of the red blood cells, and is necessary to transport the oxygen to all organs and tissues. Iron poisoning mainly occurs by overdose of iron tablets, which are used to treat anemia and alcoholism or people who engage in long-term blood donations.

The most important determinant of the severity of iron poisoning is the amount of ingested iron. The estimated amount of elemental iron that provokes fatal outcome is between 180 mg/kg and 300 mg/kg of body weight [25]. Serious toxic effects are expressed in concentrations more than 60 mg/kg of body weight of elemental iron. Because natural routes of iron excretion do not exist, the body saves excess iron and stores it for hemoglobin synthesis. Iron is stored in reticulo-endothelial cells of the liver, as well as in other organs, and it is these storage organs that are most affected by a large iron intake. The corrosive properties of iron solutions can damage the gastrointestinal tract in oral exposure. Damage to gastrointestinal mucosa further enhances the absorption and increases the amount of iron that enters systematic circulation.

Four phases of symptoms development are recognized in iron poisoning. The first phase of iron poisoning occurs in first few hours after ingestion and the symptoms involve acute gastric disturbances. These effects arise due to the corrosive action of the iron solution on intestinal mucosa. Common symptoms are vomiting, diarrhea and abdominal pain. The vomit and stools are usually black due to the disintegrating iron preparation, but later may become blood-stained. If poisoning is mild, the person may recover after this phase without progressing through the other clinical phases. In severe cases of poisoning, drowsiness, coma or convulsions may be observed and are attributed to the effects of circulating free iron, which cause the release of substances with vasodilation properties from the liver.

The second phase begins between 6 and 12 hours after ingestion and lasts between 12 and 48 hours. In this phase symptoms abate as the iron binds to reticulocites in the liver. The third phase involves more severe symptoms, including liver damage and necrosis, which may lead to jaundice and renal failure. As a consequence of the damage to these organs, metabolic acidosis, hypoglycemia and shock are usually displayed. The incidence of mortality in iron poisoning, is highest in the third phase.

In the last phase of iron poisoning, the symptoms of gastrointestinal obstruction are the most pronounced, and the intoxicated person will enter the last phase after a period of approximately 2 to 5 weeks.

When deciding how to treat a victim of acute iron ingestion, it is important to know how much iron the person has ingested. An initial iron concentration of more than 5 mg/l of blood is an indication for urgent intravenous treatment with a chelating agent of deferrioxamine [25]. Some patients may exhibit severe allergic reactions to deferrioxamine, which may range from rash to anaphylaxis [8].

LEAD

Lead is found naturally in the Earth's crust. Plants growing in soil with high lead content can readily absorb the metal and introduce it into the food chain. The toxicity of lead was recognized early in history. Even so, lead has been used extensively for both industrial and domestic applications for hundreds of years. Vehicle exhausts, wastes and industrial effluents significantly contribute to environmental contamination with lead. Tetraethyl lead was used in the past, and in some countries is still used, as a petrol (gasoline) additive; however, the use of leaded petrol has been abandoned in many countries. In the past, lead was intensively used as a constituent of paints, causing many cases of intoxication in children. In recent years, concern has arisen about the relationship between chronic exposure to lead and impairment of intelligence and behavioral difficulties, especially in children; however, this link has not been definitively verified [2]. Currently, lead is used in the production of batteries, as a component of some insecticides. In addition, cigarette smoke releases approximately 0.017-0.98 μg of lead per cigarette [12]. In the past, tap water was often contaminated with lead because of the common use of lead pipes. Fresh lead surfaces are quickly covered with a layer of insoluble lead salts, formed with ions such as carbonates, sulphates and chlorides, which normally present in the water. This protective layer prevents further lead dissolution and water contamination. However, new pipes, in which this protective layer has not yet formed, can be a source of lead contamination. Lead release also may occur in cases where pipes remain void for some time, causing the protective layer and its partial laps to crack, exposing a new layer of soluble lead. After longer periods of disuse, pipes may be cleansed by allowing water to run for a while before caching and using or drinking the water. Lead pipes are nowdays replaced with pipes of other materials because of the many health concerns associated with lead poisoning.

All lead compounds are very toxic; however, organic lead compounds are considered the most dangerous. Elemental and inorganic lead compounds pose a risk mainly through absorption after ingestion or inhalation, but organic lead compounds, such as tetraethyl lead, are efficiently absorbed by a skin contact. Pulmonary absorption of small lead particles ($d < 1$ μm) is very rapid and efficient, and occurs after the inhalation of fumes resulting from the burning of lead paints. Gastrointestinal absorption of lead varies with the age of the individual. The susceptibility of children to lead poisoning is more pronounced because they absorb around 50% of an ingested amount, while adults absorb only 10-20% [12] .

After lead is absorbed, it binds to plasma components and hemoglobin, and is transferred to all other tissues. Lead compounds, especially organic compounds, are able to pass blood-brain and placental barriers. Excess lead, which is not excreted, accumulates in tissues, such as teeth or bones, where it replaces calcium ions due to their chemical similarity. About 90% of absorbed lead is stored in the bones, with a half-life of 25 years or more. In specific physiological conditions, such as pregnancy, lactation, osteoporosis or hypothyroidism, lead can be mobilized from the storage sites, increasing its concentration in plasma and provoking intoxication. Lead is excreted from the blood with a half-life of around 25 days. In tissues, lead is retained with a half-life of about 40 days [6]. Blood lead levels are not very useful as an indicator of lead exposure, since they demonstrate only the level of recent exposures. The consequences of previous exposure to lead can be observed in bones and may be detected by x-ray. Lead is excreted from the body mainly in urine, but also in feces, and small amounts may also appear in the hair, nails, sweat, saliva and breast milk.

The toxicity of lead results from its affinity for cell membranes and mitochondria, as well as its interference with mitochondrial oxidative phosphorylation. Lead exposure also affects sodium, potassium and calcium ATP-ase pumps. The activity of calmodulin, calcium-dependent intracellular messengers, as well as the activity of the brain enzyme protein kinase C, are also severely impaired by lead exposure. Lead stimulates the formation of inclusion bodies in the cells, which may translocate the metal into the cell nuclei and alter the gene expression [1].

Acute lead poisoning is accompanied with intense thirst and metallic taste, dry throat, nausea, vomiting, diarrhea and colic. Lead influences the central nervous system, provoking muscle pain, parasthesia (feeling of tingling and numbness in extremities), fatigue, convulsions and loss of consciousness. Severe poisonings are indicated by the occurrence of convulsions and coma, and death may result from cerebral edema or renal failure.

Edema observed in acute lead poisoning results from provoked capillary injury in the brain. In the peripheral nervous system, it is assumed that lead produces degenerative changes in the supporting cells of the nerves, and secondarily in the myelin sheaths of the nerves. Impulse transmissions along nerves are disturbed due to lead's interference with calcium. The impairment of excretory function results from direct effects of lead on renal mitochondria, decreasing the energy production necessary for active transport of the essential substances. In sub-chronic poisoning, hypotension, pulse weakening and severe anemia may be display, due to the inhibition of enzymes involved in hemoglobin synthesis.

Chronic lead poisoning may be recognized by the appearance of black "lead lines" at the gingival borders of the mouth, resulting from the excretion of lead in saliva, which follows the calcium pathways. It is sometimes difficult to recognize chronic lead poisoning. Initial symptoms may occur after a few weeks, a few months, or, in some cases, even a few years after first exposure. Constant fatigue, appetite loss, headache and anemia may develop. Lead and many other metals inhibit the enzymes involved in hemoglobin synthesis, so depletion of hemoglobin also causes the skin to become slack and pale. Lead colic--sudden cramps and pain in abdomen, are characteristic of chronic lead poisoning; however, these symptoms are not very specific and may also be linked to other causes. Signs of mental retardation or deficits in language, cognitive function, balance, behavior and school performance may appear in children as a result of their exposure to lead. Hyper-excitability, confusion, convulsions and progressive lethargy appear at blood lead levels greater than 0.3 mg/l. In chronic lead poisoning, kidney damage and signs of reproductive toxicity with adverse effects on spermatozoa may be demonstrated. Other manifestations of reproductive lead toxicity include pre-term birth, reduced birth weight and reduced mental ability of the child [7]. The treatment for lead poisoning depends on the form of the lead compound. Elemental and inorganic lead may be treated effectively with chelators, but organic lead poisoning causes the compounds very quickly to form strong ligands with organic components in the body, making initial chelation ineffective. Alkyl lead is eventually converted to its inorganic form by biotransformation processes, and this portion can then be treated with chelators [8].

LITHIUM

Lithium is widely distributed on Earth; however, it does not naturally occur in elemental form due to its high reactivity. Trace amounts of lithium are present in the oceans and in some organisms, although the element serves no apparent biological function in humans. Lithium and its compounds have several commercial applications, including use in the production of heat-resistant glass and ceramics, in alloys used in aircraft industry and the production of lithium batteries, due to the element's high electrochemical potential. The neurological effect of the lithium ion makes some lithium salts, like carbonate, citrate and orotate, useful in a class of mood stabilizing drugs used to treat manic depression and bipolar disorder.

The typical daily dietary intake of lithium is from 200 to 600 μg. Rich sources of lithium include eggs, milk and milk products, processed meat, fish, potatoes and vegetables. Lithium intoxication may occur from therapeutic over dosage with lithium salts. High lithium intakes are toxic due to disturbance of potassium/sodium balance. Lithium affects the flow of sodium through nerve and muscle cells in the body. Early features of high lithium intake include nausea, vomiting and diarrhea. Adverse effects on the central nervous system may provoke convulsions. Later symptoms of intoxication with lithium include kidney and liver damage and hypothyroidism, and increased or frequent urination may be observed as a consequence of impaired kidney function. Brain and neuromuscular system effects that are provoked by lithium toxicity are often accompanied by ataxia, tremors, staggering gait, mental confusion and in severe cases coma. Other adverse effects on the nervous system include blurred vision, ringing in the ears and severe shakiness. Supplementation of magnesium intake may be recommended for the treatment of lithium poisoning.

MANGANESE

Manganese is an essential trace element that plays an important role as a cofactor for many enzyme systems [1, 26]. Manganese poisoning is rare, but serious occupational exposure may occur in mining and refining operations or foundries, and in dry battery manufacture and arc welding., Manganese in the form of ferromanganese has important applications in the iron and steel industries. Some fertilizers and pesticides contain manganese, and the element is also extensively used in chemical industries as a catalyst for many chemical processes. Manganese is released when fossil fuels are burned, and is also found in cigarette smoke (0.003 μg/cigarette) [26]. This element may occur in eight different valence states, and cationic manganese compounds are considered to be more toxic than anionic compounds. In manganese intoxication, the oral route of absorption and inhalation of the compounds in the form of dust, are most important.

Acute inhalation of manganese causes respiratory irritation and pneumonia, accompanied with fever, which does not respond to antibiotics. Chronic manganese poisoning was frequently observed in workers in manganese mines, who were usually exposed to 5-250 mg/m^3 of manganese. Symptoms of chronic manganese poisoning typically occur within several months or years of exposure. These symptoms include a mental disorder resembling schizophrenia, as well as hyperirritability, violent acts, hallucinations and difficulty in walking. In excess, manganese can cause irreversible nervous system damage. Some studies hypothesize that the mechanism of manganese neurotoxicity results from the disturbance of the depolarization processes of neural cells. Studies show that the effects of manganese on the brain result from lesions and degeneration of the basal ganglia. Manganese is also known to block calcium channels. In chronic manganese poisoning, the depletion of dopamine and the production of some neurotoxins, such as dopamine quinone and hydrogen peroxide, are observed. Long-term exposure produces symptoms similar to those of Parkinson's disease, such as a masked, expressionless face, tremors, gait disturbances and psychiatric symptoms [1]. It has also been suggested that Parkinsonism in patients with chronic liver disease may be due to manganese accumulation in their basal ganglia. In these patients, manganese accumulates in the liver due to their inability to metabolize the element. In some cases a similarity with the symptoms of Wilson`s disease also may be noticed, such as liver cirrhosis.

Severe manganese poisoning is usually treated with chelator compounds, which are not very efficient and have much higher affinity to lead or mercury [8]., Thus, calcium EDTA is recommended as a chelating agent for manganese poisoning. Levodopa also can be used as an aid to symptoms of Parkinsonism.

MERCURY

Use of elemental mercury is related mainly to industrial applications, such as liquid contact material for electric switches; in the manufacture of scientific instruments such as thermometers and barometers; and in the production of dental amalgams and zinc amalgams used in electric batteries. In the chemical industry, mercury is used as the electrode material [27] and as a catalyst for organic chemical reactions. Inorganic mercury compounds are widely used in the paint industry and in the manufacture of pigments and in anti-fouling marine paints. In the past, organic mercury compounds were widely used as fungicides and in the manufacture of explosives. Mercury salts, while very poisonous, had never been used for intentional poisonings due to their strong, metallic taste. Chemist Louis Jacques Thénard (1777-1857) accidentally drank mercury(II)-chloride during one of his lectures on mercury, but was saved by one of his students, who administrated albumin water. Albumin binds a free mercury ion and forms mercury-albuminate, which is insoluble in water. This reaction also explains the disinfectant properties of soluble mercury salts, which were widely used in the past. Mercury ions destroy bacterial cells by forming albuminates. Due to its high density, metallic mercury used to be applied for the treatment of intestinal occlusions. Elemental mercury, if orally taken, will not cause serious toxic effect, because acidic conditions are not sufficiently strong to cause its dissolution and also because its absorption from the gastrointestinal tract is very poor.

In defining mercury toxicity, the distinction between elemental, inorganic and organic mercury is much more important than the oxidation state of the element, and organic mercury compounds are generally considered the most toxic· Organic mercury compounds, such as methylated, ethylated or phenilated compounds, are volatile at room temperature. Among them, alkyl mercury compounds are the most toxic, particularly methyl mercury [26]. Mercury and its compounds are considered potent inhibitors of many enzymes, due to the binding to sulfhydryl groups. Mercury also blocks the cellular transport of potassium and sugars, due to binding to S-H groups in or on the cell membrane. Once inside the cell, the metal may be converted into inactive form, or it may react with enzymes and other cellular compounds.

Elemental mercury vapors are absorbed through lungs and are easily oxidized in red blood cells to divalent ion. The half-life of mercury ions in the blood is around 60 days. Oxidation also occurs in the brain and fetal tissue, where the toxicant remains trapped after it is oxidated by the activity of catalase and hydrogen peroxide. In brain and kidney tissues, mercury may be retained for years. Chronic exposure to mercury vapors provokes a characteristic symptom of tremor. Other symptoms include excitability, memory loss, insomnia and timidity.

Inorganic mercury may be absorbed through the gastro-intestinal tract and the skin. Symptoms of mercury poisoning with soluble salts appear immediately after ingestion. An unpleasant metallic taste is a typical sign of acute mercury poisoning and is accompanied with a feeling of burning and constriction in the throat. These symptoms soon spread to other gastrointestinal organs, followed by the development of mucilage and green and bloody vomiting. The face turns pale and the body temperature drops significantly. Urination is very difficult and in some cases complete anuria may occur. In high concentrations of mercury, damage of the mucous membranes occurs after oral intake and further absorption is enhanced. Only small amounts of inorganic mercury compound cross blood-brain barrier because inorganic compounds are metabolized to metallic mercury and Hg^{2+}, which cannot be efficiently transported through the barrier. The half-life of inorganic mercury compounds in the blood is around 40 days, but the compounds may be retained in the kidneys in the form of Hg^{2+} for a longer period. Chronic poisoning with inorganic mercury compounds is accompanied by hyper-salivation and gum inflammation, and teeth and hair fall out. Intellectual and physical abilities are usually diminished and disturbances in nervous system are usually expressed as shivering in the limbs and tongue. These symptoms are accompanied by acute renal failure, cardiovascular collapse and shock. A lower level of chronic mercury intake may cause hypertension, tachycardia and loosening of the gums.

Organic mercury compounds, particularly methyl mercury, are volatile and are easily pulmonary absorbed, but they are also well absorbed in gastrointestinal tract. The symptoms of poisoning with organic mercury start developing when more than 1.7 mg/kg of body weight of mercury has been ingested, and fatal dose is estimated to be between 10 and 60 mg/kg of body weight. The compounds are distributed to the CNS, where demethylation takes place, leading to neurological damage [28]. Acute ingestion of organic mercury compounds can cause diarrhea, blisters in the upper gastrointestinal tract and neurological symptoms, such as muscle weakness, irritability and sensations of tingling, pricking, or numbness on the skin (parasthesia). Other symptoms of effects on the nervous system include impaired peripheral vision, memory loss and depression. Chronic exposure to organic mercury may cause symptoms similar to chronic inorganic mercury poisoning, and tremor is usually accompanied with parasthesia and muscle weakness. As a highly hydrophobic substance, organic mercury compounds easily pass placental and brain barriers and are partially excreted via breast milk. They are retained in the kidneys and CNS, with a half-life of about 70 days. Methyl-mercury is considered the most toxic form of mercury. It is present in fish tissue in high concentrations, because the microbial processes that occur in anaerobic conditions cause the methylation of inorganic mercury species present in contaminated water [29, 30]. This methylated form is hydrophobic and it bioaccumulates in the tissues of aquatic organisms, biomagnifying with a factor of 10,000 or more. Meythyl-mercury further enters the food chain with fish consumption and irreversibly binds to brain tissue, provoking neurological damage, paralysis and other irreversible symptoms. A case of mass population poisoning with methyl-mercury occurred at Minamata Bay, where a local company continuously discharged about 27 tons of mercury compounds from 1932 to 1968. More than 3,000 people were affected by a strange degeneration of nervous system and started to demonstrate strange symptoms, such as uncontrolled shouting, involuntary movements and sudden unconsciousness.

The inhalation of metallic mercury should be treated with hydrocortisone, while in acute ingestion of mercury salts, gastric lavage and emesis should be induced. Oral administration of polythiol resins for binding unabsorbed mercury from the gastrointestinal tract is also recommended. Dimercaprol, succimer and penicillamine also may be administered as chelating agents,. Chronic inorganic mercury poisoning is best treated with penicillamine. For organic mercury poisoning, chelating agents are not recommended, because they can increase mercury levels in the brain. Recovery from mercury poisoning can last months or even years after exposure and even with effective chelator treatment, may often remain incomplete.

Molybdenum

The ability of molybdenum to withstand extreme temperatures without significantly expanding or softening makes it useful in applications that involve intense heat, including the manufacture of aircraft parts, electrical contacts, industrial motors and filaments. The element is also widely applied in the production of corrosion-resistant alloys. Molybdenum is an essential nutrient for animals and humans, because it is a component of more than 20 different enzymes, including sulphite oxidase, xanthine oxidase and aldehyde oxidase. Sulphite oxidase assists in the inactivation of sulphite toxins in the body, such as sulphite food preservatives and in the metabolism of sulfur-containing amino acids. The oxidation of xantine, a purine base found in most body tissues and fluids, is an important part of purine catabolism and is catalyzed by xanthine oxidase. Molybdenum-containing enzymes also have an important role in the inactivation of carcinogenic nitrosamines, as well as in normal cell function and nitrogen metabolism. Milk, beans, spinach, liver, grains, peas and other dark green leafy vegetables contain molybdenum in high concentrations and can be a good dietary source.

The tissue content of molybdenum is low, with the highest concentrations occurring in the liver, kidney, small intestine, adrenal glands and bones. The presence of dietary copper and sulfate affects the amount of molybdenum absorbed and retained by the body, and it is assumed that molybdenum interacts with these compounds in living organisms, although the mechanism of interaction has not yet been established. Absorption of molybdenum in gastrointestinal tract varies from 25% to 93%, with soluble molybdenum compounds being absorbed to a higher degree. In plasma, molybdenum is bound specifically to α_2-macroglobulin.

Molybdenum is primarily excreted in the urine and experimental data from humans suggest that up to 80% of the absorbed amount is excreted via urine. Normally, only small amounts are excreted in the feces, except in certain gastrointestinal disorders. Symptoms of molybdenum poisoning are demonstrated for element contents higher than 100 mg/kg of food or water. Signs of molybdenum toxicity include diarrhea, anemia and high levels of uric acid in the blood. Elevated uric acid levels are associated with the onset of gout and are assumed to be caused by the induction of xanthine oxidase by high molybdenum intake. Occupational exposure by inhalation to molybdenum in dusts has been associated with pneumoconiosis. In animal studies, trivalent form of molybdenum is shown to be less toxic than the hexavalent form. Most of the clinical signs attributed to molybdenum toxicity arise from impaired copper metabolism and are the same as those produced by simple copper deficiency. It is assumed that molybdenum affects the copper metabolism by replacing it from the complex with metallothionein in the liver. An excess of molybdenum forms an undissociated Cu-Mo-S complex thereby creating a picture of copper deficiency.

NICKEL

Nickel occurs in the environment naturally, but is also discharged into the environment by human activities. The element is released into the atmosphere by the burning of fossil fuels, mining and refining operations, as well as the incineration of municipal waste. In the chemical industry, nickel is used for electroplating. Cigarettes contain up to 0.51 μg of nickel per cigarette. Organic forms of nickel are much more toxic than its inorganic forms. Even though poisoning by nickel is very rare, nickel is highly irritating to the skin and many people develop allergies in contact with jewelry made with nickel.

Nickel deficiency has been linked to low blood glucose levels, abnormal bone growth, poor absorption of ferric iron and altered metabolism of calcium and vitamin B12.

Most absorbed nickel binds in plasma to albumin or nickeloplasmin. Nickel is found in highest concentrations in lung, kidney and some hormone-producing tissues. Although nickel-specific enzymes have not yet been identified, it is assumed that nickel can activate or inhibit a number of enzymes containing other elements in their cofactors. Production and activity of some hormones (prolactin, adrenalin, noradrenalin, aldosterone) are also significantly influenced by nickel concentrations. Once entered into a cell, nickel alters cell membrane properties by affecting ion channels. Significant concentrations of nickel found in RNA and DNA confirm that it reaches the chromosomes, where it interacts with these nucleic acids. In immune systems affected by nickel, response to stress and resistance to infection are diminished. Under some conditions, large amounts of nickel may decrease magnesium levels in the body and provoke accumulation of iron and zinc in certain tissues.

Approximately 10% of orally ingested nickel is absorbed and severe toxic effects are demonstrated in a dose of several grams of ingested element. Its gaseous form, nickel carbonyl, is highly toxic and is readily absorbed from respiratory tract. Most nickel is eliminated in feces and urine, and only small amount through sweat. Skin reactions such as itching, burning, redness or other rashes are the most common symptoms of nickel sensitivity. Nickel induced asthmatic attacks are less frequent and possible hypersensitivity reactions have been observed in some individuals after inhalation exposure to nickel.

The toxicity of chronic nickel exposure is reflected primarily in the cardiovascular system, where its presence is also dangerous due to carcinogenic effects which may arise, especially after inhalation exposures to nickel compounds that are known to provoke lung and nasal cancers [7]. Studies have shown that occupational inhalation of poorly water-soluble nickel compounds, such as nickel subsulfide and nickel oxide, is associated with an increased risk of lung cancer [1]. The toxicity of nickel in long-term exposure is related to the disruption of metabolic pathways and the replacement of the essential metals from their metallo-enzymes, followed by the loss of enzymes activity.

NITRITES AND NITRATES

Acute methemoglobinemia is the most important adverse health effect linked to excessive nitrate or nitrite intake. Methemoglobinemia is condition in which divalent iron from hemoglobin is oxidized to trivalent by some oxidizing agents. Trivalent iron in hemoglobin (methemoglobin) loses the ability to bind the oxygen and provokes oxygen insufficiency in all tissues, which is usually accompanied with the blue color of the skin. Besides nitrates, other oxidizing substances can provoke same effects in the body, and methemoglobinemia may also arise from systemic acidosis as a result of diarrhea and dehydration. Depending on the percentage of total methemoglobin, the symptoms of oxygen deprivation may differ and include cyanosis, arrhythmia, dizziness, lethargy, coma and convulsions [31-36]. Methemoglobin is spontaneously reduced back to hemoglobin by reductase enzymes. A certain percentage (1%) of methemoglobin normally exist in humans and is formed by endogenous reducing compounds. The toxic effects of nitrites result from their conversion into nitrates in the body. Toxicological effects of all nitrate compounds are the same, independent of whether the nitrate-containing compounds are ingested or inhaled, or whether the nitrates are produced *in vivo* after nitrite ingestion. In newborns and young children, the intestinal microflora are insufficiently developed, which limits the oxidation of nitrites in the gastrointestinal tract, and make it more susceptible to nitrite toxicity. Because of its reductive properties, ascorbic acid efficiently prevents nitrate formation in the gastrointestinal tract,.

Great health concerns about nitrates and nitrites have arisen due to wide-spread use of these compounds as preservatives and color-enhancing agents in the meat industry and because of their presence in drinking water. Cured meats are a significant source of carcinogenic nitrosamines, which form easily in the presence of nitrites. Fried bacon has the highest nitrosamine content because it is fried at very high temperatures, which are responsible for the formation of nitrosamine.. Ascorbic acid and a cheaper isomer of ascorbic acid, erythorbic acid, are normally added to preserved meat products in order to reduce the formation of nitrosamines. Other antioxidants, such as alpha-tocopherol, also may be added to cured meat products.

The amount of nitrites supplied by a typical diet is approximately 0.1 mg nitrite/day, however diet rich in leafy vegetables as well as the consumption of nitrite – rich drinking water may raice the average daily intake of nitrates up to 100-170 mg . Poisoning by nitrates, or nitrites after its conversion to nitrate, diminishes the ability of hemoglobin to carry oxygen throughout the body. Adverse symptoms occur when over 30% of hemoglobin has been converted to methemoglobin. The symptoms include arrhythmia, headache, nausea and vomiting and in severe cases may be accompanied by seizures.

Maternal exposure to environmental nitrates and nitrites may increase the risk of pregnancy complications such as anemia or premature labor. The maternal transfer of nitrates, nitrites and *N*-nitroso compounds across the placenta, and their potential effects on fetal death and malformations also have been reported [37].

Since drinking water is a significant source of exposure to nitrates, many research works are directed to the investigation of their carcinogenic activity related to long-term exposures. Elevated risk of non-Hodgkin's lymphoma and cancers of esophagus, nasopharynx, bladder and prostate have been reported in experimental animals, but have not been proven in human toxicological studies [38, 39]. An increased incidence of stomach cancer was observed in one group of workers with occupational exposures to nitrate fertilizer. However, epidemiological investigations have not yet shown an unequivocal relationship between nitrate intake and the risk of cancer.

Even though nitrates and nitrites are considered very toxic and carcinogenic, it is speculated that nitrates from the diet also may enhance host defense mechanisms against gastrointestinal pathogens by modulating platelet activity. Other beneficial effects of nitrates include the enhancement of the gastrointestinal motility and microcirculation [40-42].

Selenium

Selenium is naturally occurring element [43, 44], widely used in electronic, glass and pigment industry, as well as in metallurgy and agriculture. The element occurs in four valence states, as selenate (VI), selenite (IV), selenides (-II) and elemental selenium.

Natural sources of selenium include certain selenium-rich soils, or certain toxic plants such as locoweed (*Astragalus* and *Oxytropis* genera, family Fabaceae) which bioconcentrate the element. Anthropogenic sources of selenium include coal burning, mining and smelting of sulfide ores. Tobacco smoke contains about 0.001-0.063 μg of selenim per cigarette.

Selenium is an essential element being a component of enzymes involved in antioxidant activity and thyroid hormone metabolism. Required daily intake of selenium is 0.04-0.1 mg. Glutathione peroxidase is a selenium-containing enzyme that catalyzes the destruction of reactive peroxides in the body and protects cells and membranes from oxidative damage [45]. The enzyme can be found in blood platelets and white blood cells, and it plays an important role in the body's immune system and clotting mechanisms. Reduced glutathione peroxidase activity in a selenium-deficient diet increases the risk of cardiovascular diseases and is implicated in a higher risk of liver, lungs, breast, skin, colon, rectal and prostate cancer.

In several intra- and extra-cellular glutathione peroxidases, iodothyronine 5H-deiodinases, and in thioredoxin reductase, selenium is incorporated into the enzyme`s active center as selenocysteine. As selenomethionine, selenium is incorporated non-specifically into many different proteins, as it competes with methionine in general protein synthesis.

Vitamin E and selenium act synergistically and tend to enhance each other's effects. Vitamin E prevents the formation of peroxides, while glutathione peroxidase provokes its destruction. Even though acting synergistically, selenium and vitamine E compete in some biochemical processes.

Selenium absorption in gastrointestinal tract is mainly dependent on the solubility of the ingested compound, its chemical form and on the dietary ratio of selenium and sulfur. Selenium absorption from food is influenced by many factors, including the nature of the food source and the method of food processing. Seafood, for example, is high in selenium, but only the small portion of selenium is absorbed, possibly due to interaction with other elements present in the food. It is known that selenium forms stable selenides with mercury and other heavy metals and that these metals are almost always present in high concentrations influencing selenium biovailability. Organic forms of selenium (selenomethionine, selenocysteine) are better absorbed from the gastrointestinal tract in comparison to inorganic forms such as sodium selenite. In general, the absorption of selenium compounds from food varies from 55% to 68% [46]. After absorption, different chemical forms of the element are transformed differently into biologically active forms and these processes most strictly limit the selenium's bioavailability. The lowest bioavailability is attributed to elemental selenium. Both selenomethionine and selenite are considered as biologically active forms of the element, even though their biochemical pathways are different [47].

Once absorbed, selenium incorporates in enzymes, such as glutathione peroxidase, replacing the sulfur-containing amino acids cysteine and methionine. Selenium is also incorporated to some extent into hemoglobin and myoglobin.

Besides being essential and their significant physiological role, all selenium compounds are toxic. Elementary selenium is not highly toxic if orally taken, but it can provoke serious damage to the skin in dermal exposure. In the body, selenate is reduced to selenite which is considered a biologically toxic form and which provokes anemia and increased level of blood billirubin. Toxic effects in oral exposures are manifested in selenium intake higher than 4 mg for a person of average weight. Signs of selenium toxicity may resemble the symptoms of arsenic poisoning to some extent. Acute selenium poisoning can be recognized by the characteristic garlic-like odor of pulmonary excreted dymethylselenide, which is accompanied with vomiting and muscle cramps. Histopathological changes are observed as liver and kidney necrosis and in changes in the heart muscle. Estimated LD_{50} for sodium selenite 2.3-13 mg/kg. Inorganic selenium compounds have demonstrated mutagenic effects in toxicological studies, as well as the provocation of cataracts at high doses. In conducted studies, organic selenium has shown lesser toxicity and did not express mutagenic or oxidizing activity. Inhalation of selenium containing particles, such as selenium-dioxide, hydrogen-selenide or selenium dust, provokes severe damage in the respiratory tract, as well as the impairment of liver and kidney functions, depending on exposure duration. In the selenium photoelements industry, of occupational exposures to the element usually occur due to inhalation of the dust, which may contain 0.8-16% of selenium.

Excess selenium is excreted in urine as trimethylselenium ion and via the lungs in the form of volatile dimethyl selenide, which gives the breath a characteristic garlic-like odor. Mass selenium poisoning has been reported in some regions of the world that are naturally rich in selenium. Endemic selenium-induced diseases have been identified in China, where daily exposure to selenium from foods and water is estimated to be 0.75-5.0 mg [48]. Early signs of chronic selenium poisoning include nausea, weakness and diarrhea. With continuation of selenium intake, deformed fingernails and hair loss may be demonstrated, as well as the symptoms of adversely affects to the nervous system, such as coordination difficulties. Excessive long-term intake of selenium increases the potential risk of developing osteoporosis, an enlarged prostate and dermatitis. Symptoms are frequently accompanied with reduced glucose tolerance and peripheral neuropathy which leads to numbness.

SILVER

The principal sources of silver are the ores of copper, copper-nickel, lead and lead-zinc. Silver has been known since ancient times and has long been valued as a precious metal. Metallic silver is used in jewelry and silverware, for alloys and electroplating, and in the processing of food and beverages. Silver nitrate is used in photographic industry, ink manufacturing, porcelain coloring, and in water treatment, where it is widely used as an antiseptic. Colloidal silver is often recommended as a dietary supplement to protect from infections; however, the benefit of such products in relation to health risks is a subject of controversy. The antiseptic properties of silver arise from oligodynamic effects inhibiting the expression of enzymes and other proteins essential to ATP production. Oligodynamic effect is the toxic effect of metal ions on living cells and microorganisms, that occurs due to binding to reactive groups of the enzymes that result in their precipitation and inactivation. Besides silver, many other heavy metals, such as mercury, silver, copper, iron, lead, zinc, bismuth, gold and aluminium express oligodynamic effect.

The accidental or intentional ingestion of large doses of silver nitrate primarily provokes corrosive damage to gastrointestinal tract, followed by abdominal pain, diarrhea and vomiting. Dehydration and secondary effects may provoke convulsions and shock, and in very large doses, death. The estimated lethal dose of silver nitrate for humans is approximately 10 g.

In sub-chronic exposures to either metallic silver or silver compounds in small doses over periods of months to years, a discoloration of the skin, very specific for silver (argyria) is observed. Argyria is a gray or blue-gray discoloration of the skin that results from silver deposition in the tissues in the form of insoluble silver salts, such as silver chloride and silver phosphate. Damage to the skin is irreversible and silver deposits are also formed in internal organs, mucous membranes, eyes and the body's elastic fibers. Deposited insoluble silver salts are further oxidized in the tissues into metallic silver, which is subsequently photoreduced into black silver sulfide, which is responsible for characteristic skin color. Small dissolved quantities of soluble silver sulfide may form albuminates, which tend to form complexes with amino or carboxyl groups of proteins, RNA or DNA. Albuminate complexes may be reduced with ascorbic acid or catecholamines. Following oral or inhalation exposure to silver compounds, humans excrete silver primarily in the feces (about 90%) with only very minor amounts excreted in urine.

Chronic inhalation exposure to silver oxide or silver nitrate dusts results in upper and lower respiratory irritation and deposition of granular silver-containing deposits in eyes, disturbing night vision. Silver nitrate solutions are highly irritating to skin, mucous membranes and eyes. Mild allergic responses have been observed in dermal contact with silver. and cases of immediate and delayed contact dermatitis have been reported.

THALLIUM

Thallium is found in trace amounts in the Earth's crust and is emited into the atmosphere by anthropological industrial activities, such optical glass, semiconductors, low-temperature thermometers and switches. The element is also a constituent of green fireworks and tobacco smoke, and it is frequently used in catalytic chemical processes. Some thallium salts are strong rodenticides, but have been replaced with a new generation of pesticides [7].

Thallium salts are extremely poisonous and ingestion of more than 10-15 mg/kg of the compound can be lethal. Thallium poisoning most frequently occurs due to ingestion of its salts, which may be confirmed by X-rays. Inhalation of dust or fumes from smelting is another possible way of intoxication with thallium compounds [2]. Within the first few hours after absorption, thallium distributes throughout the vascular space. After 48 hours the element starts entering CNS and other tissues. Similar to other toxic metals, within first 12-48 hours, signs of thallium poisoning are related to disturbances to gastrointestinal tract, provoking nausea, vomiting, diarrhea, gastritis and damage to the pancreas and salivary glands. After 2-5 days, the nervous system responds with parasthesia or hyperesthesia (an abnormal increase in sensitivity to stimuli, such as sound, taste or texture), as well as muscle pain. Accompanying symptoms almost always include headaches and respiratory depression. Severe pain, loss of reflexes, convulsions, coma, dellirium, dementia and psychosis also are commonly displayed. Involuntary scanning movement of the eyes also is a typical sign of thallium poisoning and occurs due to damage to optic nerves. Subsequent arrhythmia, tachycardia and skin rash may appear, and after 2-3 weeks the hair starts falling out. In humans, lethal oral doses of thallium ranging from 0.9 mg/kg to 9.4 mg/kg have been reported, which are equivalent to inhalation exposure of a male of the average weight to concentrations ranging from 40 mg/m^3 to 450 mg/m^3 for 30 minutes, assuming a breathing rate of 50 liters per minute and 100% absorption [49, 50].

The mechanism of thallium toxicity is due mainly to binding to sulfydryl groups in the body, disturbing many biochemical processes. By size and charge thallium resembles potassium and affects bodily processes related to flux of potassium across membranes. Only one-third of plasma thallium content is excreted by urinary tract. Remaining amounts are excreted mainly by intestinal mucosa; however, the majority of the intestinally excreted amount is reabsorbed significantly, prolonging the half-life of the element.

In cases of thallium ingestion, further absorption of the element is stopped by inducing emesis and performing gastric lavage. Potassium ferrihexacyanoferrate (Prussian blue or Berlin blue) should be administrated in order to trap the thallium ions in the form of complex in gastrointestinal tract to prevent further absorption. In contrast to poisoning with other metals, activated charcoal is useful in treating thallium poisoning and aids in reducing the severity of symptoms by disrupting the enterohepatic circulation. Other treatments that may be helpful include forced diuresis, treatment with potassium chloride and peritoneal dialysis [1]

TIN

Tin is found naturally in the Earth's crust, but it is also released into environment by mining and coal and oil combustion, as well as during production and use of tin products. Tin is used mainly in tinplated containers, but it is also extensively used in solders, in alloys such as bronzes and in more specialized alloys such as dental amalgams or titanium alloys used in aircraft engineering. Tin-containing antifouling paints and agents were extensively used in the past until it was proven that they are persistent environmental pollutants. Tin compounds are used as stabilizers in the plastics industry and in the production of food containers. Inorganic tin compounds have important applications in a variety of industrial processes, such as glass strengthening, color bases, catalysts, stabilizers in perfumes and soaps and in dentistry as anticariogenic agents.

The typical daily dietary intake of tin is from 1 mg to 40 mg. After absorption, all forms of tin are metabolized to other, less toxic forms. It is estimated that oral absorption of tin from the diet is less than 5%, but may be as much as 20% because is highly influenced by the presence of anion complements and by its own oxidation state.. Divalent compounds are more readily absorbed than tetravalent. Tin distributes mostly to the bones, lungs, kidneys and liver. Penetration through the blood-brain and placental barriers appears to be very slight for tin compounds. Absorbed inorganic tin is excreted mainly in urine. The fraction excreted with bile varies with the type of the compound and is estimated to be below 15%.

Toxicity of tin compounds may be explained partially by mitochondrial damage and disturbance of oxidative phosphorilation in cells due to inhibition of oxygenase synthesis. Organic tin compounds are much more toxic than inorganic tin salts and are bioaccumulated by plants, fish and animals. High intake of inorganic tin compounds can cause abdominal pain, anemia and liver and kidney problems [7]. After tin absorption, the production of catecholamines may be stimulated and this may lead to hyperglycemia, changes in blood pressure and damage to immune system. In poisoning with organotin compounds, psychomotor disturbances are observed, including tremor, convulsions, hallucinations and psychotic behavior. In some cases cerebral edema may occur. Organotin compounds (triethyl tin) are well absorbed dermally, causing severe irritation and burning to the skin. Technological operations associated with the processing of tin are known to cause excessive occupational inhalation exposure to tin oxide, which may result in a benign pneumoconiosis (stannosis), many cases of which have been reported in the past. The symptoms of tin hydride inhalation are similar to that of arsine gas, but the effects are less severe, and hemolysis is omitted.

VANADIUM

Vanadium is a trace element that is present at nanogram levels in most plant and animal tissues. The element enters the environment naturally from processes such as weathering and volcanic eruptions, as well as from human activities. Vanadium is also released during the burning of fossil fuels and is a constituent of tobacco smoke.

The most common valence states of vanadium are III, IV and V. The pentavalent form (VO^{3+}) predominates in extracellular body fluids, whereas the tetravalent form (VO^{2+}) is the most common intracellular form. The highest concentrations of vanadium in mammalian tissues are in the kidneys, spleen, liver, bone, testes and lungs. Total human body content of vanadium is believed to be between 0.1mg and 1 mg. *In vitro* vanadium is shown to have regulatory effects on numerous enzymes, including protein tyrosine phosphatases and kinases, and to mimic insulin effects to a substantial degree, enhancing the sensitivity of insulin receptors on membranes. *In vivo* studies demonstrate that vanadium is essential with respect to normal iodine metabolism and thyroid function. In developed countries, vanadium deficiency most often occurs due to the increased intake of antagonistic elements, such as chromium, rather than from inadequate diet. The bioavailability of orally administered vanadium is very low, usually less than 1%, and most of the ingested element is excreted unabsorbed, indicating its low oral toxicity.

Pentavalent compounds are the most toxic and the toxicity of vanadium compounds usually increases with the valence state. Most of the toxic effects of vanadium compounds result from local irritation of the eyes and upper respiratory tract rather than systemic toxicity. Vanadium is generally more toxic when inhaled than when taken orally. The lungs absorb soluble vanadium compounds (V_2O_5) well, but the absorption of vanadium salts from the gastrointestinal tract is poor. The excretion of vanadium by the kidneys is rapid with a biological half-life of 20-40 hours in urine. Animal studies have shown that toxic levels vary considerably, depending on the age and species of test animals, as well as on the content of proteins and other trace elements in the diet. Symptoms of acute vanadium toxicity demonstrated on experimental animals include dehydration, diarrhea, breathing difficulties and cardiac irregularities. The LD_{50} in rats has been determined to be approximately 50 mg/kg expressed as sodium metavanadate. In prolonged exposures, renal function is impaired and weight loss and depressed growth are observed. In human studies, relatively large vanadium doses of 50 mg and more may lead to the occurrence of green tongue and gastrointestinal disturbance [51]. It is believed that high doses of vanadium inhibit the synthesis of cholesterol and acetylcholinesterase in humans. The inhalation of vanadium compounds may produce both acute and chronic respiratory effects. Inhalation of vanadium pentoxide provokes pulmonary irritation, as well as irritation to the eyes and the upper respiratory system, leading to cough, wheezing and dyspnea.

ZINC

The human body contains between 1.5 g and 2.5 g of zinc, concentrated in specialized areas of brain, pancreas, eyes, prostate, liver, kidneys and adrenal glands. The element is present in nearly all cells, particularly in the nucleus. Zinc is an essential element that has structural, enzymatic and regulatory roles. More than 60 enzymes require zinc for their activity, including RNA polymerases and superoxide dismutase, which protect the cells from free radical attack. The metal also supports the neuronal activity and memory functions, since it is actively taken up by synaptic vesicles. Zinc plays a vital role in protein synthesis, and in the development of reproductive organs, prostate functions and male hormone activity, by maintaining the normal serum testosterone. It is also required for the transport of vitamin A from the liver. It has been shown that zinc enhances white blood cell function and increases the number of T-lymphocytes, and is therefore contraindicated in leukemia. Zinc metabolism may be affected by physical stress, hormones, cytokines and toxins.

The bioavailability of zinc from food depends on the presence of other elements, such as copper, iron, calcium, cadmium and lead. In some foods, zinc bioavailability is low because of binding to other food components, such as o phytates or oxalates in grains and vegetables. Zinc deficiency has been implicated as a factor in birth defects and low birth weight, as well as delayed sexual development and impaired learning.

High levels of zinc exposure occur mainly from contaminated food and water. Ingestion of large amounts of zinc may occur when acidic food or drink are consumed from galvanized containers. Excessive zinc intake eventually affects the balance of other important nutrients such as iron, calcium, selenium, nickel, phosphorus and copper, as well as vitamins A, B1 and C. Emetic effects and other gastrointestinal problems occur with zinc intake of more than 150 mg/day. In chronic exposures to elevated zinc concentrations, hair loss, anemia and depressed immune functions may be observed. Adverse effects affecting reproductive system include loss of libido, impotence, prostatitis, ovarian cysts and menstrual problems. In some cases, muscle spasms and dizziness are quite prominent. Long-term high oral doses also affect the level of HDL cholesterol by decreasing it and as well as serum ceruloplasmin activity. Inhalation of zinc fumes can cause dyspnea, flu-like symptoms and neurological damage.

REFERENCES

[1] Hu, H. Heavy metal poisoning. In: Fauci, AS; Braunwald, E; Isselbacher, KJ; Wilson, JD; Martin, JB; Kasper, DL; Hauser, SL; Longo, DL. *Harrison's principles of Internal medicine*. 14th ed. New York: McGraw-Hill; 1998.

[2] Baldwin, DR; Marshall, WJ. Heavy metal poisoning and its laboratory investigation. *Ann.Clin. Biochem.* 36, 1999, 267-300.

[3] Švarc-Gajić, JV; Suturović, ZJ; Marjanović, NJ; Kravić, SZ. Determination of As(III) And As(V) in Waters By Chronopotentiometric Stripping Analysis. *Acta Period. Technol. (APTEFF)* 37, 2006, 105-115.

[4] Eric, O; Uthus, EO. Arsenic essentiality: A role affecting methionine metabolism. *J. Trace Elem. Exp. Med.* 16(4), 2003, 345-355.

[5] Švarc-Gajić, JV; Suturović, ZJ; Marjanović, NJ; Kravić, SZ. Determination of As(III) and As(V) in oilseeds by chronopotentiometric stripping analysis: Development of a method. *Mol. Nutr. Food Res.* 49, 2005, 337-342.

[6] Bartolome, B; Cordoba, S; Nieto, S; Fernandez-Herrera, J; Garcia-Diez, A. Acute arsenic poisoning: clinical and histopathologic features. *Brit. J. Dermat.* 141, 1999, 1106-1109.

[7] Williams, F; Robertson, R; Roworth, M. Scottish Centre for Infection and Environmental Health. *Detailed profile of 25 major organic and inorganic substances.* 1st ed. Glasgow, SCEIH, 1999.

[8] Clowes, W. Emergency treatment of poisonings: Heavy metals. In: *British National Formulary,* British Medical Association, Royal Pharmaceutical Society of Great Britain. 37th ed. Beccles (Suffolk); 1999.

[9] Arena, JM; Thomas CC. *Poisoning. Toxicology. Symptoms. Treatments,* ed. 4, Illinois: Springfield; 1979.

[10] Beliles, RP. The Metals. In: Clayton, GD; Clayton, FE. *Patty's Industrial Hygiene and Toxicology.* 4th ed., New York: John Wiley and Sons; 1994.

[11] Brenniman, GR; Levy, PS. In: Calabrese, EJ. *Advances in Modern Toxicology IX.* Princeton: Princeton Scientific Publications; 1984.

[12] Timbrell, JA. *Introduction to toxicology.* 3d edition. London: Taylor and Francis; 2001.

[13] Goldwater, LJ; Clarkson, TW; Smith, RG. Five of potential significance. In: Lee, DHK; *Metallic contaminants and human health.* 1st ed. New York: Academic Press; 1995.

[14] Mertz, W. Chromium occurrence and function in biological systems. *Physiol. Rev.* 49, 1969, 163-239.

[15] Mertz, W. Chromium in human nutrition: a review. *J. Nutr.* 123, 1993, 626-33.

[16] Mertz, W. Interaction of chromium with insulin: a progress report. *Nutr. Rev.* 56, 1998, 174-177.

[17] Hopkins, JLL; Ransome-Kuti, O; Majaj, AS. Improvement of impaired carbohydrate metabolism by chromium(III) in malnourished infants. *Am. J. Clin. Nutr.* 21, 1968, 203-211.

[18] Jeejeebhoy, KN; Chu, RC; Marliss, EB; Greenberg, GR; Bruce-Robertson, A. Chromium deficiency, glucose intolerance, and neuropathy reversed by chromium supplementation in a patient receiving long-term total parenteral nutrition. *Am. J. Clin. Nutr.* 30, 1977, 531-538.
[19] Anderson, R. Chromium. In: Mertz, M. *Trace Elements in Human and Animal Nutrition.* San Diego, CA: Academic Press, 1987.
[20] Lukaski, HC. Chromium as a supplement. *Annu. Rev. Nutr.* 19, 1999, 279-302.
[21] Stoecker, BJ. Chromium. In: Bowman, B; Russell, R. *Present Knowledge in Nutrition* Washington, DC: ILSI Press; 2001.
[22] Li, XS; Zhi, JL; Gao, RO. Effects of Fluoride Exposure on the Intelligence of Children, *Fluoride* 28, 1995, 182-189.
[23] Proudfoot, AT. *Acute poisoning: diagnosis and management.* 2nd ed. Oxford: Butterworth-Heinemann; 1993.
[24] Caravati, EM; Richard C. Dart, RD. *Medical toxicology.* Lippincott Williams and Wilkins; 2003.
[25] Matthew, J; Lawson, AAH. *Treatment of common acute poisonings.* 4th ed. Edinburgh: Churchill Livingstone; 1979.
[26] Lee, JW. Manganese intoxication. *Arch. Neur.* 57(4), 2000, 597-599.
[27] Švarc-Gajić, JV; Suturović, ZJ; Marjanović, NJ; Kravić, SZ. Selenium determination in sunflower seed using mercury film electrode, *Cent. Europ. J. Occup. Envir. Med.,* 9(4), 2003, 296-303.
[28] Health and Safety Executive. *Mercury and its organic divalent compounds.* 1st ed., London: HSE books; 1996.
[29] Švarc-Gajić, JV; Suturović, ZJ; Marjanović, NJ; Kravić, SZ. Determination of mercury by chronopotentiometric stripping analysis using glassy carbon vessel as a working electrode. *Electroanalysis* 18, 2006, 513-516.
[30] Švarc-Gajić, JV; Stojanović, ZJ, Suturović, ZJ; Marjanović, NJ; Kravić, SZ. Direct mercury determination in natural waters by chronopotentiometric stripping analysis in macroelectrode process vessel. *Desalination*, 2009, in press.
[31] Bradberry, SM. Occupational methaemoglobinaemia: mechanisms of production, features, diagnosis and management including the use of methylene blue. *Toxicol. Rev.* 22(1), 2003, 13-27.
[32] Osterhoudt, KC. Methemoglobinemia. In: Ford, MD; Delaney, KA; Ling, LJ; Erickson, T; *Clinical toxicology.* 1st ed., Philadelphia: WB Saunders Company; 2001.
[33] Tabacova, S; Balabaeva, L; Little, RE. Maternal exposure to exogenous nitrogen compounds and complications of pregnancy. *Arch. Environ. Health* 52(5), 1997, 341-347.
[34] Bruning-Fann, CS; Kaneene, JB. The effects of nitrate, nitrite and *N*-nitroso compounds on human health: a review. *Vet. Hum. Tox.* 35(6), 1993, 521-38.
[35] Cantor, KP. Drinking water and cancer. *Cancer Caus.Cont.* 8, 1997, 292-308.
[36] Eichholzer, M; Gutzwiller, F. Dietary nitrates, nitrites, and *N*-nitroso compounds and cancer risk: a review of the epidemiologic evidence. *Nutr. Rev.* 56, 1998, 95-105.
[37] Fan, AM; Steinberg, VE. Health implications of nitrate and nitrite in drinking water: An update on methemoglobinemia occurrence and reproductive and development toxicity. *Reg. Tox. Pharm.* 23(11), 1996, 35-43.

[38] Barrett, JH; Parslow, RC; McKinney, PA; Law, GR; Forman, D. Nitrate in drinking water and the incidence of gastric, esophageal, and brain cancer in Yorkshire, England. *Cancer Caus.Cont.* 9, 1998,153-159.

[39] Ward, MH; Mark, SD; Cantor, KP; Weisenburger, DD; Correa-Villasenor, A; Zahm SH. Drinking water nitrate and the risk of non-Hodgkins lymphoma. *Epidemiology* 7(5), 1996, 465-71.

[40] McKnight, GM; Duncan, CW; Leifert, C; Golden, MH. Dietary nitrate in man: Friend or foe? *Br. J. Nutr.* 81(5), 1999, 349-358.

[41] Lundberg, JO; Weitzberg, E; Cole, JA; Benjamin, N. Nitrate, bacteria and human health. *Nat. Rev. Microb.* 2(7), 2004, 593-602.

[42] Hill, MJ. Nitrate toxicity: myth or reality? *Br. J. Nutr.* 81(5), 1999, 343-4.

[43] Švarc-Gajić, JV; Suturović, ZJ; Marjanović, NJ; Kravić, SZ. Development of a chronopotentiometric stripping method for the determination of selenium in mixed diets. *Food Chem.* 92, 2005, 771-776.

[44] Švarc-Gajić, JV; Suturović, ZJ; Marjanović, NJ; Kravić, SZ. Chronopotentiometric stripping analysis of selenium in feed: Development of a method. *Asian J. Anim. Vet. Adv* 1, 2006, 66-75.

[45] Papp, LV; Lu, J; Holmgren, A; Khanna, KK. From selenium to selenoproteins: synthesis, identity, and their role in human health. *Antiox. Redox. Signal* 9(7), 2007, 775–806.

[46] Zhang, J; Wang, X; Xu, T. Elemental selenium at nano size (Nano-Se) as a potential chemopreventive agent with reduced risk of selenium toxicity: comparison with se-methylselenocysteine in mice. *Tox. Sci.* 101(1), 2008, 22–31.

[47] Zhang, J; Wang, H; Yan, X; Zhang, L. Comparison of short-term toxicity between Nano-Se and selenite in mice. *Life Sci.* 76(10), 2005, 1099–109.

[48] Tan, J; Zhu, W; Wang, W; Li, R; Hou, S; Wang, D; Yang, L. Selenium in soil and endemic diseases in China. *Sci. Total Envir.* 284(1), 2002, 227-235.

[49] Tanaka, J; Yonezawa, T; Ueyama M. Acute thallotoxicosis: neuropathological and spectrophotometric studies on an autopsy case. *J. Tox. Sci.* 3, 1978, 325-334.

[50] Venugopal, B; Luckey TD. Metal toxicity in mammals. Vol. 2. Chemical toxicity of metals and metalloids. New York, NY: Plenum Press; 1978.

[51] Dimond, EG; Caravaca, J; Benchimol, A. Vanadium: Excretion, toxicity, lipid effect in man. *Am. J. Clin. Nut.* 12, 1963, 49-53.

Chapter VIII

TOXICOLOGICAL EFFECTS OF ORGANIC TOXICANTS

PESTICIDES

Since the time of the earliest agrarian activities, people have attempted to improve yields and quality of the crops they harvested. In early civilisation (1000 BC), sulphur was used in wine technology to destroy bacteria and fungus. Beginning in the 15th century, metal compounds such as arsenic, mercury and lead were applied in agriculture.. The pesticidal properties of nicotine, the first natural insecticide, were discovered in 18th century and nicotine was extracted from tobacco leaves and sprayed on foliage. Soon thereafter, the pesticidal properties of many other natural compounds were revealed, such as rotenone, isolated from the root of tropical leguminous (*Lonchocarpus nicou)*, or pyrethrin, an alkaloid, isolated from perennial plant (*Chrysanthemum cinerariaefolium*). Today, synthetic pesticides are widely used to improve crop yields and enhance agricultural production by reducing a wide range of pests. Because of their extensive use, humans are continually exposed to persistent residues of these compounds directly by applying them in agriculture, or producing or transporting them, and indirectly, through contaminated food and water.

Exposure to pesticides occurs orally, dermally or by inhalation. Oral exposure usually occurs accidentally and as a result of carelessness, for example, by blowing out a plugged nozzle, smoking or eating without washing the hands, or consuming fruit that has been recently sprayed. Dermal exposure accounts for about 90% of all pesticide poisonings. Intoxication from exposure to liquid pesticide formulations through the skin is much more rapid from pesticide powders and granules; however, pesticide dust also may remain on clothing for long periods of time, prolonging skin contact. The severity of dermal exposure depends on the dermal toxicity of the pesticide, the rate of absorption through the skin (skin humidity, injury), the exposed area, duration of the contact, pesticide concentration and body part that is in contact with the pesticide (Table 8.1.). Absorption through the skin in genital area is rapid enough to approximate the effects of injecting the pesticide directly into the bloodstream.

Inhalation exposure results from breathing pesticide vapors, dust or spray particles and is most serious in fumigant pesticides which are very active gases. Absorption through respiratory tract also occurs by breathing the smoke from burning pesticide containers or pesticide fumes during application, as well as fume inhalation while mixing and pouring pesticide formulations. Another means of inhalation exposure is by smoking tobacco products containing pesticide residues.

Table 8.1. Rates of parathion absorption through the skin on various body regions

Body region	Relative absorption (%)
Forearm	8.6
Palm of hands	11.8
Ball of foot	13.5
Abdomen	18.4
Scalp	32.1
Forehead	36.3
Ear canal	46.5
Genitalia	100

Table 8.2. Classes of pesticides according to target species

Pesticide group	Target organism
Acaricides	Mites
Algicides	Algae
Avicides	Birds
Bactericides	Bacteria
Fungicides	Fungus
Herbicides	Plants/weeds
Insecticides	Insects
Molluscides	Molluscs
Nemacides	Parasitic nematodes (roundworms)
Piscicides	Fish
Rodenticides	Mice/rats
Virucides	Viruses

Pesticides are typically categorized on the basis of the target organism (Table 8.2.). Different natural or synthetic classes of substances that disturb mating behaviour (mating distrupters), plant growth (plant growth regulators) or reproduction (chemosterilants) are also used in pest control. Antifeedants are chemicals that inhibit feeding, due to their toxicity or a bad taste. Herbicide safeners selectively protect crop plants from herbicide damage without reducing their activity in target weed species. Biopesticides are natural pesticides derived from animals, plants, bacteria or certain minerals, and may be divided into three major classes: microbial pesticides, plant incorporated protectants and biochemical pesticides. Microbial pesticides is are certain bacterium, fungus, virus or protozoa, specifically affecting a certain pest (weed, insect). Plant incorporated protectants are pesticidal substances that are produced by a plant that was modified using recombinant DNA technology. by which a certain gene, responsible for the production of active substance, is introduced into genetic pool of the plant. Biochemical pesticides, as opposed to conventional chemical pesticides which are toxic and synthetic substances, are naturally occurring compounds that control pests by non-toxic mechanisms. They include substances such as insect sex pheromones that interfere with mating, as well as various scented plant extracts that attract or repel insects. Pesticide synergists do not themselves possess the capability of a pesticide, but significantly enhance and promote the effectiveness of other pesticides, thereby reducing effective pest concentrations, as required.

Because pesticides pose risks, the World Health Organization (WHO) has recommended the classification of synthetic pesticides according to their risk, from extremely hazardous (class Ia), highly hazardous (class Ib), moderately hazardous (class II), and slightly hazardous (class III) [1]. The lethal toxicity of certain active compounds of different classes of pesticides is presented in Appendixes II, III and IV.

The pesticides are also conveniently classified in different groups on the basis of their chemical structure. Organochlorine and organophophorous pesticides, carbamates, pyretroides, phenoxy compounds, triazines and organometallic compounds are the major classes within which it is possible to establish and define some toxicological, biochemical and analytical similarities.

Organochlorine Pesticides

Organochlorine pesticides used used between 1940 and 1960 in agriculture and forestry. dichlorodiphenyltrichloroethane (DDT) was first synthesized in 1874 by Zeidler and rediscovered in 1939 by Paul Muller. Even though DDT and other organochlorine pesticides were banned after their biocumulative properties and adverse effects on animal reproduction were recognized, other organochlorine pesticides remain in use, such as lindane and endosulfan. Compounds belonging to the organochlorine class have low vapor pressure, high liposolubility and high persistency in the environment. All cyclodien organochlorine pesticides (aldrin, dieldrin and endrin) have a low water solubilities (high log K_{ow}) and tend to persist in the environment, being strongly sorbed to suspended solid particles and sediments. Organochlorine pesticides are highly toxic to aquatic organisms. The majority of the remaining environmental burden of these substances appears to be in sediments and marine organisms, adversely affecting their reproduction, development and endocrine status.

Even though they are considered acutely not very toxic, organochlorine pesticides express potent estrogenic and enzyme-inducing effects, influencing the reproduction. These compounds may induce convulsions even in low concentrations, without any previous symptoms. The symptoms of acute exposure to organochlorine pesticides include headache, nausea, vertigo and motor hyperexcitability. Chronic poisoning may provoke persistent tremors and seizures, which can be triggered by sound stimuli. The toxicity of aldrin and dieldrin results from the inhibition of the flux of chloride ions by antagonistic action on gamma-aminobutyric acid (GABA) receptors. The disturbance of neurotransmitter activity results only in partial repolarization of neurons and the state of uncontrolled excitation. Calcium transport across neural and other membranes is also impaired by inhibiting Ca^{2+}-ATPase, and results in accumulation of calcium ions in synapses. Accumulated calcium ions further induce the release of neurotransmitters from vesicles, depolarisation of neurons and propagation of stimuli through the CNS.

Dermal toxicity of aldrin is very close to its oral toxicity, indicating extensive dermal absorption. After absorption, aldrin is rapidly metabolized to dieldrin in biotranformation reactions involving the cytochrome P450 monooxigenase system of the liver and lungs. Dieldrin is further metabolized by liver microsomal enzymes to less toxic metabolites *cis*-aldrinol and 9-hydroxydieldrin and photoconverted to more toxic 2-ketodieldrin. More hydrophilic aldrin metabolites are excreted through feces, urine and to lesser extent through milk. Endrin is hydroxylated by liver microsomal enzymes to more water soluble metabolites *syn*-12-hydroxyendrin and anti-12-hydroxyendin. Oxidative metabolite 12-ketoendrin is significantly more toxic than the parent compound and more than 90% of the compound is biliary excreted. High concentrations of 12-ketoendrin can be stored in liver and adipose tissue. Animal studies of long-term exposure demonstrate increased incidence of liver tumors as a result of exposure to aldrin and dieldrin, but these findings have not been confirmed in human studies.

Organophosphosphorus and Carbamate Pesticides

The application of relatively safe, but persistent, organochlorine pesticides was superseded in the 1970s by the use of more acutely toxic but degradable organophosphorus pesticides. Organophosphorous pesticides are among the most acutely toxic of all pesticides, acting as nerve toxicants like nerve gases. In the environment, they are not persistent but break down relatively quickly by hydrolysis and exposure to sunlight, although small amounts can be detected in food and drinking water. Agricultural exposure is the most common cause of poisoning with organophosphorus and carbamate pesticides. Some organophosphorous compounds are, besides being used in agriculture, also encountered in military applications as nerve agents such as sarin, VX or soman, or are used as industrial chemicals as lubricants, plasticizers and fuel additives. In addition to their use as insecticides, carbamates are used to treat certain medical diseases, such as glaucoma and the neuromuscular illness, myasthenia gravis.

Commonly used organophosphorous pesticides include parathion, malathion, methyl parathion, chlorpyrifos, diazinon, dichlorvos, phosmet, tetrachlorvinphos and azinphos methyl, but there are more than 100 commercially used organophosphorous compounds.

This group of pesticides is considered acutely very toxic and according to their acute toxicity, three subgropus are recognized: the most toxic group with the LD_{50} in the range 1-30 mg/kg (e.g., chlorfenvinphos), the medium toxicity group with the LD_{50} values in the range 30-50 mg/kg (e.g. dichlorvos) and the least toxic group with the LD_{50} values 60-1300 mg/kg (e.g. malathion).

Symptoms of acute poisoning with organophosphorous pesticides include excessive sweating, salivation and lacrimation, nausea, vomiting, diarrhea and abdominal cramps. Intoxication manifestations are accompanied with the general weakness, headache, poor concentration and tremors. In serious cases, respiratory failure and death may be expected. Low occupational exposure results in cumulative poisoning, causing increased susceptibility to further exposures and producing progressive effects on the nervous systems [2]. Some research reports suggest that chronic exposure to organophosphorous pesticides produces depression and disturbed heart function [3]. Mutagenic and carcinogenic effects in mammals, with the exception of dichlorvos which EPA classifies as a possible human carcinogen, have not been confirmed [4].

Carbamates are esters of carbamic acid with the general structure -NH(CO)O-. Some of commercial products include aldicarb, carbofuran, methomyl, oxamyl, thiodicarb, furadan, fenoxycarb, carbaryl, sevin, 2-(1-methylpropyl)phenyl-N-methylcarbamate and ethienocarb. Although organophosphorous pesticides and carbamates are structurally distinct, they have a similar mode of toxicity by binding to active sites of acetylcholinesterase and inhibiting this enzyme by means of steric influence [5]. After absorption, the compounds bind to erithrocytes and first affect the activity of plasma acetylcholinesterase. Acetylcholinesterase hydrolyzes acetylcholine to choline and acetic acid. The inhibition of acetylcholinesterase causes an excess of acetylcholine in synapses and neuromuscular junctions. Both muscarinic and nicotinic acetylcholine receptors are affected by acetylcholine excess. Accumulation of acetylcholine in the synapses provokes neurological symptoms, specific for poisoning with these chemicals. Accumulation of acetylcholine at postganglionic muscarinic synapses leads to increased parasympathetic activity of the smooth muscles in lungs, provoking bronchospasm, in the gastrointestinal tract, causing diarrhea, and in the heart, resulting in bradycardia. The functions of the bladder and secretory glands are impaired and the activity in postganglionic sympathetic receptors for sweat glands is enhanced. The impact of acetylcholine on nicotinic receptors provokes persistent depolarization of skeletal muscles resulting in progressive weakness, hypotonicity and paralysis. The compounds belonging to organophosphate and carbamate pesticides are able readily to cross the blood-brain barrier, and this is one of the reasons for their adverse effects on the CNS expressed in the form of induced seizures and CNS depression.

The main difference in the mechanisms of action between organophosphorous pesticides and carbamates is that carbamates spontaneously hydrolyze from the active sites of acetylcholinesterase within 24 hours, causing the symptoms to reverse. In the case of organophosphorous pesticides, the protein molecule of acetylcholinesterase becomes irreversibly inactivated due to phosphorylation and loss of the alkyl side chain. The result of metabolic biotransformation of carbamates is ethylenethiourea, a mutagenic, teratogenic, carcinogenic and antithyroid substance, which can also be present as an impurity in a technical products or can be formed during processing of foodstuffs containing traces of carbamates.

The rate of absorption of organophosphorous pesticides and carbamates varies depending on the route of absorption and the type of compound. Symptoms usually occur within few hours after oral ingestion, but appear almost immediately in inhalation exposure. The severity of poisoning depends on a number of factors, including the type of compound, the amount and route of exposure, and the time until initial treatment.

In patients who recover from cholinergic crisis, an intermedial syndrom occuring after 24 – 96 hours of exposure and a new wave of adverse reactions may be experienced. In this phase the risk of respiratory paralysis is high with continuation of disturbance in presynaptic and postsynaptic neuromuscular transmission. Some organophosphorous pesticides can induce delayed neurotoxicity manifested as a disorder of peripheral nerves that typically occurs 9-14 days after exposure. Phosphorylation of other esterases in nervous tissues and impairment in their function causes esters to accumulate in nerve cells, and is responsible for burning and tingling sensations that may progress to paralysis of the lower limbs. Complete recovery from the poisoning with organophosphorous compounds occurs within 12-15 months, but permanent effects have been reported.

Organophosphorous and carbamate poisoning can be confirmed by measuring serum acethylcholine activity or its activity in erythrocytes. Further absorption through the gastrointestinal tract can be reduced by administrating activated charcoal within the first half-hour of ingestion. In the case of poisoning, an atropine may be administrated intravenously as an antidote. Atropine is a muscarinic antagonist that competes with acethylcholine at the muscarinic receptors. However atropine it does not bind to nicotinic receptors, and therefore, it is ineffective in the treatment of neuromuscular toxicity and muscular weakness [6, 7]. Oximes are effective reactivators of acethylcholinesterase activity. They enhance the activity of acetylcholinesterase at muscarinic sites and decrease the requirement for atropine [8]. The activity of acetylcholinesterase is refreshed by oximes attachment to phosphorus atom and formation of oxime-phosphonates which later dissociates from acetylcholinesterase molecule. The most effective oxime antidote is pralidoxime [9]. Benzodiazepines are commonly used to treat seizures that have severe effects on the CNS.

Pyrethroids

Pyrethroids are class of insecticides that are least acutely toxic to mammals due to their poor absorption and rapid biotransformation. Acute toxicity of some pyrethroid pesticides is listed in Table 8.3.

This group of compounds is naturally occurring and the first pyrethroidal insecticides were derived from chrysanthemum flowers. At present, their application is widespread for the control of mosquitoes and other insects, and in pet sprays and shampoos. Natural pyrethroides are chemically unstable and brake down rapidly upon exposure to air and sunlight. At the beginning of the 1970s, many more stable synthetic pyrethroids were introduced to market because they demonstrated enhanced stability and rapid insecticidal action. Some of common synthetic pyrethroids include permethrin, resmethrin and sumithrin. Synthetic pyretroides are effective against a wide range of insects and pests and are also often mixed with other pesticides to achieve a broader range. Piperonyl butoxide, a synergist, enhances the effects of a pesticide's active ingredients and is frequently included in commercial products.

The toxicodynamic action of pyretroide pesticides is interference with the transmission of nerve impulses by inhibiting Ca- and Mg-ATPases by blocking GABA receptors and sodium ion channels [10, 11]. In high concentration exposures, dizziness, headache, nausea and diarrhea may occur. Aggressive behavior and tremor, as well as a feeling of itching or prickling without any exterior signs may be experienced. To date, there is no firm evidence that pyrethroids are reproductive toxicants and that they may cause birth defects in humans. Possible carcinogenic effects that may occur in humans due to long-term exposures to pyretroides remain to be examined.

Table 8.3. Acute toxicity of pyrethroid pesticides

Common name	LD_{50} (rat, oral) (mg/kg)	LD_{50} (rabbit, dermal) (mg/kg)
Allethrin	860	11 332
Bifenthrin	375	>2000
Cyfluthrin	869 - 1271	>5000 (rat)
Cyhalothrin	79	632 (rat)
Cypermethrin	250	>2000
Deltamethrin	31 - 139 (female)	>2000
Esfenvalerate	451	2500
Fenpropathrin	70.6 -164	>2000
Fluvalinate	261 - 282	>20 000
Permethrin	430 - 4000	>2000
Resmethrin	1244 - >2500	>2500
Tefluthrin	969	>2000 (rat)
Tetramethrin	>5000	>2000
Tralomethrin	284	>2000

Table 8.4. Chemical classes of herbicides

Chemical group	Chemical subgroup
Amide herbicides	Anilide herbicides
	Arylalanine herbicides
	Chloroacetanilide herbicides
	Sulfonanilide herbicides
	Sulfonamide herbicides
	Thioamide herbicides
Aromatic acid herbicides	Benzoic acid herbicides
	Pyrimidinyloxybenzoic acid herbicides
	Pyrimidinylthiobenzoic acid herbicides
	Phthalic acid herbicides
	Picolinic acid herbicides
	Quinolinecarboxylic acid herbicides
Arsenic herbicides	
Benzoylcyclohexanedione herbicides	
Benzofuranyl alkylsulfonate herbicides	
Benzothiazole herbicides	
Bipyrydiium herbicides	
Carbamate herbicides	
Carbanilate herbicides	
Cyclohexene oxime herbicides	
Dinitroaniline herbicides	
Dinitrophenol herbicides	
Diphenyl ether herbicides	Nitrophenyl ether herbicides
Dithiocarbamate herbicides	

Table 8.4. (Continued)

Chemical group	Chemical subgroup
Halogenated aliphatic herbicides	
Iimidazolinone herbicides	
Inorganic herbicides	
Organophosphorus herbicides	
Phenoxy herbicides	Phenoxyacetic herbicides
	Phenoxybutyric herbicides
	Phenoxypropionic herbicides
Pyrazole herbicides	Benzoylpyrazole herbicides
	Phenylpyrazole herbicides
Pyrimidinediamine herbicides	
Quaternary ammonium herbicides	
Thiocarbamate herbicides	
Thiocarbonate herbicides	
Thiourea herbicides	
Triazine herbicides	Chlorotriazine herbicides
	Methoxytriazine herbicides
	Methylthiotriazine herbicides
Triazinone herbicides	
Triazole herbicides	
Triazolone herbicides	
Triazolopyrimidine herbicides	
Uracil herbicides	
Urea herbicides	Phenylurea herbicides
	Sulfonylurea herbicides Pyrimidinylsulfonylurea herbicides
	Triazinylsulfonylurea herbicides
	Thiadiazolylurea herbicides

Established toxicity of pyretroides after inhalation and dermal absorption is shown to be quite low; however, in sensitive persons either an asthmatic condition or a skin rash and inflammation may be observed after single exposure. Some individuals, especially children, with the history of allergies or asthma, are particularly sensitive. Symptoms of acute toxicity due to inhalation include sneezing and nervous disturbance. Affected nervous system leads to incoordination, tremors, convulsions, facial swelling and burning and itching sensations. Ingestion of high doses of pyretroides first provokes diarrhea and then, excitation. The state of excitation may further develop and proceed to convulsions and paralysis. The symptoms of poisoning with pyretroides last about 48 hours. Lethal outcomes in poisoning with pyrethroides are rare and if they occur they are caused by respiratory arrest. Many pyrethroids have been linked to disruption of endocrine system, which adversely affects reproduction and sexual development. Pesticides also interfere with the immune system and increase the chances of breast cancer [12].

Herbicides

Herbicides are primarily synthetic chemicals used to fight weeds. Many natural products such as salt or soap also possess herbicidal properties and can provoke a plant`s death. A wide array of different compounds (Table 8.4.), varying in their chemical structure and toxicity, can be effective in weed control through different mechanisms.

Most very toxic herbicides, such as mercury and arsenic salts, have been withdrawn from use and replaced with less toxic compounds, which express low mammalian toxicity and have even higher LD_{50} values than many commonly used chemicals (Table 8.5.).

Even though herbicides encompass a great number of different chemical compounds, they are generally characterized with relatively low mammalian toxicity in comparison to other groups of pesticides, because they target biochemical processes specific to plants, such as photosynthesis. In addition, they are very rapidly eliminated from the body, due to the absence of specific binding components in mammalian tissues that could retain herbicide molecule it in the body. Not all herbicides are characterized with low toxicity; paraquat and endothal are an exception and are highly toxic and should be used with care.

Table 8.5. Comparison of oral (rat) acute toxicity of some herbicides and common substances

Herbicide	LD50 (mg/kg)	Common consumer chemical	LD50 (mg/kg)
Paraquat	~100	Nicotine	9
Triclopyr	630	Caffeine	192
2,4-D	666	Bleach	192
Pendimethalin	1050	Tylenol	338
Atrazine	3090	Ammonia (10%)	350
Glyphosate	4900	Codeine	427
Imazaquin	>5000	Table salt	3000

Table 8.6. Acute toxicity of bipyridylium herbicides

Herbicide	LD_{50} (rat, oral) (mg/kg)	LD_{50} (rabbit, dermal) (mg/kg)
Diquat	600	260
Paraquat	150	236 - 325

Bipyridylium Herbicides

Bipyridylium herbicides, an older class of herbicides, contain two pyridyl rings and act as contact herbicides; they may also be used as desiccants for cotton, sugarcane or sunflower. This class of herbicides is nonselective, causing a rapid destruction of plant cell membranes, particularly in sunlight. This action may occur in only a few hours following application. Bipyridylium herbicides are mostly absorbed by the plant`s leaves and are very limited in translocation to other tissues due to plants quick death. Paraquat (1,1'-dimethyl-4,4'-bipyridylium) and diquat (1,1'-ethylene-2,2'-bipyridylium) are the two most widely used bipyridylium herbicides today.

The target organ of adverse health influence with the bipirydylium herbicides diquat and paraquat are the lungs, where they are quickly transfer after absorption. Diquat and paraquat are structurally similar to diamines and poliamines, which are normally transported to the alveoli by specific transport mechanisms involving active transport. Due to its organ-specific accumulation, exposure to paraquat is accompanied by the proliferation of fibrous tissue in the lungs. It is speculated that fibrosis occurs due to lipid peroxidation of alveoli membranes as a result of a redox cascade of the paraquat. Diquat is considered less acutely toxic than paraquat (Table 8.6) and typically results in great damage to kidneys besides also injuring the lungs. The most obvious symptoms of diquat poisoning are neurologic effects, including nervousness, irritability, restlessness and disorientation. Intoxicated individuals are often unable to recognize familiar persons and shows diminished reflexes.

Phenoxy Herbicides

A wide variety of phenoxy herbicides used today includes organic acids which are typically applied in an ester or salt form. Further, they may be grouped into phenoxyacetic, phenoxybutyric and phenoxypropionic subtypes, the latter itself containing the aryloxyphenoxypropionic subtype, which includes a great number of commercial products.

Phenoxy herbicides are used to control broadleaf weeds and are considered relatively safe pesticides in terms of acute, chronic and long-term effects. Most of the substances belonging to this group are systematic pesticides, being absorbed by plant leaves and translocated to the roots, where they affect enzymatic activity and plant growth. Phenoxy herbicides are degraded relatively quickly in the environment with a half-life of 10 days in sandy soils, and about 30 days in clay soils.

=Aa member of phenoxy family of herbicides, 2,4-dichlorophenoxyacetic acid (2,4-D) was the first successful selective herbicide developed. 2,4-D and other chlorophenoxy herbicides are not persistent in the environment and have low acute toxicity (Table 8.7). Symptoms of acute toxicity of 2,4-D usually start as fatigue, weakness and nausea. Manifestations vary with different commercial products due to variability in the amounts and types of specific additives, surfactants and solvents. 2,4,5-trichlorophenoxyacetic acid (2,4,5-T) is usually contaminated with 2,3,7,8-tetrachlorodibenzo-para-dioxin due to specific technological process of production, and is banned in most countries. Dioxin, as an impurity of the product, was the cause of mass poisoning and great number of children being born with malformations during the Vietnam War, when the herbicide was extensively used. 4-(2,4-dichlorophenoxy) butyric acid (2,4-DB) is a slightly toxic compound in oral exposures. However, preliminary evidence suggests that the compound may be weakly carcinogenic. In the soil, the compound is broken down by the action of soil microorganisms to the product 2,4-D. The half-life of the parent compound, 2,4-DB, in the soil is seven days.

Potassium 2-(2-methyl-4-chlorophenoxy)propionate (MCPP) is highly irritating to the skin and eyes, and contact with it causes redness and swelling and may provoke cloudy vision. Studies show that MCPP also may be mutagenic at very high doses.

Table 8.7. Acute mammalian toxicity of chlorophenoxy herbicides

Common name	Chemical name	LD_{50} (rat, oral) (mg/kg)	LD_{50} (rabbit, dermal) (mg/kg)
2,4-D	2,4-dichlorophenoxyacetic acid	500 - 949	-
2,4-DB	4-(2,4-dichlorophenoxy) butyric acid	>2000	>10 000
MCPA	2-methyl-4-chlorophenoxyacetic acid	900 - 1160	>4000
MCPP	2-(2-methyl-4-chlorophenoxy)propionate	1000 - 1166	>2000

Table 8.8. Acute toxicity of triazine pesticides

Common name	LD_{50}, (rat, oral) (mg/kg)	LD_{50}, (rabbit, dermal) (mg/kg)
Atrazine	1780	750
Cyromazine	3387	-
Hexazinone	1690	5278
Metribuzin	2000	>20 000
Prometon	2980	>2000
Prometryn	5235	>2020
Simazine	>5000	>3100

Triazine Herbicides

The triazine class of herbicides encompasses compounds that were among first used to selectively beat weeds. This class of chemicals has a heterocyclic structure, and compounds are composed of carbon and nitrogen atoms. With the exception of metribuzin, most of them are symmetrical molecules. Typical representatives of this group of compounds include atrazine, hexazinone, metribuzin, prometon, prometryn and simazine. Acute toxicity in this group of herbicides is not very pronounced (Table 8.8.) [13]. Some of the triazines are moderately irritating to eyes, skin and respiratory tract. Atrazine is considered slightly to moderately toxic to humans. Long-term exposure to high levels of atrazine has been shown to cause adverse health effects in animals, including tremors, changes in organ weights, and damage to the liver and heart. Hexazinone is not considered to be highly acutely toxic, but can cause serious and irreversible eye damage. Prometon has been shown to pose a risk for substantial but temporary eye injury. Prometon, prometryn, or simazine are not likely to cause serious toxic effects. As a family of pesticides, triazines are one of the least toxic to wildlife and only prometon is considered moderately toxic to fish.

Organometallic Pesticides

In the past, many metallic compounds were widely used for pest control; however more recently, their application in agriculture is significantly reduced due to their high toxicity.

Table 8.9. Different classes of organometallic pesticides

Arsenic herbicides Use	
Dimethylarsinic acid	Herbicide
Calcium bis(hydrogen methylarsonate)	Herbicide
Disodium methylarsonate	Herbicide
Methylarsonic acid	Herbicide
Ammonium/sodium hydrogen methylarsonate	Herbicide
Potassium arsenite	Herbicide
	Insecticide
	Rodenticide
Copper pesticides	
Copper(II) oleate	Fungicide
	Insecticide
Bis(quinolin-8-olato-O,N)copper(II)	Fungicide
Tricopper dichloride dimethyldithiocarbamate	Fungicide
Copper(II) silicate	Fungicides
Copper(I) oxide	Fungicides
Dicopper chloride trihydroxide	Bird repellent
	Fungicides
Copper(II) tetraoxosulfate	Algicide
	Fungicides
	Herbicides
	Molluscides
Copper(II) hydroxide	Bactericide
	Fungicide
[μ-[carbonato(2−)-kO:kO′]]dihydroxydicopper	Fungicide
Copper(II) acetate	Fungicide
Organotin pesticides	
Decyltriphenylphosphonium bromochlorotriphenylstannate(IV)	Fungicide
Triphenyltin	Fungicide
	Molluscide
Bis(tributyltin) oxide	Fungicide
Organomercury compounds	
(3-ethoxypropyl)mercury(II) bromide	Fungicide
Ethylmercury(II) acetate	Fungicide
Ethylmercury bromide	Fungicide
Ethylmercury chloride	Fungicide
Ethylmercury 2,3-dihydroxypropyl mercaptide	Fungicide

Ethylmercury phosphate	Fungicide
N-(ethylmercury)-p-toluenesulphonanilide	Fungicide
μ-(2,2′-binaphthalene-3-sulfonyloxy)bis(phenylmercury)	Bactericide Fungicide
2-methoxyethylmercury chloride	Fungicide
Methylmercury benzoate	Fungicide
1-cyano-3-(methylmercurio)guanidine	Fungicide
Phenylmercury(II) quinolin-8-olate	Fungicide
Methylmercury(II) pentachlorophenolate	Fungicide
(Phenylmercurio)urea	Fungicide
Phenylmercury(II) chloride	Fungicide
Phenylmercury(II) acetate	Fungicide Herbicide
o-(phenylmercuriooxy)phenol	Fungicide
Phenylmercury(II) nitrate	Fungicide
Phenylmercury(II) salicylate	Fungicide
Sodium 2-(ethylmercuriothio)benzoate	Fungicide Bactericide
Tolylmercury(II) acetate	Fungicide

Both inorganic and organic compounds of copper, arsenic, mercury and tin have very potent pesticidal properties (Table 8.9.). Organic mercury pesticides are very lipophilic and are easily absorbed through the skin. After absorption organic mercury compounds are accumulated in the brain tissue, reaching concentrations even 10 times higher there than in other organs and tissues. It is not uncommon for an intoxicated person first to experience noncharacteristic symptoms, such as headacke, nausea and insomnia, which develop into more serious neurological damage, which usually starts with tremor, confusion and disturbed vision. Copper compounds are used alone or in combination with other pesticides. Inorganic copper sulfate is widely applied as a fungicide used in agriculture, but also in the treatment of swimming pool water to prevent the growth of algeae. Systematic effects in the accidental oral ingestion of copper pesticides are often omitted due to the strong emetic properties of copper. Inorganic arsenic compounds have been used for a long time in agriculture as insecticides and as wood and fur preservatives, but are nowadays replaced by less toxic organic compounds (Table 8.10). Among arsenic herbicides, dimethylarsinic acid, mono- and di-sodium methanearsonate and calcium bis(hydrogen methylarsonate) are used extensively in combination with other compounds (Figure 8.1.). The mode of action of arsenic pesticides is to inhibit plant growth by uncoupling phosphorylation. If organoarsenic pesticides are ingested, the symptoms of acute poisoning usually appear within one hour.

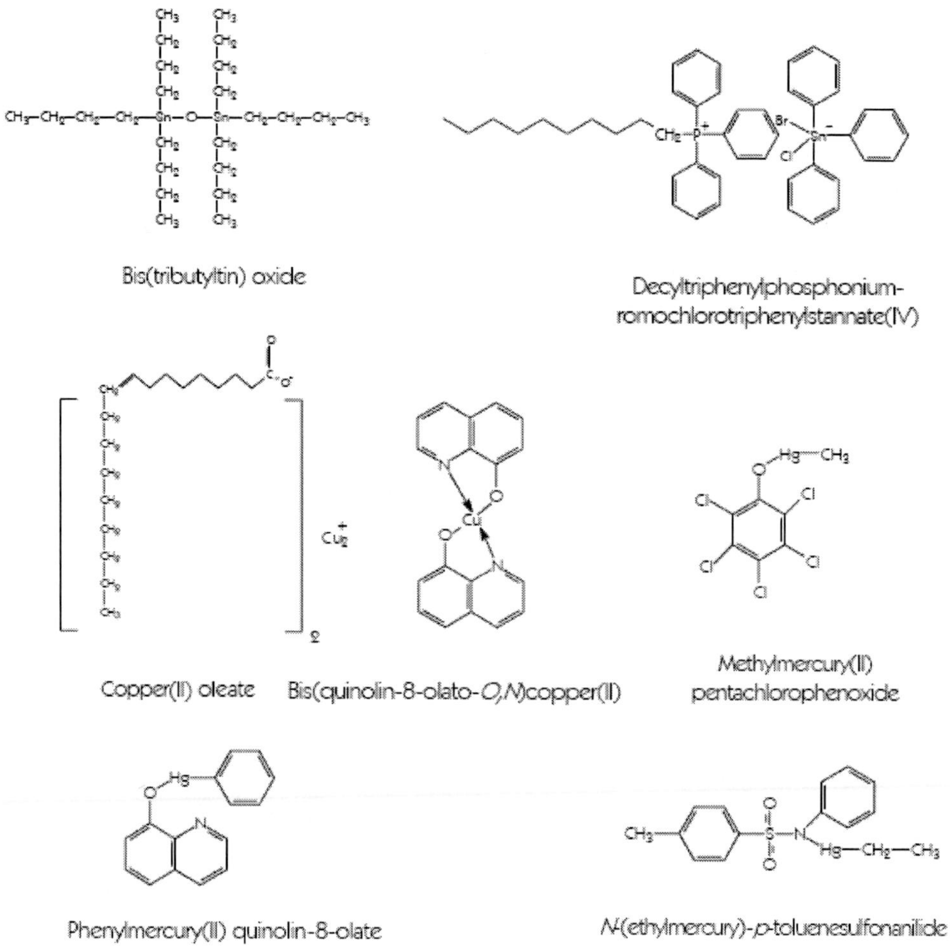

Figure 8.1. Organometallic pesticides.

Table 8.10. Acute toxicity of arsenic herbicides

Common name	LD_{50} (rat, oral) (mg/kg)	LD_{50} (rabbit ,dermal) (mg/kg)
Dimethylarsinic acid	2756	-
Disodium methanearsonate	600	236 - 325
Monosodium methanearsonate	1700	>2500

Dimethylarsinic acid is best absorbed by inhalation;in dermal and oral exposures toxic blood level is reached slower and symptoms development is much more gradual. In severe poisonings with arsenic pesticides, garlicky odor of the breath and feces are noticeable as well as a salty, metallic taste in the mouth. Effects on central nervous system are manifested as dizziness, headache and confusion. Symptoms may progress with weakened muscles, spasms, coma and convulsions. Death usually occurs in one to three days after exposure and is usually a result of circulatory failure. A sign of chronic exposure to organic arsenic pesticides is the formation of white bands across the nails.

Table 8.11. Commonly used fumigants

Compound	Chemical structure	Use
Acrylonitrile	CH_2=CH–C≡N	Insecticide
Carbon disulfide	CS_2	Insecticide
Carbon tetrachloride	CCl_4	Insecticide
Chloroform	$CHCl_3$	Insecticide
Chloropicrin	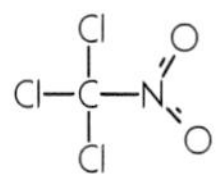	Insecticide Fungicide Rodenticide
1,2-dichloropropane	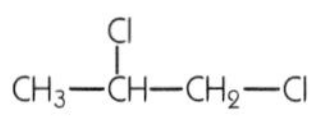	Insecticide Nematicide
Ethyl formate	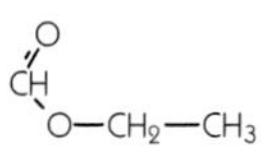	Insecticide
Ethylene dibromide	$CH_2Br–CH_2Br$	Insecticide Nematicide
Ethylene dichloride	$CH_2Cl–CH_2Cl$	Insecticide
Hydrogen cyanide	HCN	Insecticides Rodenticide
Iodomethane	CH_3I	Fungicide Herbicide Insecticide Nematicide
Naphthalene	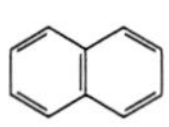	Insecticide
Para-dichlorobenzene	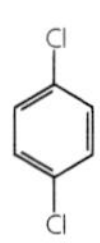	Insecticide Fungicide
Phosphine	PH_3	Insecticide
Sulfuryl fluoride	SO_2F_2	Insecticide Rodenticide Rodenticide

Table 8.12. The types of rodenticides

Chemical group	Representatives
Natural rodenticides	Scilliroside
	Strychnine
Coumarin rodenticides	Brodifacoum
	Bromadiolone
	Coumachlor
	Coumafuryl
	Coumatetralyl
	Difenacoum
	Difethialone
	Flocoumafen
	Warfarin
Indandione rodenticides	Chlorophacinone
	Diphacinone
	Pindone
Inorganic rodenticides	Arsenous oxide
	Phosphorous potassium arsenite
	Sodium arsenite
	Thallium sulphate
	Zinc phosphide
Organochlorine rodenticides	Gamma-1,2,3,4,5,6-hexachlorocyclohexan (Gamma HCH)
	1,2,3,4,5,6-hexachlorocyclohexane (HCH)
	Lindane
Organophosphorus rodenticides	Phosacetim
Pyrimidinamine rodenticides	Crimidine
Thiourea rodenticides	Antu
Urea rodenticides	Pyrinuron
Unclassified rodenticides	Bromethalin
	Chloralose
	α-chlorohydrin
	Ergocalciferol
	Fluoroacetamide
	Flupropadine
	Hydrogen
	Norbormide
	Sodium fluoroacetate

Fumigants

Fumigants are highly toxic substances that are used in gaseous form to kill insects, fungus or worms (Table 8.11.). Some of these gases, such as chloropicrin, hydrogen cyanide and sulfuryl fluoride, are efficiently used in rodent control. Commercial fumigant products are availabe in the form of compressed gases or the gas may be generated at the required site from liquid (ethylendibromide, formaldehide) or solid substances (aluminiumphosphide). The fumigant chloropicrin is a strong pulmonary irritant and is commonly added to other fumigants to prevent abundant inhalation of the active compound. This gas is also a very intense lacrimator which can severely irritate the eyes and throat in concentrations as low as 1 mg/m^3. Breathing chloropicrin fumes even for a very short period can lead to severe lung injury. Methyl bromide is a colorless, odorless, and tasteless respiratory poison. Unlike phosphine and chloropicrin, which become gaseous and effective only in contact with air, methyl bromide is effective immediately and is sold as compressed gas. In skin contact, the gas may cause severe burns. Like methyl bromide, phosphine is also an odorless and colorless gas. Phosphine is considered less toxic than other fumigants; however, exposure to this gas can lead to serious illness or even death at concentrations as low as 0.3 mg/m^3. It is produced in reaction of aluminum, zinc or magnesium phosphide pellets with the air.

Rodenticides

Various chemicals, such as phosphorus paste or barium carbonate, were used in the past to exterminate rodents. Wide use and availability of very toxic inorganic salts, such as arsenic trioxide, thallium sulfate, zinc phosphide and calcium cyanide, have often led to accidental and intentional poisoning with rodenticides. Natural alkaloids like strychnine or strychnine sulfate, can also be used for efficient rodent control. Today, different groups of pesticides are used to fight rodents (Table 8.12.). Fumigants such as sulfur dioxide, carbon monoxide, hydrogen cyanide, and methyl bromide are also effective for this purpose.

The most commonly applied group of rodenticides is the anticoagulant class, which includes coumarin and indandione compounds, which act by preventing normal blood clotting. Disturbance of the coagulation process is accomplished by interference with the epoxide-reductase enzyme necessary for recycling of vitamin K (warfarin, coumafuryl), an important participant in coagulation process; or by preventing the production of blood clotting factors (difenacoum). Brodifacoum, a very toxic anticoagulant of a second generation, acts by producing hypoprothrombinemia. In order for prothrombin factors to be activated, oxidation of vitamin K and the carboxylation step in hepatocytes must be coupled. Brodifacoum toxicity results from the inhibition of coupled carboxylation reaction, which prevents blood from clotting. Other coumarin compounds include warfarin, coumafuryl, and bromadiolone. In rodents, these compounds have a half-life up to 55 hours. In contrast, indandione compounds have a half-life of 4 to 5 days, but may affect hemostasis for a longer period of approximately 30 days. In addition, indandione compounds may interfere with the endocrine pancreatic function, resulting in reduced intestinal absorption of vitamin K. Because of marked difference in the duration of action of coumarin and indandione compounds, identification of the ingested substance is important for planning of the poisoning treatment.

Warfarin, a common representative of coumaril group of rodenticides, is a odorless, tasteless substance, effective in very low doses. Oral LD_{50} for sodium warfarin in rats is 323 mg/kg for males and 58 mg/kg for females. Average or large doses of warfarin in humans may cause hemorrhage. In humans the compound is considered highly toxic by inhalation and ingestion and moderately toxic by dermal absorption. Warfarin has been established as a human teratogen, because it has been shown to cause birth defects in the offspring of women who receive clinical doses of the compound during any trimester of pregnancy. Therapeutic use by pregnant women has resulted in fetus malformations and mental retardation in infants. Bromadiolone, another coumarin rodenticide acting similary as warfarin, is absorbed well through gastrointestinal tract, skin, and respiratory system. The major route of elimination in different animal species after oral administration is via feces. The liver is the main organ of accumulation and storage of the compound. Elimination from the liver is performed with an initial rapid phase (2-8 days) and a slower phase with a half-life of 170 days. Bromadiolone has a highly acute oral (LD_{50} = 1-3 mg/kg, rodents) and dermal (LD_{50} = 9.4 mg/kg, rabbits) toxicity. In less severe cases of difenacoum poisoning, excessive bruising, nose and gum bleeding, as well as blood in the urine and feces may be observed. Bleeding from internal bodily organs may not be evident until a few days after ingestion. In cases of less severe poisoning with difenacoum, vitamin K_1 may be given as an antidote, together with fresh, frozen plasma, in order to rapidly restore blood clotting factors. Gastric lavage and repeated administration of charcoal is recommended for recent poisonings.

Chlorophacinone, an Indandione rodenticide, can be used for warfarin-resistant rodents. In addition to its anticoagulant action, it also uncouples oxidative phosphorylation in mammal tissues. If ingested by humans, it may reduce clotting ability. In cases of poisoning with chlorophacinone, intravenous and oral administration not of vitamins K3 or K4, but specifically of vitamin K1, as well as blood transfusion, are indicated.

Other classes of rodenticides act by mechanisms other than disturbing blood coagulation. A poisonous alkaloid originating from the seed of the plant of *Strychnos* genus, strychnine was used in the past, and is still used today, as a potent rodent poison. Strychnine is orally rapidly absorbed and symptoms of poisoning usually appear within 20 minutes. The substance increases the irritability of the spinal cord, causing severe contractions of the muscles. The symptoms begin with cramps and soon culminate in powerful convulsions that soon subside, but recur at a touch, a noise or some other minor stimulus. Death usually occurs due to asphyxiation resulting from continuous spasms of respiratory muscles.

Some rodenticides, such as chloralose, are designed only to immobilise the target species at sublethal levels. A primarily narcotic compound, chloralose, is a central nervous system depressant which is efficiently metabolized, resulting in recovery within a few hours from ingestion. It can also be used for surgical procedures on animals. Rodenticides belonging to vitamine D family (calciferol, ergocalciferol) act by raising serum calcium levels, which affects the muscles and provokes general weakenes and decreased heart rate. Clinical signs of intoxication arise within 18 to 36 hours after ingestion and include depression, increased urination and increased water intake.

Sodium fluoroacetate is rapidly absorbed from the gastrointestinal tract and exposure by oral route is the most danderous in poisonings, as the substance is not absorbed to any significant extent through the intact skin. The mechanism of toxicity arises by blocking the Krebs cycle during cell respiration by formation of fluorocitric acid and citric acid accumulation in the cells. After ingestion, a latency period lasting from 30 minutes to two hours occurs, followed by nausea and mental apprehension with facial twitching, numbness and convulsions. Zinc phosphide, an inorganic rodenticide, upon digestion reacts with stomach acid, generating phosphine gas which enters into blood circulation. Chronic exposure to phosphine causes dizziness, nausea, vomiting, diarrhea, coughing, headache and chest pain. Damage to the kidneys, liver and stomach have been reported in humans, but only at high acute doses of zinc phosphide.

ETHANOL

The word alcohol derives from the Arabic "al-kuhul," which denotes a fine powder of antimony used as eye makeup. The word alcohol originally referred to any fine powder, but medevial alchemists later applied the word to refined products of destillation. In narrow sense, alcohol usually denotes ethanol which is a flammable, colorless, slightly toxic substance with a distinctive perfume-like odour. Alcohol has been identified as intoxicating ingredient in beverages since prehistory. The use of alcoholic beverages was indicated during Neolithic period on the basis of 9,000-year-old artefacts. The isolation of ethanol as a relatively pure compound was first achieved by Persian alchemists who developed destillation during the Abbasid caliphate. Absolute ethanol was first obtained in 1796 by Johann Tobias Lowitz, by filtering distilled ethanol through charcoal. Denaturated ethanol is a mixture of ethanol and various additives that prevent human consumption. These additives are either toxic, such as methanol, or have an unpleasant taste or odor, like denatonium benzoat.

In contact with the tongue and mucose membranes, ethanol produces a heat-like sensation. Some individuals have less effective alcohol and acetaldehyde dehydrogenases, the enzymes responsible for ethanol biotransformation, and can experience more severe symptoms from ethanol consumption than individuals with normal enzymatic status. Since ethanol inhibits the production of antidiuretic hormone, overconsumption can lead to dehydratation. Ethanol consumption activates the dehydrogenases responsible for alcohol metabolism and reduces susceptability in subsequent exposures. Persons with increased levels of enzymes metabolize ethanol more rapidly. When taken orally, ethanol is first transfered to tissues with high water content, such as blood and brain; however, cell membranes are generally highly permeable to ethanol, influencing efficient alcohol transfer from the bloodstream to almost all tissues. Women become intoxicated with ethanol more quickly than men, because their body water content is lower and after alcohol absorption, alcohol concentrations occur in their blood. In addition, the, the activity of alcohol dehydrogenases is lower in women and man have a more active first pass metabolism of alcohol.

As a small molecule, alcohol is able to cross the blood-brain barrier. In synapses, it acts on GABA receptors and keeps the synapses open for prolonged periods, enabling the entrance of charged molecules. Other targets of ethanol impact have been suggested, such as ion channels and intracellular signaling molecules.

Table 8.13. Symptoms of intoxication at different ethanol blood level

Blood concentration (g/l)	Symptoms
0.5	Euphoria, talkativeness, relaxation
1	Central nervous system depression, impared motor and sensory function, impared cognition
>1.4	Decreased blood flow to brain
3	Unconsciouness
4	Possible death
>5.5	Very high probability of death

After absorption, ethanol is diluted approximately 5000 fold in the blood. Small doses of ethanol generally produce euphoria and relaxation (Table 8.13.). People tend to be talkative and may exibit poor judgment. At concentrations higher than 1 mg/l of blood, ethanol acts as a central nervous system depressant, producing at progressivly higher doses impaired sensory and motor funtions, slower cognition and then unconsciouness. Ethanol acts to surpress brain functions and causes unsteady walking, slurred speech, disturbed sensory perception and an inability to react quickly. Even though ethanol is generally thought to be a depressant, at low levels it can stimulate certain brain areas, such as N-methyl-D-aspartate receptors, making them more receptive to neurotransmitters. Stimulated areas are responsible for thinking and pleasure sensation. Neurological euphoria induced by alcohol has been associated with rapid release of dopamine in limbic areas of the brain. Body relaxation can be explained by heightened alpha waves in the brain, which can be observed by EEG. At very high concentrations, ethanol produces general anesthesia. It is assumed that alcohol suppresses glucose metabolism in 29% of the brain, as well as in the part of the brain responsible for receiving visual inputs, causing blurred vision.

The product of ethanol metabolism, acetaldehyde, is more toxic than ethanol itself. The body quickly detoxifies acetaldehyde partly by conjugation with glutathione and other thiol containing molecules. When glutathione is depleted and completely engaged in bonding to acetaldehyde, acetaldehyde further concentrates in the blood and partially oxidizes to acetic acid. Alcohol metabolism further proceeds by conversion of acetaldehyde into harmless acetic acid. The reaction is catalized by class of enzymes belonging to aldehyde dehydrogenases. Some people, especially those of East Asian origin, have a genetic mutation in acetaldehyde dehydrogenase gene, resulting in a less potent enzyme. This leads to the formation of high concentrations of acetaldehyde in the blood after alcohol intake and intolerance to even low concentrations of alcohol, with symptoms such as flushing, nausea and dizziness soon after alcohol consumption. Most of the symptoms of alcohol poisoning, such as headacke and nausea, arise from the combination of dehydratation that occurs within the body and acetaldehyde accumulation in the blood. Similar symptoms provoked by the same mechanisms can also be artifically induced by the drugs used to treat alcoholism. Many pathological conditions associated with chronic alcohol abuse, including liver cirrhosis and some forms of cancer, have been directly linked to acetaldehyde impact. Drug therapy with some medications, including paracetamol, as well as exposure to certain organochlorine pesticides, can deplete the glutathione supply, enchancing both acute and long-term risks associated with ethanol consumption. Ethanol has been shown to increase the growth of

Acinetobacter baumannii, a causative bacteria of pneumonia, meningitis and urinary tract infections. International Agency for Research on Cancer of the World Health Organisation has classified ethanol as a carcinogen, while the National Institute on Alcohol Abuse and Alcoholism reports that alcohol may act as cocarcinogen. Studies suggest that the mechanism by which carcinogenesis may be provoked by ethanol are linked to the effects of acetaldehyde on DNA. In the body, acetaldehyde appears to react with naturally occurring polyamines creating mutagenic DNA adducts. The risk for mouth, tracheal and esophageal cancer is 35 times greater in individuals who consume alcohol and smoke. Prolonged ethanol consumption also has been associated with liver cirrhosis. The formation of scar-like fibrous tissue in the liver usually starts as a fatty liver, i.e., fat accumulation in hepatocytes. Besides alcohol, other factors can provoke this condition, such as diabetes, obesity and poor diet. Certain drugs such as corticosteroids can also induce fatty liver. Usually fatty liver has no particular external symptoms, but elevation in some liver enzymes can indicate possible liver pathology. Histopathologically, the condition is characterized by the presence of numerous large lipid vacuoles in the liver that compress and displace the cell nucleus into the periphery of cells.

Liver pathology in heavy, prolonged alcohol consumption can be categorized histologically into three phases. Steatosis or fatty liver, is the first stage, which is reversible and is associated with a variety of metabolic disturbances. Acetic acid that is produced by metabolizing acetaldehyde, a primary biotransformation product of alcohol, can be converted either to fats or to carbon-dioxide and water, depending on a metabolic pathway. In impaired liver function, the metabolic pathway converts acetic acid into fats and fat accumulation leads to inflammation of the liver cells. Produced fats are mostly deposited locally in the liver, building up plaques around the hepatocyte capillaries and causing their death, which initiates the second stage of the liver damage, alcoholic hepatitis or fibrosis. The final stage, alcoholic cirrhosis, is irreversible and is associated with cell death due to buildup of fibrous tissue. One of the key alterations occurring at the early stages of impairment of the liver function is the change in the structure and function of mitochondrial ribosomes. Altered ribosomes have an increased susceptibility to dissociation and decreased ability to participate in translation. The exact mechanism responsible for altered ribosome function is unknown; however, preliminary studies suggest that oxidative stress and aldehyde adduct formation may be directly involved [14]. Another suggested mechanism of liver pathology in alcohol consumers is disturbed cellular oxidation-reduction equilibrium. Alcohol and aldehyde dehydrogenases, which are responsible for alcohol metabolism in the liver, require NAD cofactor, which is converted to NADH in oxidation reactions. The capacity of the cells to convert NADH back to NAD is exceeded during alcohol metabolism, and as a result, cells accumulate an excess of NADH. Most affected is the citric acid cycle. Many NAD-mediated enzyme reactions in the liver are disturbed and abnormal metabolism in the hepatocytes takes place. Accumulation of NADH is considered to be the major cause of alcoholic fatty acid.

Alcohol is able to cross the placenta from maternal circulation. Fetal alcohol syndrome (FAS) is a disorder of permanent birth defects that occur in the offspring of women who consume alcohol during pregnancy. This disorder was clinically described in humans about 30 years ago. Similar effects, without obvious alterations to appearance, but with effects primarly on the nervous system, are known as Fetal Alcohol Effects (FAE). It is not exactly known whether it is the amount, frequency or timing of alcohol consumption during pregnancy that causes differences in the degree of fetal damage. FAS is characterized by variable degrees of birth defects and mental retardation, initially identified by reduced head size and distinctive facial features. Altered facial features occur due to induction of cell death in the craniofacial region during the gestation period [15]. Besides microcephaly (small head size), typical facial features include short eye openings, a flat midface, a low nasal bridge, a thin upper lip and a small jaw. Usually, vertical grooves between nose and mouth are indinstinct, part of the outer ear is underdeveloped and skin folds may be formed at the inner corner of the eyes [16]. In addition to distinct facial effects, there may also be abnormalities associated with the fingers or toes, such as permanent flexion and deformities. Other conditions may commonly co-occur with FAS, such as heart murmur, joint anomalies, involuntary eye movements, a cleft lip with or without a cleft palate, as well as many other symptoms.

Neurological effects of alcohol on the fetus arise due to high incidence of cell death in the outer layer of the developing neural tube and can be manifested as significantly reduced IQ. The risk of brain damage exists during the whole gestation period, since the fetal brain develops throughout the entire pregnancy. Primarily, cognitive functions are affected, promoting poor memory and attention deficit, as well as impulsive behavior. Mental health problems and drug addiction that occur later in the life also have been linked to prenatal exposure to alcohol [17, 18].

Ethylene Glycol

In its pure form ethylene glycol is an odorless, colorless, dense liquid with a sweet taste. Due to its low freezing point, the compound is widely used as automotive antifreeze, coolant and cleaning liquid for car windshields and aircraft exteriors. This alcohol is used by plastic industry to produce polyester fibres and resins. A high boiling point and high affinity to water also makes it a good dehydrator for natural gas. Due to its sweet taste, children and animals are at high risk of ingesting large quantities of ethylene glycol. An estimated lethal dose in humans is about 1.4 ml/kg. Poisoning with ethylen-glycol usually progresses through three phases. The first phase starts with neurological symptoms, such as dizziness, slurred speach and confusion. Over time, the body transforms ethylene-glycol into other toxic metabolites, glycolic acid, glycoxylic and oxalic acid. The second phase is a result of the accumulation of metabolic acids, and this provokes tachycardia, hypertension, hyperventilation and acidosis. The third stage of ethylene-glycol poisoning occurs due to kidney failure provoked by the formation of crystalline calcium-oxalate.

In cases of acute poisoning with ethylene glycol, ethanol is used as an antidote to compete for alcohol dehydrogenases and limit the formation of toxic metabolites. Usually a 5-10% ethanol solution in water containing 5% dextrose is used for the treatment of poisoning.

Propylene glycol, whose most abundant application is as a moisturizer for cosmetics, food and tobacco products, is considered practically non-toxic and is used as a non-toxic antifreeze. In biotransformation processes, it is metabolized to lactic acid, which is also produced naturally by muscle activity.

ISOPROPANOL

Isopropanol has a very wide application as a disinfectant, gasoline additive, solvent and good cleaning substance. Vapors of isopropanol are heavier than air and highly flammable in a wide concentration range. This alcohol is more toxic than ethanol and acute toxicity of isopropanol is approximately twice that of ethanol. In the liver, it is oxidized to acetone by the activity of alcohol dehydrogenase. In intoxicated persons, the biotransformation metabolite is responsible for a fruity, acetone-like odor of the breath. Symptoms of poisoning with isopropanol start with flushing, headacke and dizziness. Manifestations progress with nausea and vomiting and may end with coma due to CNS depression.

METHANOL

Methanol is a colorless, volatile and flammable liquid that is readily miscible with water. It has a light odor that is distinctly different from that of ethanol. Originally distilled from wood, methanol is now manufactured synthetically from carbon oxide and hydrogen. Methanol is primarily used in the manufacture of other chemicals and as a solvent. It is also added to a variety of commercial and consumer products such as windshield washing fluid, paint remover, embalming fluid, lacquer and ink.

The majority of available information on methanol toxicity in humans is related to acute rather than chronic exposure. Toxic effects after repeated or prolonged exposures to methanol are believed to be qualitatively similar, but less severe than those induced by acute exposure. These effects include primarly CNS and visual disturbances such as headache, dizziness, nausea and blurred vision [19].

Methanol is readily absorbed from the gut, skin and lungs. The highest serum concentration is usually reached 30-60 minutes following oral ingestion. Due to its miscibility with water, methanol distributes well to all body tissues. In the liver, methanol is metabolized slowly and it follows zero-order kinetics. Approximately 3% of a methanol dose is excreted through the lungs or excreted unchanged in the urine. Like ethylene glycol, methanol itself is relatively non-toxic; however, it is metabolized to highly toxic compounds that are responsible for acidosis and blindness characteristic of methanol poisoning. In biotransformation reactions methanol is converted to highly toxic formic acid responsible for acidosis, blindness and potentially death. Formic acid inhibits cellular respiration leading to lactic acidosis. A lethal dose of pure methanol in humans is estimated to be 1-2 ml/kg. Permanent blindness has been reported in exposures to as little as 0.1 ml/kg in children and 6-10 ml for adults of average weight. Initial symptoms of methanol poisoning may appear 12 hours post-ingestion, but usually develop after 24 hours. They may ressemble symptoms of ethanol intoxication and consist of drowsiness, confusion and ataxia. The condition also is accompanied by general weakness, headache, nausea, vomiting and abdominal pain. Some of the symptoms that may mimic an alcohol hangover and are due to mild intoxication caused by the methanol itself. As methanol metabolism proceeds, a severe metabolic acidosis develops. Severe metabolic acidosis in conjunction with visual effects are the hallmarks of methanol poisoning. Patients usually describe blurred vision, double vision or changes in color perception. There also may be a constricted visual field and, occasionally, a total loss of vision. Other characteristic visual dysfunctions include pupillary dilation and loss of pupillary reflex. Further signs and symptoms include shallow respiration, cyanosis, seizures and electrolyte disturbances, as well as various hemodynamic changes including profound hypotension and cardiac arrest. There may be mild to profound loss of memory, confusion and agitation, which may progress to stupor and coma as the severity of acidosis increases. In severe cases of poisoning with methanol, death is highly possible. Individuals who survive serious methanol poisoning may suffer from permanent blindness or other neurological damage. In methanol poisoning, coingestion of ethanol may produce a confusing clinical picture as the toxic effects of methanol may be masked or delayed.

Therapy for methanol poisoning include the treatment of metabolic acidosis, inhibition of methanol metabolism and enhanced elimination of unmetabolized compound and formed toxic metabolites. Sodium bicarbonate should be administered in order to treat acydosis and to correct serum pH. Inhibition of further methanol biotransformation is performed by supplying ethanol, which competes for alcohol and aldehyde dehydrogenases. Elimination of methanol may be enhanced by administrating folic acid, a cofactor in conversion of formic acid to carbon dioxide, and by performing hemodialysis.

The signs of methanol toxicity in laboratory animals are quite different from those observed in humans, and the differences in toxicity are attributed to the ability of experimental animals to metabolize formate more efficiently than humans [21, 20]. The major effect of methanol in rodents, dogs, cats and other experimental animals, is CNS depression similar to that produced by other alcohols. In experimental animals metabolic acidosis and ocular toxicity are not observed.

Organic Solvents

Organic solvents are liquid volatile compounds or their mixtures that are relatively stable chemically and include aliphatic, cyclic, aromatic and halogenated hydrocarbons, ketones, amines, esters, alcohols, aldehydes and ethers. The use of organic solvents is widespread in many industries and for various applications. They are used for extraction, dissolution or suspension of compounds not soluble in water. Organic solvents are used in paints, adhesives, glues, coatings and cleaning agents, and in production of dyes, polymers, plastics, textiles, printing inks, agricultural products and pharmaceuticals. Workers employed in listed industries and those involved in paintingprofessions, are chronically exposed to organic solvents and are at high risk of developing adverse health effects. Toxicological studies have demonstrated that these effects can be both reversible and irreversible, depending on the exposure time and total amount of the solvent that is absorbed. Inhalation and percutaneous absorption are the primary routes of solvent transfer into the peripheral blood. Solvent absorption by inhalation depends on its concentration in inhaled air, the blood/air partition coefficient of the solvent, alveolar ventilation rate and duration of exposure. Increased levels of physical exercise increase pulmonary ventilation and cardiac output and leads to increased pulmonary absorption. Dermal solvent absorption depends on duration of the contact, skin area, degree of skin hydration and skin injury. Presence of skin cuts, abrasions or skin diseases enchance dermal absorption significantly.

Some common health effects can be defined for different organic solvents. In epidemiologic studies oriented toward defining the toxicity of solvents, chronic changes in peripheral nerve function were observed. Sensory and motor nerve conduction velocities were impaired and these symptoms persisted for months to years following cessation of exposure.

To date, most experimental animal studies with organic solvents have been directed to determination of their acute neurotoxic effects rather than potential chronic neurotoxic effects from long-term exposure. Acute toxic effects of solvent inhalation noted in animals reflect those seen in humans. Exposure to lower concentrations first provokes narcosis and psychomotor impairment. In higher amounts of inhaled solvents, CNS depression may lead to respiratory arrest, unconsciousness and in severe cases, death. Solvent inhalation by workers may cause effects ranging from an alcohol-like intoxication to narcosis and death from respiratory failure, with a spectrum of intermediate symptoms that include drowsiness, headache, dizziness and nausea.

Human studies of acute exposure to methyl chloroform, styrene and toluene demonstrate impaired psychomotor functions of the CNS, impaired reaction time, as well as impaired perceptual and sensory motor speed. Impact on manual dexterity and coordination was also profound [22-24]. Exposure to xylene for 6 hr/day over a period of 6 days at varying concentrations from 90 ppm to 200 ppm caused an impairment in body balance, manual coordination and in choice reaction times [25]. Most of the observed neurological effects of xylene disappeared a few days after exposure, suggesting the development of tolerance.

Three categories of chronic effects of organic solvents can be defined, varying from minimal and reversible, to pronounced and irreversible. Chronic effects are often correlated with changes in the structure of the nervous tissue and its function. Effects on nervous system provoked by long-term poisoning with organic solvents have high probability of being irreversible. Peripheral nerves have the capacity to regenerate, but damage to CNS is more often permanent. The mildest type of disorder is characterized by fatigue, memory impairment, irritability, difficulty in concentrating and mild mood disturbance. The second level of disorder involves both symptoms of neurotoxicity and abnormalities of performance on neuropsychological testing. The condition can be characterized by sustained personality or mood changes, emotional instability and diminished motivation. Diminished concentration, memory, and learning capacity are usually observed. The third and the most pronounced level of disorder is described as severe chronic toxic encephalopathy. The condition is characterized by global deterioration in intellectual and memory functions that are irreversible, or only poorly reversible. This type of disorder is observed in persons who abusively inhale organic solvents.

Following absorption, organic solvents undergo biotransformation and accumulate in lipid-rich tissues such those in the nervous system. Sometimes the outcome of the biotransformation leads to more toxic metabolites. which are often capable of producing additional toxic effects by covalently binding to essential macromolecules, such as proteins, RNA and DNA. For example, n-hexane and methyl n-butyl ketone are both metabolized to 2,5-hexanedione, which has been shown to have greater neurotoxic potency than either parent compound.

Many factors influence the metabolism and toxic potential of organic solvents. Acute ethanol ingestion increases blood levels of toluene and xylene through competition for active sites of enzymes involved in biotransformation, whereas chronic ethanol ingestion induces solvent-metabolizing enzymes and thereby lowers blood solvent concentrations. Workplace exposure to several solvents simultaneously or to solvent mixtures may result in similar metabolic interactions.

Occupational exposure to organic solvents also pose a risk with respect to carcinogenic or reproductive hazards. Examples of carcinogens are benzene, carbon tetrachloride, trichloroethylene and 1,1,2,2-tetrachloroethane. Recognized reproductive hazards include 2-methoxyethanol, 2-ethoxyethanol and methyl chloride.

Aniline

Aniline is a colorless liquid of specific smell which turns dark in contact with air. It is a very important chemical, used in the production of a wide array of pharmaceutical products and synthetic dyes. In chronic intoxication, related mostly to occupational exposures of workers imployed in the industry of aniline dyes, the disturbance of kidney and digestive functions is observed. As with most organic solvents, aniline is hematotoxic and in chronic exposures causes anemia due to oxidation of hemoglobin to methemoglobin. Methemoglobinemia is indicated by blue skin (cyanosis). Blue coloration appears first in skin areas around nails, nose and cheeks. In addition, aniline acts on the blood by provoking hemolysis, which contributes to severe anemia. Particularly dangerous is chronic, long-term exposure to aniline, where the latent risk may evolve much later, and in some cases serious symptoms may start developing several years after first exposure.

Benzene

Benzene is a colorless liquid of specific smell that differs depending on inherent impurities such as carbon disulfide or thiophen. Benzene vapors concentrate in lower layers (near to the ground) because they are more dense than air. This chemical is commonly applied in the production of synthetic organic products, such as paints, drugs, fragrances and explosives. Benzene is a very good solvent and is used in the rubber industry, in the production of synthetic fabrics and material, in car painting and others.

In inhalation, exposures benzene pose the most serious risk. Orally taken, 6-7 g of benzene can cause death, while in exposure via inhallation same effect would occur by inhaling 64 μg/l of benzene for a short time. Chronic inhalation of contaminated air containing as low as 0.5 μg/l of benzene, which cannot be detected by smell, during prolonged time, carries a risk of serious health impairment. Persons with impaired lung, kidney or liver functions, as well as children, are more susceptible to adverse effects of benzene inhalation.

Symptoms of benzene poisoning vary significantly for chronic and acute exposure. Acute poisonings are mostly accidental and can occur in unexpected situations. Cases of benzene poisoning which ended with lethal outcome were reported in workers doing tar coating. The workers were using a benzene solution of tar for coating the inner surfase of a wooden tank. Because the tank was large, two workers were performing coating job by entering the tank.

Benzene vapors concentrated in the inner space and both workers died in a short time [26]. Manifestations of acute poisoning with benzene vapors usually starts with a feeling of excitation, which develops into shivering, depression and lost of consciousness. In mild cases, only headache and vertigo are observed. These symptoms are accompanied by nausea, vomiting and potentially convulsions. Chronic exposure can be recognized by anemia, breathing difficulties, disturbances in digestion and increased body temperature. Erytrocyte countd below 1,000,000 can also point at possible benzene intoxication, but that is unspecific indicator and may be caused by many other factors and also by other organic solvents.

Carbon Disulfide

Pure carbon disulfide is a colorless liquid; however, technical product is usually yellow colored. The smell of the pure carbon disulfide resembles that of chloroform, but can differ if the chemical contains impurities. Carbon disulfide vapors are 2.5 times heavier than the air and is an consideration in occupational exposures when defining ventilation parameters and when sampling the air. Carbon disulfide is a very good solvent for fats, resins and caoutchouc (India-rubber), phosphorous and iodine. Workers in artificial silk, celluloid, parafine, rubber products industries, as well as workers involved in sulphur production, vulcanization, dry cleaning and fumigant application, are at higher risk of intoxication.

Poisoning with carbon-disulfide is usually occupational and occurs mostly via inhalation and less frequently via dermal routes. Concentrations of 0.1-0.2 μg/l of carbon-disulfide in air can provoke serious health symptoms in chronic exposure, while inhalation of 1.5 μg/l of carbon-disulfide in air can cause the severe effects of acute poisoning. Symptoms of acute poisoning with carbon disulfide are similar to symptoms of poisoning with other organic solvents and are characterized with nausea and vomiting. In chronic poisoning, neurological symptoms start with headache and proceed to speech disorder and impaired vision. Laughter and angry behavioral episodes in individual are recognized and these usually end in dementia. Hallucinations and impulses toward murder or suicide are characteristic in poisoning with carbon disulfide and in some cases may be quite pronounced [26].

Chloroform

Chloroform is a colorless, dense liquid with characteristic sweet smell and taste. The liquid has a high vapor pressure. It is poorly miscible with water but mixes well with ethanol and ether. Chloroform is unstable when exposed to light and moisture and easily decomposes to various products, including very toxic phosgene:

$$2CHCl_3 + O_2 \rightarrow 2COCl_2 + 2HCl$$

To avoid decomposition reactions in the bulk chemical, chloroform is stored in dark bottles with an addition of 0.5 % - 1% of alcohol.

Data concerning oral toxicity of chloroform in humans are inconsistent. The cases where several grams of chloroform taken by mouth caused lethal outcomes, but cases also were reported in which recovery occurred after ingestion of 90 g of chloroform. Variability of excretion rates and different amounts of chloroform eliminated by vomiting might be the reason for this data inconsistency. Air levels of chloroform higher than 100 μg/l cause serious health effects in acute inhalation exposure. Chronic inhalation of air containing 2-3 μg/l of chloroform can lead to jaundice and to other symptoms related to neurological disturbances.

A symptom of oral poisoning with chloroform are strong burning feelings in all internal organs and vomiting, by which majority of the substance is expelled. In the later stages, circulatory failure may provoke cardiac arrest and coma. Poisoning with chloroform should be treated with pure oxygen, and the body temperature of the patient should be increased by warming.

Diethyl ether

Diethyl ether, or ether, is a flammable liquid with high vapor pressure and a very specific odor. The boiling point of this solvent is 35 °C. Miscibility with water is limited but it mixes well with other organic solvents, such as ethanol, benzene, chloroform, petrolether and others. Diethyl ether is widely used in industry as a solvent, for the production of artificial silk, collodium and various polymers. In the past, it had medical application as a general anesthetic and narcotic. The side effects in such applications were frequent, especially concerning lungs injury, but less pronounced than in the case of chloroform, which was also used in the past as general anesthetics. Ether abuse using inhalation or oral intake aimed to intentionally provoke narcosis was frequent in the past. In ether abusers, the enzymatic system responsible for ether biotransformation was activated as a result of frequent consumption, and many abusers developed a certain resistance to consumption of up to half a liter of ether, mixed with ethanol. In susceptible persons, ether can cause a lethal outcome in amounts of 30 g by mouth. The symptoms of diethyl ether intoxication are similar to those of alcohol poisoning.

Formaldehyde

Formaldehyde is a gas with characteristic smell and lacrimating properties. It is usually used in the form of a concentrated solution (40%), known as formalin. Formaldehyde solutions cannot be further concentrated, because the polymerization takes place. Today formaldehyde is one of the most important substances used in chemical industry. It is used in the manufacture of resins, leather goods, paper, pharmaceuticals and other products. It is also used as a disinfectant and in chemical industry for polymer production. Cigarette smoke is a source of indoor formaldehyde and tobacco smoke may contain up to 40 ppm of formaldehyde. The substance binds strongly to proteins, and this property is responsible for antiseptic activity of formalin. Namely, the substance reacts with all amino groups of the protein part of the microorganisms, provoking their death. Similarly, formalin also is a very good embalming fluid and is used to temporarily preserve human remains.

Serious injuries in humans occur for several milliliters of formalin taken by mouth. Formaldehyde is readily absorbed through the gastrointestinal tract and lungs and to a much lower extent, through the skin. After inhalation, the substance is usually absorbed in the upper airways and nose due to good solubility in water. Formaldehyde is also produced in the body and is normally found in the blood. It has been estimated that human body produces approximately 50 grams of formaldehyde per day.

Health effects associated with formaldehyde exposure are most likely to occur at body sites that have direct contact with this chemical, i.e., the eyes, nose and throat. Since the aldehyde is a strong irritant it provokes sneezing, bronchitis and conjunctivitis. In persons exposed to environmental air containing between 0.1 and 1 mg/l of formaldehyde, eye and throat irritation, fatigue, headache, and nausea were reported. Formaldehyde can also cause allergies and trigger asthma symptoms. Inhalation of high levels of formaldehyde, approximately 15 mg/l, have been associated with the appearance of cancer of the nose in rats and mice. A large number of occupational epidemiologic studies have been performed to assess whether formaldehyde is a human carcinogen. Although some studies suggest that there may be an increased risk of cancer, particularly of the nose, others have not confirmed this finding. Positive studies suggest an association with cancer only in groups exposed to high formaldehyde concentrations for a prolonged period .

Nitrobenzene

Nitrobenzene is a light-yellow liquid that turns dark in contact with air. It is not soluble in water, but in alcohol and ether can be easily dissolved. It has very specific odor, which resembles that of almonds, and this property is widely exploited in pharmaceutical and food industry. It is also used in the chemical industry for the production of synthetic dyes, explosives and other items. Generally, the symptoms of poisoning with benzene derivatives containing nitro or amino groups are similar, and consequently the symptoms of poisoning with nitrobenzene are very close to those of aniline poisoning.

Nitrobenzene is an organic solvent of high acute toxicity. Lethal outcome in humans can be observed with only 10-20 drops of nitrobenzene taken orally. Occupational exposure most frequently occurs via inhalation of contaminated air or via dermal absorption. In both cases the absorption is very efficient and toxicity is very high. The cases of intoxication after application of soaps containing nitrobenzene fragrance illustrate the compound's high dermal toxicity [26].

Phenol

At room temperature phenol is a crystalline substance of a specific smell. It is poorly dissolved in water, but dissolves well in ethanol, ether and glycerin. One of phenol's features is a strong caustic effect and because of this, concentrated solutions denaturate proteins very efficiently. Longer dermal contact with phenol may cause serious gangrenous changes. Phenol possesses very strong antiseptic properties, and in the past it was used as an universal disinfectant. The chemical is widely used in pharmaceutical and chemical industries and for the production of various polymers. Lethal oral doses for humans are between 8.5 g and 60 g, but as much as 1 g of phenol administrated to open wound may cause lethal outcome. If taken orally, concentrated solutions of phenol, as with all caustic substances, cause a sensation of burning in the mouth and the other organs of digestive tract. In the first phase of phenol poisoning, the intoxicated person experiences alcohol-like symptoms, after which the pulse weakens, the pupils contract and hypothermia develops. Death comes quickly after oral intake, and for high phenol concentrations, a lethal outcome may come within a few minutes. Characteristic symptoms of phenol poisoning are the dark color of the blood and urine, due to hydroquinone formation in metabolic transformation. In chronic poisoning with phenol, atypical symptoms of headache, cough and skin lesions may be observed.

Phthalates

Phthalates are lipophilic, colorless liquid esters of phtalic acid of low volatility and high boiling point. They were developed during the last century and most phtalates produced are used as plasticizers, contributing to the strength, flexibility and durability of plastics. This group of compounds is widely used in the production of polyvinyl chloride plastic materials, such as plastic bags, food packaging, toys, blood-storage containers and intravenous tubing, as well as in some pesticide formulations. Phthalates are also used as industrial chemicals added to many consumer products such as vinyl flooring adhesives, detergents, lubricants and personal care products. They are extensively used in the automobile industry, home furnishings and plastic clothing.

Once they enter into environment, phthalic acid esters adsorb to particulate matter or bind to fulvic acid in the soil and water and become dispersed. Their estimated half-life in water under aerobic conditions is from 2 days to 2 weeks, but if adsorbed onto solid particles they can persist in soil sediments for over a century. Phthalates are not bioaccumulated in the tissues of animals because they are readily metabolized and excreted. Thus, biomagnification in the food chain is not specific for phtalates. In the atmosphere, phtalates photodegradate quickly and their half-life is from 0.2 day to 4 days [27].

In animal studies, the toxic effects of different phtalates vary significantly in their ability to cause testicular and liver damage, anti-androgenic effects, teratogenicity and liver cancer. The mechanism of carcinogenesis in liver tissue involves peroxisomal proliferation [28-30]. Some of the manifestations in animals cannot be extrapolated to humans because the body mechanisms differ, such as that of peroxisome proliferation, which is pronounced in animal species and is responsible for hepatic changes. The testes are shown to be a primary target tissue for phtalate toxicity, resulting in decreased testicular weight and tubular atrophy. Additional research concerning the toxicity of phtalates in humans should be conducted;, however, according to existing data, phtalates are not classified as very toxic and are readily excreted from the human body. As primary urinary metabolites, phthalate monoesters are useful biomarkers of a recent phthalate exposure [31]. In analytical practice, the determination of phtalates in environmental and biological samples, as well as in various products, is performed by gas chromatography with mass spectrommetric detection after appropriate extraction [32, 33].

REFERENCES

[1] The WHO Recommended Classification of Pesticides by Hazard. Guidelines to Classification; 2004.

[2] Stevens, R; Sturgeon, A. Neuropsychological effects of long-term exposure to OP sheep dips. *Lancet* 345, 1995, 122-129.

[3] Davies, DR. Organophosphates, affective disorders and suicide. *J. Nutr. Envir. Med.* 5, 1995, 367-374.

[4] Minton, NA; Murray, VSG. A review of Organophosphate Poisoning. *Med. Tox.* 3, 1988, 350-375.

[5] Worek, F; Koller, M; Thiermann, H; Szinicz, L. Diagnostic aspects of organophosphate poisoning. *Toxicology* 214(3), 2005, 182-189.

[6] Johnson, MK; Jacobsen, D; Meredith, TJ. Evaluation of antidotes for poisoning in organophorus pesticides. *Emerg. Med.* 12(1), 2000, 22-37.

[7] Thiermann, H; Szinicz, L; Eyer, F; Worek, F; Eyer, P; Felgenhauer, N. Modern strategies in therapy of organophosphate poisoning. *Toxicol. Lett.* 107(1-3), 1999, 233-239.

[8] Worek, F; Kirchner, T; Backer, M; Szinicz, L. Reactivation by various oximes of human erythrocyte acetylcholinesterase inhibited by different organophosphorus compounds. *Arch. Tox.* 70(8), 1996, 497-503.

[9] Willems, JL; De Bisschop, HC; Verstraete, AG; Declerck, C; Christiaens, Y; Vanscheeuwyck, P. Cholinesterase reactivation in organophosphorus poisoned patients depends on the plasma concentrations of the oxime pralidoxime methylsulphate and of the organophosphate. *Arch. Tox.* 67(2), 1993, 79-84.

[10] Tomlin, C. A World Compendium. *The Pesticide Manual.* Incorporating the agrochemicals handbook.10th ed; 1994.

[11] Vettorazzi, G. *International regulatory aspects for pesticide chemicals.* Bungay, Suffolk, Crop Protection Publications. Toxicity Profiles: Florida, Boca Raton; CRC Press, Inc.; 1979.

[12] Garey, J; Wolf, M. Estrogenic and Antiprogestagenic Activities of Pyrethroid Insecticides. *Biochem. Res. Comm.* 251(3), 1998, 855-859.

[13] Fishel; FM. Pesticide Toxicity Profile: Triazine Pesticides. Institute of Food and Agricultural Sciences, University of Florida, 2006.

[14] Marsano, LS. Diagnosis and treatment of alcoholic liver disease and its complications. *Alc. Res. Health* 27(3), 2003, 247–256.

[15] Sulik, KK. Genesis of alcohol-induced craniofacial dysmorphism. *Exp. Biol. Med.* 230(6), 2005, 366-375.

[16] Wattendorf, DJ, Muenke, M. Fetal alcohol spectrum disorders. *Am. Fam. Physician* 72(2), 2005, 279-282.

[17] Sowell, ER; Mattson, SN; Kan, E; Thompson, PM; Riley, EP; Toga, AW. Abnormal cortical thickness and brain-behavior correlation patterns in individuals with heavy prenatal alcohol exposure. *Cereb. Cortex* 18(1), 2008, 136-44.

[18] Riley, EP; McGee, CL. Fetal alcohol spectrum disorders: an overview with emphasis on changes in brain and behavior. *Exp. Biol. Med.* 230(6), 2005, 357-65.

[19] Kavet, R; Nauss, KM. The toxicity of inhaled methanol vapors. *Crit. Rev. Toxicol.* 21, 1990, 21-50.

[20] Gilger, AP; Potts, AM. Studies on the visual toxicity of methanol: V. The role of acidosis in experimental methanol poisoning. *Am. J. Opht.* 39, 1955, 63-86.

[21] Tephly, TR. The toxicity of methanol. *Life Sci.* 48, 1991, 1031-1041.

[22] NIOSH. Criteria for a recommended standard: occupational exposure to 1,1,1-trichloroethane (methyl chloroform). Center for Disease Control, National Institute for Occupational Safety and Health, DHEW (NIOSH) Publication No. 76-184; 1976.

[23] NIOSH. Criteria for a recommended standard: occupational exposure to styrene. National Institute for Occupational Safety and Health, DHHS (NIOSH) Publication No. 83-119; 1983.

[24] NIOSH. Criteria for a recommended standard: occupational exposure to toluene. National Institute for Occupational Safety and Health, HSM 73-11023, NTIS No. PB-222-219/A06; 1973.

[25] NIOSH. Criteria for a recommended standard: occupational exposure to xylene. National Institute for Occupational Safety and Health, DHEW (NIOSH) Publication No. 75-168; 1975.

[26] Mokranjac, M. *Toksikološka hemija*, Naučna Knjiga, 1949 (in Serbian).

[27] Johnson, I; Thomas, B; David, L; Stalling, M; James, W; Schoetter, A. Dynamics of Phthalic Acid Easters in Aquatic Organisms, Fate of Pollutants in the Air and Water Environments Part 2: Chemical and Biological Fate of Pollutants in the Environment. *Adv. Envir. Sci. Tech. Ser.* 8, 1977, 283 – 298.

[28] Schmezer, P; Pool, BL; Komitowski, D. Various short-term assays and two long-term studies with the plasticizer di(2-ethylhexyl)phthalate in the Syrian golden hamster. *Carcinogenesis* 9(1), 1988, 37-43.

[29] Roth, B. Di-(2-ethylhexyl)-phthalate as Plasticizer in PVC Respiratory Tubing Systems: Indications of Hazardous Effects on Pulmonary Function in Mechanically Ventilated, Preterm Infants. *Eur. J. Ped.* 147, 1988, 41-46.

[30] Klimisch, HJ; Hellwig, J; Kaufmann, W. Di-(2-ethylhexyl)phthalate (DEHP): Investigation of inhalation toxicity in rats after repeated exposure. *Human Exp. Tox..* 10, 1991, 68-75.

[31] Blount B. Levels of Seven Urinary Phthalate Metabolites in a Human Reference Population. *Env. Health Pers.* 108(10), 2000, 85-93.

[32] ISO 18856: Water quality. Determination of selected phthalates using gas chromatography/mass spectrometry; 2004.

[33] Chen, H; Wang, C; Wang, X; Hao, N; Liu, J. Determination of phthalate esters in cosmetics by gas chromatography with flame ionization detection and mass spectrometric detection. *Inter. J. Cos. Sci.* 27(4), 2005, 205-210.

Chapter IX

PERSISTENT ORGANIC POLLUTANTS

The presence of different chlorinated organic compounds has been detected in distinct areas, far away from their sources. Identification of traces of pesticides in polar ice and in tissues of arctic animals initiated research on the environmental distribution of stable organic compounds at a global scale. After atmospheric emission, substances with high Henry's Law constants (0.1-50 Pa·ml/mol) evaporate at latitudes of warmer climates, followed by their condensation and fall-out closer to the poles. The compounds are carried as vapors or are adsorbed onto solid particulate matter and are returned to the groud by precipitation events. Generally, the concentrations of stable organic compounds in the Antarctic are lower than the levels in Arctic regions because most of the industrialized regions are located in the Northern hemisphere. Many organic pollutants, which are mostly generated by humans, are not readily biodegraded and are very persistant in the environment. These compounds may circulate in the environment for years and, eventually reach humans from one source or another. Only small amounts of stable circulating substances are affected by oxidative processes and UV irradiation. Mostly being highly chemically stable, such compounds are commonly denoted as Persistent Organic Pollutants (POPs). Due to their lipophilic character these compounds tend to bioaccumulate in tissues of the organisms in the food chain, reaching the concentration factors up to 100.000 and severely influencing biodiversity. Concentrations of these compounds increase along the food chain, dispersing from plankton to whales and birds. This fact marks this group of compounds as toxicologically very important, since they are highly toxic and in bioconcentrated amounts, their impact on human health can be dramatic. Transboundary contamination is important factor in the global distribution of POPs via economic pathways and international trade, which stress the issue of persistant organic polutants as an international problem.

Persistent organic pollutants encompass different classes of substances that have some common characteristics, such as global transport through various environmental compartments, persistancy in the environment, high chemical stability, bioaccumulation and high toxicity. The POPs class partially consists of compounds that are relatively new to human awareness, such as dioxins, and most of them are products of the 20th century, for example polychlorynated byphenyls, organotin compounds, short chain paraffins, perfluorinated compounds and polybrominated flame retardants. Polycyclic aromatic hydrocarbons (PAHs) are not intentionally made products, but have environmental impacts that are similar to other industrial POPs. Some pesticides, like those of the organochlorine group, also persist in the environment for very long periods and can be classified on the basis of this and other features as POPs.

Industrial development initiated the production of many new products, whos toxic effects and environmental impacts were not completely recognized and considered at the time. Subsequent scientific developments have demonstrated their adverse influence on a global scale and revealed their toxicity, contributing to awareness of the risk of the spontaneous introduction of new compounds. Many of the compounds belonging to POPs were present in the environment before the development of analytical methods enabling their detection and determination; however, with further progress in industry, their concentrations started to increase rapidly in both the environment and humans, and modern analytical techniques have since enabled the reliable confirmation of such emerging evidence. The experience from the past should impose a willingness in future technological development to consider all aspects of the introduction of new chemicals, with particular emphasis on toxicological concerns.

Brominated Flame Retardants

Brominated flame retardants (BFRs) have routinely been added to consumer products for several decades in a successful effort to reduce fire-related injuries and property damage. This group of compounds reduce the likelihood of ignition and hinder the spread of the fire in manufactured items, providing valuable extra time for remedial action in the early stages of a fire. Brominated flame retardants are unstable in flame and they act by releasing HBr, which suppresses the fire and slows down the initial burn rate. The compounds are used as additives in a variety of polymers, such as polystyrene foams and epoxy resins. They are also extensively used in the production of computers, electronic and electrical equipment, textiles, furniture and building materials.

The application of BFRs is justified with respect to saving lives and reducing costs and damages from fires, but after their persistancy and adverse health effects were recognized, new considerations have emerged. Concern about this class of chemicals has arisen because of their occurrence in the environment and in human tissues. In addition, by their burning they can generate other persistant environmental contaminants such as polyhalogenated dibenzo-p-dioxins and polyhalogenated dibenzo-p-furans, which are known to have extremly high potential to adversely affect both human health and the environment.

There are more than 175 different types of flame retardants (Figure 9.1.), encompassing halogenated organic compounds, phosphorus-containing and nitrogen-containing substances, as well as inorganic flame retardants. Some of commonly applied representatives of brominated flame retardants that have been investigated for basic toxicological information, include hexabromocyclododecane, tetrabromobisphenol A and commercial mixtures of polybrominated diphenyl ethers. However, the toxicological database for brominated flame retardants is very limited and the current literature is incomplete and often conflicting. This is also the case for many other newly introduced chemical compounds, i.e., the compounds whose application is linked to past century.

Hexabromocyclododecane is a nonaromatic, brominated, cyclic highly lipophilic alkane (log K_{ow} = 5.6), with low water solubility (0.0034 mg/l, 25 °C). The commercial product is composed of three diastereomers: α-, ß- and γ-hexabromocyclododecane. Recent studies have shown that the compound has strong bioaccumulative properties and that it persists in air with a half-life of 3 days and in water with a half-life of 2-25 days [1]. The results of mutagenicity test performed on yeast and *Salmonella*, as well as lack of chromosomal aberrations in human peripheral blood lymphocytes, indicate the absence of genotoxic character for hexabromocyclododecane. *In vitro* studies have shown that under the influence of hexabromocyclododecane, the uptake of dopamine is disturbed and that neurobehavioral alterations may be expected [2]. In order to gain more informations on hexabromocyclododecane regarding developmental toxicity, endocrine disruption, and long-term effects, including carcinogenesis, additional toxicological studies are needed.

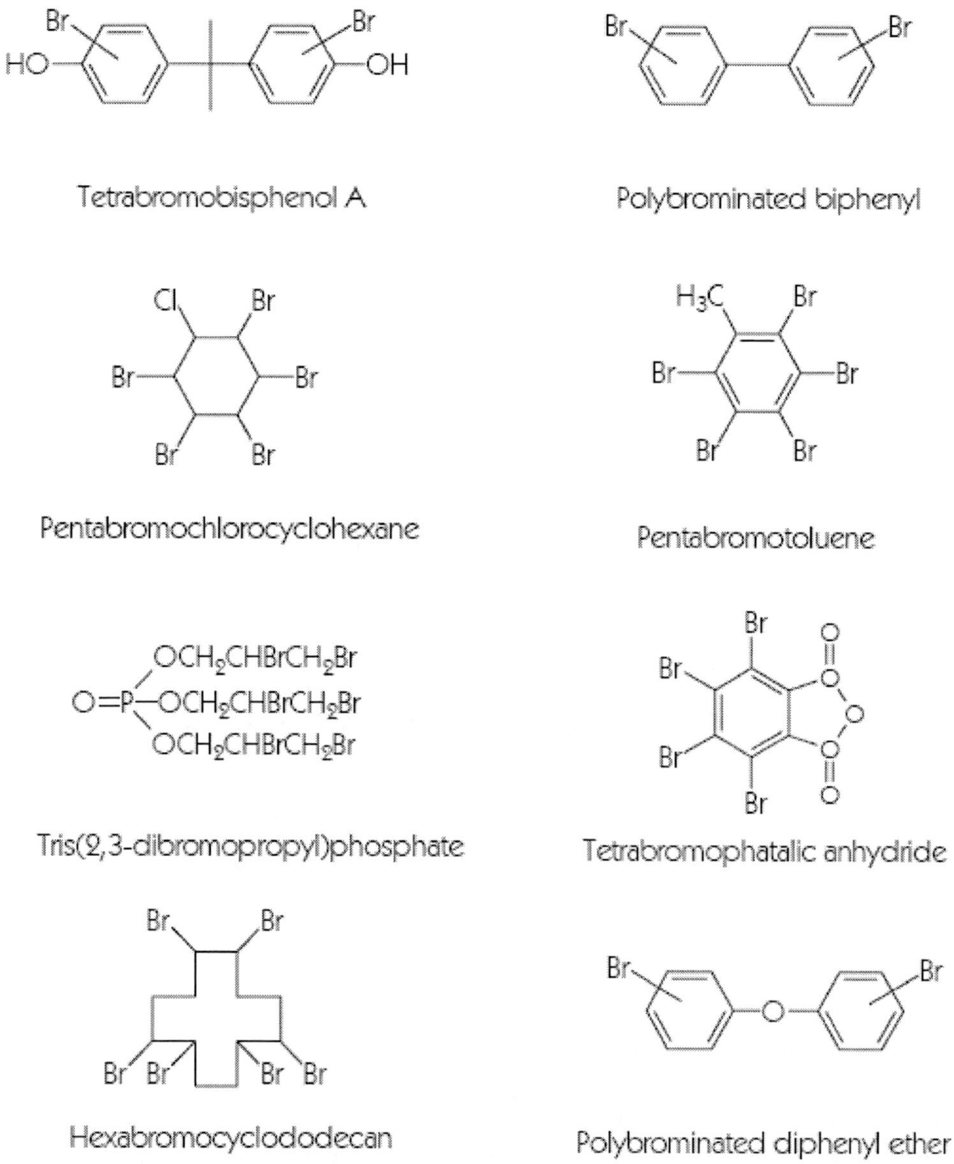

Figure 9.1. The most common brominated flame retardants.

Tetrabromobisphenol A is highly lipophilic (log K_{ow} = 4.5) and therefore poorly water soluble (0.72 mg/l, 25 °C). After a single oral dose, 51-95 % of the dose is rapidly excreted in feces [3]. It is assumed that after rapid excretion in bile, debromination occurs in intestines by the acitivity of gastrointestinal flora. The presence of conjugates in bile suggests that enterochepatic circulation and reabsorbtion slows the elimination rate of the compound. Animal *in vivo* studies have shown that tetrabromobisfenol A is not potent as an inhaled or dermal toxicant, nor as a skin and eye irritant [4]. In conducted studies, the substance did not express teratogenic effect. On the basis of reduced defense ability against bacteria and viruses, as well as neurotoxic effects observed in experimental animals, the inhibition of immunological response, as well as the inhibition of dopamine have been suggested for tetrabromobisphenol A. Generation of free radicals induced by tetrabromobisphenol A metabolism may be implicated in its carcinogenesis [5]. Some of the most recent concerns regarding the toxicity tetrabromobisphenol A focus on the possibility of the compound to act as an endocrine disruptor. its structural similarity to bisphenol A, a known weak environmental estrogen, suggests that tetrabromobisphenol A might have the ability to bind to estrogen receptors and disrupt signaling. In conducted experiments, it was demonstrated that the compound also acts as thyroid hormone agonist [6].

Polybrominated diphenyl ethers potentially involve 209 different congeners, varying in both number and position of bromination. Commercial products represent the mixture of different congeners, but contains pentabromodiphenyl ether, octabromodiphenyl ether and decabromodiphenyl ether in the highest concentrations. All congeners are persistent in environment, with the half-lives in water and soil/sediment of 2 months and 6 months, respectively. These compounds are considered stable, but under the influence of UV light debromination occurs relatively easily. Photodecomposition seems to be dependent on the degree of bromination. The lower brominated congeners degrade in the air with the half-life > 1 day, whereas the octa- and deca- congeners decompose rapidly (5-6 h). In water, lower brominated congeners containing four to seven bromines are more bioaccumulative and persistant, with bioconcentration factors > 5000 [7]. Because of the differences in bioaccumulation and persistence, the congener patterns in biota are different from those in commercial products. In general, lower brominated mixtures are more toxic than the higher congeners. Pentabrominated diphenyl ether is more toxic than octabrominated diphenyl ether, whereas decabromodiphenyl ether is, according to present knowledge, considered nontoxic to invertebrates.

In oral exposure to polybrominated diphenyl ethers, approximately 10% of the dose is eliminated in the bile as hydroxy/methoxy metabolites with 5 to 7 bromine atoms [8]. Lipophilic tissues are the preferred site for deposition, with highest tissue concentrations in adipose tissue, the adrenal glands, the gastrointestinal tract and skin. In general, the concentration of polybrominated diphenyl ethers in human breast milk shows a rapid increase over the past two decades [9]. In animal studies that were directed to defining acute oral toxicity of polybrominated diphenyl ethers, effects such as the impairment of spontaneous motor behavior, neurobehavioral disturbance and alteration of the susceptibility to cholinergic transmiters were observed [10]. Multiple studies and some *in vitro* tests demonstrate that polybrominated diphenyl ethers can perturb the thyroid system in animal models [11,12] and inhibit estrogen sulfotransferase [13]. Future toxicological studies directed to defining the toxicity of polybrominated diphenyl ethers should consider various congeners, as well as their metabolites.

Brominated flame retardants are a structurally diverse group of compounds that include aromatic and aliphatic compounds, cyclic aliphatic substances, phenolic and phthalic anhydride derivatives. The reliable determination of every individual compound in environmental and biological samples poses different analytical requirements. Some of the compounds are thermolabile, demanding the application of liquid chromatography for their separation, usually combined with mass detector to achieve sufficient selectivity [14, 15]. Most individual brominated flame retardants and their metabolites can be determined by gas chromatography since they are sufficiently volatile. The problems of coelution of different congeners, as well as the potential for interference from other halogenated organic compounds are probable and should be solved by adequate choice of experimental parameters. Mass detector and electron capture detector should provide sufficient sensitivity for the determination of trace levels of brominated flame retardants [16].

Even though the concentrations of brominated flame retardants are rapidly increasing in human tissues, they are still lower than those occurring in animal samples and in comparison to PCBs level. Nevertheless, great precaution should be taken in allowing the further production and application of brominated flame retardants, and all toxicological aspects recognized to date should be taken into consideration.

Chlorinated Short-Chain Paraffins

Chlorinated paraffins are commercially produced chlorinated linear hydrocarbons containing between 10 and 30 carbon atoms. Commercial products are viscous yellowish liquids or low-temperature melting solids with a chlorine content of 40% - 70%. Their properties vary depending on the properties of the raw material they were produced from, the temperature of chlorination and the chlorine content. Chlorinated paraffins have properties similar to polychlorinated biphenyls (PCBs) and have replaced them in various uses, depending on their chain length [17]. Short-chain chlorinated paraffins containing between 10 and 13 carbon atoms are used as lubricants, additives and sealing fluids, while medium-length paraffins containing between 14 and 17 carbon atoms are used as PVC plasticizers tailored to replace phtalates due to cost considerations. Chlorinated paraffins containing between 20 and 30 carbon atoms are flame retardants, components of paints, lubricants and additives in cutting oils used to improve the surface roughness of blades [18].

The major source of environmental contamination with chlorinated paraffins most likely occurs via the use or disposal of these products, release of effluents from sewage treatment plants, as well as via direct emissions from the industries dealing with their production. Movement of chlorinated paraffins through various environmental compartments is determined by their low volatlity and low water solubility, i.e., high log K_{ow} values (5 – 13). Due to such properties, the compounds are adsorbed on solid particles and sediments and their availability for uptake by biota is reduced. The content of chlorinated paraffins in natural waters depends on their use in nearby regions and reaches μg/l level in both marine and freshwaters, while in sediments their concentration can reach up to 10 mg/l. Even though under neutral conditions, hydrolysis of chlorinated paraffins is very slow, they can be degradated by aquatic microorganisms. Microbiological degradation is a slow process, requiring microorganism first to aclimate to the environment in the vicinity of the effluent source.

Reported research results of different groups of authors concerning the bioacumulation of chlorinated paraffins in biota are inconsistent. The results of some authors indicate that no significant bioaccumulation in aquatic organisms occurs and that tissue contents are similar to sediment contents [18]. However, other research results suggest that low-molecular weight chlorinated paraffins may accumulate to a greater extent than paraffins with longer chains, and that concentration factors in mussles can reach up to 6,000 [19]. For medium-sized chlorinated paraffins, the concentration factor is estimated to be around 1,000 and for high molecular chlorinated paraffins around 50 . It is considered that bioaccumulation potential of high molecular weight compounds is limited due to their low water solubility and strong tendency to adsorb onto suspended particles. It is estimated that the uptake is insignificant for chlorinated paraffins with molecular mass more than 600 g/mol, corresponding to carbon chain length of 24 and chlorination level around 50%. Variability in bioaccumulation properties of chlorinated paraffins of different lengths is the reason that short-chain chlorinated paraffins are toxicologically more interesting compared to high-molecular weight compounds. Also, the toxicity of chlorinated paraffins increases with the decrease of the carbon chain length [20].

Although short chain paraffins are considered not very toxic, recent findings related primarly to the impact of, short-chain paraffins on human health have evoked attention of toxicological community. Animal toxicological studies demonstrate carcinogenic and necrotic effects on the liver under the influence of short-chain compounds, but it is confirmed that the mechanisms by which short-chain chlorinated paraffins cause tumours in rodents are specific for animal species and, therefore, are of no relevance for human health [21]. Furthermore, chlorinated paraffins have been clearly shown to be non-genotoxic. *In vitro* experiments of percutaneous absorption of 0.037 mg/kg/day, conducted on the culture of human skin, demonstrate the cancer risk of approximately 1 in 20 000 [22]. In human experiments, cases of skin irritation and sensitization have been observed. Another early warning sign related to toxic effects arise from the fact that short-chain chlorinated paraffins induce the phase II detoxification enzyme glutathione S-transferase, even at very low concentrations [23].

Determination of chlorinated paraffins in environmental and biological samples is very difficult owing to the presence of many congeners, which are difficult to separate. The compounds may also undergo hydrolysis or dehydrochlorination in aqueous solutions and consequently many derivatives may be present in real samples. Determination of the level of contamination requires water and sediment samples to be stored at ambient temperatures, and their analysis within a month of sampling, avoiding plastic containers, such as PVC containers, which may contaminate them. In analysis of short-chain paraffins, non-specific methods of thin-layer chromatography on aluminium oxide plate [24] and gas chromatographic methods, in which the peaks are usually poorly resolved and where other halogenated compounds interfere [25], were improved by introducing a more selective mass detector with negative ionization [26, 27]. Prior to application of analytic techniques, chlorinated paraffins should be extracted from the samples and the extract must be purified by applying adsorption or gel permeation chromatography. Purifed extract may be directly introduced into the mass spectrommeter, due to the high sensitivity enabled by high ion concentrations in the ion source [28, 29]. The method also requires an extensive clean-up procedure of the samples; however, the risk of interference is still high. With the development of new analytical techniques, both the selectivity and the sensitivity of measures of chlorinated paraffins have improved; however, the obtained results represent only estimates of real concentrations, as it remains difficult to detect individual substances by applying the current analytical techniques.

DIOXINS

The dioxin family consists of 75 different compounds that include a chlorine atom attached in any position on two aromatic rings. Dibenzofurans (Figure 9.2) are structurally and toxicologically very similar as a class, and they accompany dioxins in most of the samples. In pure form, dioxins are colorless solids or crystal compounds. In the environment, they tend to be associated with ash, soil, or any surface with a high organic content, such as plant leaves, because their solubility in water is very low (~0.0002 mg/l, 25 °C). Dioxins are naturally produced from incomplete combustion of organic material in forest fires or volcanic activity. Industrial, municipal and domestic incineration and combustion activities are the predominant source of man-induced environmental pollution. These toxic compounds can also be generated in processes that do not use the thermal energy, such as the chlorine bleaching of paper, if the source of the chlorine atoms and the organic matter are present. Dioxins are common contaminants of many industrial products, such as chlorinated phenols, used as antiseptics, or 2,4,5-trichlorophenoxyacetic acid (2,4,5-T), a herbicide that is a component of Agent Orange and extensively used in Vietnam War as a defoliant. It is assumed that herbicides highly contaminated with dioxins were the reason for the birth of a large number of children with deformities and that these children were exposed transplaccntally and via mothers milk to dioxins during the Vietnam War. Dioxins have also been detected at low concentrations in cigarette smoke and car engine exhaust.

2,3,7,8-tetrachlorodibenzodioxine (2,3,7,8,-TCDD) (Figure 9.2.) is considered the most toxic representative of this group of compounds and its toxicity is adopted to be 1. The toxicity of other organic pollutants of similar structure (dibenzofurans, chlorinated byphenyls) is expressed relatively to 2,3,7,8-TCDD (Table 9.1.) as toxic equivalent factors (TEQ).

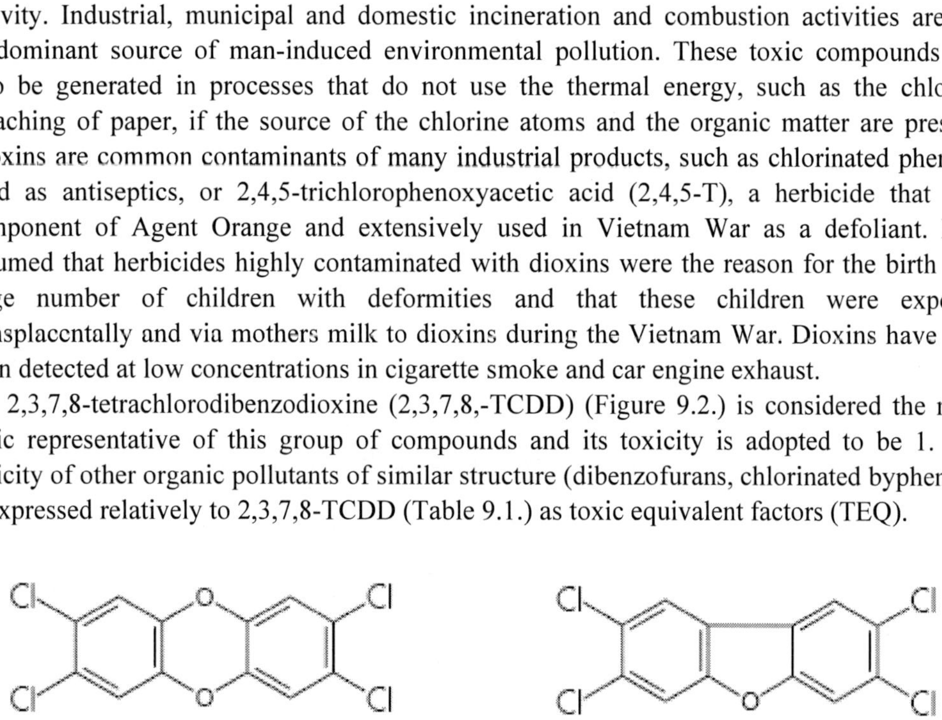

2,3,7,8-TCDD 2,3,7,8-TCDF

Figure 9.2. The most toxic dioxin 2,3,7,8-tetrachlorodibenzodioxin (2,3,7,8-TCDD) and related 2,3,7,8-tetrachlorodibenzofuran (2,3,7,8-TCDF).

In industrialized countries, the daily intake of chlorinated dibenzodioxins and chlorinated dibenzofurans is estimated to be 50-200 pg of TEQ. Being highly lipophilic (log K_{ow} = 6.8) after absorption, dioxins accumulate in fat tissue with a half-life of up to seven years.

Table 9.1. International toxic equivalent factors of dioxins and dibenzofurans for humans and mammals

Dioxins	TEQ
2,3,7,8-tetrachlorodibenzodioxin (2,3,7,8-TCDD)	1
1,2,3,7,8-pentachlorodibenzodioxin (1,2,3,7,8-PeCDD)	1
1,2,3,4,7,8-hexachlorodibenzodioxin (1,2,3,4,7,8-HxCDD)	0.1
1,2,3,7,8,9-hexachlorodibenzodioxin (1,2,3,7,8,9-HxCDD)	0.1
1,2,3,6,7,8-hexachlorodibenzodioxin (1,2,3,6,7,8-HxCDD)	0.01
1,2,3,4,6,7,8-heptachlorodibenzodioxin (1,2,3,4,6,7,8-HpCDD)	0.0001
Furans	TEQ
2,3,7,6-tetrachlorodibenzofuran (2,3,7,6-TCDF)	0.1
1,2,3,7,8-pentachlorodibenzofuran (1,2,3,7,8-PeCDF)	0.05
2,3,4,7,8-pentachlorodibenzofuran (2,3,4,7,8-HxCDF)	0.5
1,2,3,4,7,8,-hexachlorodibenzofuran (1,2,3,4,7,8-HxCDF)	0.1
1,2,3,7,8,9-hexachlorodibenzofuran (1,2,3,7,8,9-HxCDF)	0.1
1,2,3,6,7,8-hexachlorodibenzofuran (1,2,3,6,7,8-HxCDF)	0.1
2,3,4,6,7,8-hexachlorodibenzofuran (2,3,4,6,7,8-HxCDF)	0.1
1,2,3,4,6,7,8-heptachlorodibenzofuran (1,2,3,4,6,7,8-HpCDD)	0.01
1,2,3,4,5,7,8-heptachlorodibenzofuran (1,2,3,5,6,7,8-HpCDD)	0.01
Octachlorodibenzofuran	0.0001

The toxicokinetic behavior of dioxins depends on their lipophilicity, which increases with the degree of their chlorination. Absorption and tissue partitioning of dioxins, besides being directly dependent on lipophilicity of the specific compound, depends also on the dose, body composition, age and gender. Estimation of exposure level to dioxins on the basis on lipid concentration is not recommended and steady-state models using first-order kinetics should be used instead.

In most experimental animals, determined lethal levels of most toxic 2,3,7,8-TCDD are shown to be on the order of magnitude of several μg/kg, classifying this compound as very toxic. However, this measure differs in some ways in comparison to other very toxic substances, such as strychnine or polytoxin. 2,3,7,8-TCDD provokes lethal a effect within several days, while for other very toxic substances, a lethal effect is demonstrated soon after intoxication.

Typical manifestation of acute poisoning with high concentrations of dioxins, similar to that of other very lipophilic organic toxicants, such as chlorinated byphenyls, include chloracne. Chloracne is a severe skin disease with acne-like lesions that occur mainly on the face and upper body. Other skin effects noted in people exposed to high doses of 2,3,7,8-TCDD include skin rash, discoloration and excessive body hair. Changes in blood and urine composition observed in exposed people indicate possible liver damage. Most, if not all, of these responses are dependent on the binding of dioxins to aryl hydrocarbon receptors, inducing changes at biochemical, cellular and tissue levels. Extraordinarily potent 2,3,7,8,-TCDD has been shown to affect a wide range of organ systems in many animal species. The most sensitive endpoints of 2,3,7,8,-TCDD exposure are related to immunotoxicity and reproductive and neurobehavioral effects. A number of genes encoding biotransformation enzymes, such as cytochrome P450 monooxygenase, glutathione S-transferase and UDP-glucuronosyltransferase, are targeted by aryl hydrocarbon-receptors, and by this mechanism dioxins are able to induce biotransformation enzymes. Enzyme induction, changes in hormonal level, and reduced glucose tolerance are examples of subtle changes that may occur at comparatively low dioxin exposures. These are also other common symptoms of exposure to other chemicals that have a similar structure and that also bind to arylhydrocarbon receptors [30].

Risks from dioxin and furan exposure are not related to acute toxicity, and the primary health risk is from chronic, long-term exposure at much lower levels. According to conducted studies, 2,3,7,8-TCDD has been classified as a proven human carcinogen, contributing to lung, larynx and tracheal cancer, Hodgkin's disease, non-Hodgkin's lymphoma and leukemia. Long-term exposure to lower levels of dioxin can cause a variety of effects in animals, such as weight loss, liver damage, disruption of the endocrine system and weakening of the immune response. In animal studies, prolonged exposure to 2,3,7,8-TCDD has been shown to cause reproductive damage and birth defects, even though firm confirmation requires more toxicological research. Severe teratogenic effects of dioxins were also demonstrated very obviously in humans in Agent Orange accidents. In animal studies, different animal species exposed to dioxins during pregnancy had miscarriages or offspring with severe birth defects including skeletal deformities, kidney defects and weakened immune responses.

Since dioxins are shown to be toxic at very low concentrations, their reliable determination requires very sensitive and selective analytical methods. Gas chromatography with high-resolution mass spectrommetry is the current standard for analysis of dioxins and related compounds [31], because it is difficult to differentiate compounds that are structurally so similar using detectors of lower resolution. High-resolution spectommetry can discriminate compounds that differ in molecular wight to the 4 decimal places. In order to adequately define the risk of dioxins in a specific sample, the content of each individual isomer should be determined and multiplied by toxic equivalent factors. The overall toxicity is then estimated by resuming different toxic equivalents of different isomers and expressing the overall toxicity through the toxicity of 2,3,7,8-tetrachlorodibenzodioxin. Sample preparation for dioxin determination is usually time consuming and tedious, because many purification and preconcentration steps are required to reduce interferences from other organic substances and to reach detectable levels. Sample preparation can require up to 24h, but the use of automated extraction and cleanup procedures significantly simplifies the process and contributes to better reproducibility and accuracy. Bioanalytical methods for dioxin determination offer advantages in speed, expense, and field portability [32, 33]. An enzyme-linked immunosorbent assay (ELISA) method enables on-site screening of soil, sediments or human blood for dioxins. Currently, the only commercially available antibody that can be used in bioanalytical tests is anti-2,3,7,8-tetrachlorodibenzodioxin monoclonal antibody, which shows high reactivity and selectivity with 2,3,7,8-tetrachlorodibenzodioxin, but can not be used for the determination of other isomers.

Organotin Compounds

Organometallic compounds containing tin are a very important group of substances used for various purposes and most of them contain tetravalent tin. The only well-established compound with divalent tin is the cyclopentadienyl compound ($C_{10}H_{10}Sn$). Industrially important organotin compounds are mono- ($RSnX_3$), di- (R_2SnX_2), tri- (R_3SnX) or tetrasubstituted (R_4Sn) with R representing usually butyl, octyl or phenyl group, and X being chloride, fluoride, oxide, hydroxide, carboxylate or thiolate. Monosubstituted organotin compounds have a limited application and are used as stabilizers in poly(vinyl chloride) films. Disubstituted organotin compounds are used mainly in the plastics industry, particularly as stabilizers and as catalysts in the vulcanization and production of polyurethane foams. A class of organotin compounds called estertins ($(R\text{-}O\text{-}CO\text{-}CH2\text{-}CH_2)_2SnX_2$) have been developed for use as stabilizers in poly(vinyl chloride) [34, 35]. Trisubstituted organotin compounds have strong biocidal properties and are implemented as fungicides, bactericides, herbicides, molluscicides, insecticides, nematocides, rodent repellents and antifouling agents in boat paints. Tetrasubstituted organotin compounds are mainly used as intermediates in the preparation of other organotin compounds.

In natural waters, tin content is usually below 1 μg/l and organotin compounds are primarily adsorbed on suspended particles and sediments [36]. The persistency of organotin compounds in the environment varies considerably depending on the conditions and the type of the compound, being strongly influenced by UV light, oxidation and microbal processes [37]. Under usual environmental conditions, the half-life of triphenyltin compounds is between a few days and 140 days [38]. Diorganotin compounds have shorter half-lives [37]. Organotin compounds may enter water from antifouling paints and may be subject of biotransformation (methylation) by Pseudomonas, species which influence their concentraton in aquatic organisms. The concentration factor varies for different organometallic species and different marine organisms and due to significant bioacumulation of methylated forms, content in tissues of some marine organisms may reach 0.2-20 mg/kg [39].

The application of tin pesticides and antifouling agents, and migration of tin from poly(vinyl chloride) materials contribute to tin content in food, which is generally below 2 mg/kg. Organotin compounds are more readily absorbed from the gastrointestinal tract than inorganic compounds and the absorption of compounds with short chains is the most efficient. After absorption, the compounds are biotransformed in the liver and intestinal mucosa by dealkilation reactions. The mode of excretion of organotin compounds largely depends on the specific compound and occurs in urine, bile and milk.

Some butyltin compounds are known to produce gastrointestinal irritation when administrated orally, while dermal application provokes skin irritation with severe local demage, as well as eye irritation. Dermal LD_{50} values of various trimethyltin derivatives are 50-100 mg/kg (mice) [40] and for bis(tributyltin) is 200 mg/kg (rats and mice) [41]. Appendix V may be consulted for a comparison of the acute toxicity of different organotin compounds. .

Systemic effects of monosubstituted, disubstituted and trisubstituted organotin compounds differ. In general, mono- and di-organotin compounds are less toxic than triorganotin compounds, and the toxicity of trialkyltin compounds decreases as the number of carbon atoms in the alkyl chain increases. According to toxicological research results, only trisubstituted compounds have a specific effect on the central nervous system and produce cerebral edema [42], whereas disubstituted compounds do not produce this effect, but are potent irritants that induce inflammatory reactions in the bile duct and necrotic changes in the liver [43]. Tetrasubstituted compounds toxicologically resemble trisubstituted compounds.

In experimental animals, acute poisoning with trialkyltin compounds cause characteristic lesions in the central nervous system, manifested as cerebral edema. Cerebral damage causes insomnia and convulsions, which disappear quickly upon cessation of administration. Neurological problems are accompanied with bradycardia and constipation. Tetraalkyltin compounds may acutely produce muscular weakness, tremors and hyperexcitability, which may end with paralysis and respiratory failure. There is no evidence that organotin compounds are carcinogenic or teratogenic. The mechanism of toxic effects is not yet clarified for all organotin compounds, but for trialkyltin compounds, the disturbance of mitochondrial respiration is suggested. Trialkyltin compounds are shown to provoke discharge of a hydroxyl-chloride gradient across mitochondrial membranes and to interfere with adenosine triphosphate (ATP) synthesis [44, 45].

For the treatment of poisoning with dialkyltin compounds, dimercaprol has been suggested as an effective antidote because it prevents the accumulation of produced alpha-keto acids during their biotransformation [46, 47]. While effective in poisoning with dialkyltin compounds, dimercaprol is ineffective for the treatment of poisoning caused by triethyltin compounds, possibly because trialkyltin compounds do not react with sulphydyl groups [48, 49]. Steroid therapy with dexamethasone for poisoning with triethyltin compounds diminishes the severity of brain edema due to enchanced excretion and catabolism of the substance [50].

Organotin compounds in environmental and other samples can be determined with the application of chromatographic techniques coupled with spectrommetric detection, such as atomic absorbtion spectrometry [51], flame photometric detection [52, 53] or mass spectrometry [54]. Prior gas chromatographic separation tin compounds must be derivatized in order to enchance their volatility. Use of an atomic emission detector enables very specific detection of separated compounds; however, mixtures containing both inorganic and organic tin compounds that may cooccur in various media may still pose a problem. Furthermore, in order to improve reliability of the overall analytical method, the procedures of quantitative extraction and sample preparation must be improved. Liquid chromatography can also be used for separation and determination of different tin organometallic species [55, 56]; however, in practice it is less frequently applied.

PERFLUORINATED COMPOUNDS

Fluorinated organic compounds are man-made compounds in which the hydrogen atom of carbon-hydrogen bond is replaced with a fluorine atom. The chemical bond between carbon and fluorine is extremely stable, making these molecules unbiodegradible. Such compounds have the capability to repell the water and oil and to dramatically decrease surface tension. These properties contribute to their wide use in the production of food containers and pans, as well as in many different cosmetic products. The use of fluoride polymers in fabrics impregnation protects clothing from moisture. Perfluorinated compounds have been applied for more than a half of century as refrigerants, surfactants, adhesives and insecticides. They are also components of paper coatings and may be used as fire retardants. Most commonly applied representatives of this group of compounds are perfluorooctane sulfonate (PFOS), perfluorooctanoate (PFOA) and perfluorooctane sulfonylamide (PFOSA). These compounds are important chemicals used mainly in the manufacture of plastics, electronics, textiles and construction materials. PFOA is used in the production of chemicals and the aircraft industry. The toxic potential of these compounds is not well characterized and in order to underline the mechanisms of their adverse effects, basic toxicological studies should be conducted.

Recent reports indicate the presence of perfluorinated compounds in the environment, classifying them as environmental pollutants [57]. Despite the fact that biomagnification of these compounds has not been detected in biota, perfluorinated compounds persist in the environment and have a potential role in increasing the risk of endocrine, reproductive and developmental dysfunction in humans [58]. Biopersistancy of PFOA in the environment is estimated to be 98 days. Occupational studies have revealed that perfluorinated compounds influence liver and pancreatic functions, the endocrine system and lipid metabolism. Data on human toxicity are not yet available since the awareness of the toxicity of perfluorinated compounds and their persistency is relatively new; however, some epidemiological studies offer valuable information. In workers exposed to PFOA with blood levels of 5 mg/l, elevated levels of cholecystokinin and estrogen were detected. Animal studies performed on monkeys show that PFOS is well absorbed orally but poorly metabolized and excreted, with a long half-life of 200 days [59]. In conducted animal studies, the compound has demonstrated hepatic toxicity and alteration of thyroid hormone levels [60]. The potential mechanism of the toxicity is that PFOS interferes with mitochondrial biochemical processes and intercellular communication [61, 62].

Results of environmental monitoring of perfluorinated organic compounds in different compartments indicate that perfluorinated organic compounds are present in ppb levels in water and air, and in ppm levels in soil. So far, serious environmental threats from this class of stable compounds has not been recognized; however, considering adverse endocrine effects of perfluorinated organic compounds in humans, they should be characterized as components of emerging interest and should be completely toxicologically defined and characterized.

Liquid chromatography with mass detector applied after extraction procedures enables the determination of fluorinated organic compounds in different samples [63-65]. However, their separation and determination with the application of gas chromatography is not suitable for these compounds because of their extremly low volatilility.

POLYCHLORINATED BIPHENYLS (PCBS)

Polychlorinated biphenyls (PCBs) were first manufactured in 1929, but their use was banned or severely restricted three decades later because of their recognized risks to human and environmental health. Commercial products were manufactured and sold as mixtures of several congeners with a variety of trade names, including Aroclor, Pyranol, Pyroclor (USA), Phenochlor, Pyralene (France), Clophen, Elaol (Germany), Kanechlor, Santotherm (Japan), Fenchlor, Apirolio (Italy) and Sovol (USSR). Depending on the number and location of chlorine atoms in the two phenyl rings, 209 of different congeners of polychlorinated byphenyls are possible (Figure 9.3.). About 10% of totally produced amounts since 1929 still circulates in the environment. These synthetic compounds have been used as an insulating material in electric equipment, such as transformers and capacitors, and also as heat transfer fluids and lubricants. PCBs had an important application in wide range of products such as plasticizers, surface coatings, inks, adhesives and paints. Trace levels of PCBs are found even in very remote areas, far away from the source, proving that they are globaly transported through all environmental compartments.

Polychlorinated byphenyls are highly lipophilic, stabile compounds that accumulate in the tissue,, and the highest concentrations are found in animals at the top of the food chain and in humans. Biotransformation of these lipophilic molecules starts with hydrolysis in phase I. Chlorine content, the substitution pattern and the presence of certain isoenzymes of the cytochrom P450 system, are important factors in the biotransformation rate of PCBs [66]. In general, metabolism rate of PCBs decreases with the increased number of chlorine atoms.

Polychlorinated byphenyls are very stable in the environment and in general, bacteria cannot use chlorinated aromatic hydrocarbons as a substrate. However, some microorganisms (e.g., *Acinetobacter sp*, *Achromobacter sp*, *Acetobacter*, *Alcaligenes*, *Pseudomonas*) are capable of using lower chlorinated PCBs as a carbon source. An increase in number of chlorine substituents decreases the biodegradation of PCBs, and chlorine atoms in *ortho*-position on either of the aromatic rings of the biphenyl molecule significantly inhibit biodegradation [66]. Burned at high temperatures, polychlorinated byphenyls can be converted to more toxic dioxins.

All humans are exposed to greater or lesser extent to PCBs through food, water and air. Average daily dietary intake of PCBs in humans is estimated to be < 0.5 μg, and these concentrations are not likely to cause adverse health effects. People who consume large amounts of fish and other aquatic organisms may be exposed to higher dietary levels of PCBs, and accumulated PCBs in human body remain there for years.

The toxicity of polychlorynated byphenyls depends on multiple variables, including the type of PCB, the dose and route of exposure. The mechanism of toxicity of PCBs is related to both dependent and independent mechanisms on aryl hydrocarbon-receptors and is somewhat similar to other organic pollutants such as dioxins. The toxicity of coplanar and non-coplanar byphenyls, expressed as toxic equivalent factors, are presented in Tables 9.2 and 9.3.

The aryl hydrocarbon receptor is an intracellular receptor that has a role as a transcription factor; it is normally inactive in cytosol. Upon binding to a ligand, the factor translocates into the nucleus, leading to changes in gene transcription [67]. Ligands for these intracellular receptors can be both synthetic and naturally occurring. Synthetic representatives include dibenzo dioxins, dibenzofurans, polycyclic aromatic hydrocarbons and polychlorinated biphenyls. Naturally occurring compounds that have been identified as ligands of aryl hydrocarbon receptors include derivatives of tryptophan, billirubin, arachidonic acid metabolites, several dietary carotinoids and others [68, 69].

Table 9.2. International toxic equivalent factors of coplanar biphenyls for humans and mammals

Biphenyls	TEQ
3,3`,4,4`- tetrachlorobiphenyl (3,3`,4,4`-TCB)	0.0001
3,4,4`,5`- tetrachlorobiphenyl (3,4,4`,5-TCB)	0.0001
3,3`,4,4`,5- pentachlorobiphenyl (3,3`,4,4`,5-PeCB)	0.1
3,3`,4,4`,5,5`-hexachlorobiphenyl (3,3`,4,4`,5,5`-HxCB)	0.01

Table 9.3. International toxic equivalent factors of non-coplanar biphenyls for humans and mammals

Biphenyls	TEQ
2,3,3`,4,4`- pentachlorobiphenyl (2,3,3`,4,4`-PeCB)	0.0001
2,3,4,4`,5- pentachlorobiphenyl (2,3,4,4`,5-PeCB)	0.0005
2,3`,4,4`,5- pentachlorobiphenyl (2,3`,4,4`,5-PeCB)	0.0001
2`,3,4,4`,5- pentachlorobiphenyl (2`,3,4,4`,5-PeCB)	0.0001
2,3,3`, 4,4`,5`- hexachlorobiphenyl (2,3,3`,4,4`,5`-HxCB)	0.0005
2,3,3`, 4,4`,5- hexachlorobiphenyl (2,3,3`,4,4`,5-HxCB)	0.0005
2,3`,4,4`,5,5`- hexachlorobiphenyl (2,3`,4,4`,5,5`-HxCB)	0.00001

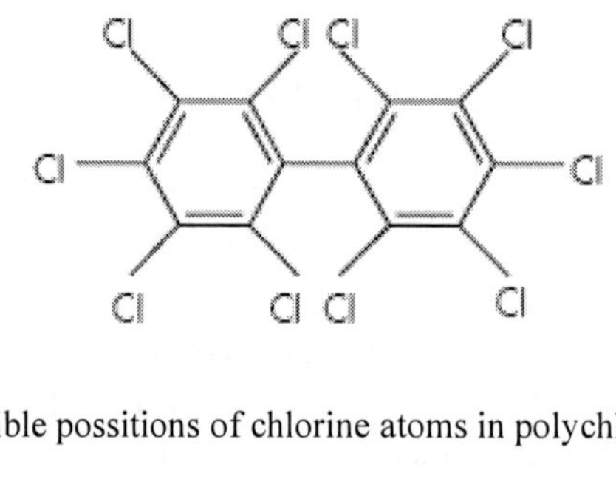

Figure 9.3. Structure and possible possitions of chlorine atoms in polychlorinated byphenyls.

It is believed that metabolites of PCBs may exert estrogenic effects by inhibiting the metabolism of estradiol [70]. Anti-estrogenic activity of PCBs seems to depend somewhat on the pattern of chlorine substitution on the parent molecule, as well as on the formation of hydroxylated metabolites. Results of animal studies demonstrate that PCBs have potential toxic effects on the male reproductive system. It was also shown that compounds induce hepatic phase I (cytochrome P450 enzymes) and phase II (UDP glucuronyltransferases, glutathione transferase) enzymes [71]. PCBs can disrupt both the production and distribution of thyroid hormones by variety of different mechanisms, such as stimulation of the thyroid gland, decreasing the activity of iodothyronine sulfotransferases in the liver, disturbing the transport mechanisms and other means.

The current state of knowledge suggests that low-level exposure to PCBs is unlikely to cause adverse health effects to humans, but some evidence of prolonged low-level exposures shows subtle effects on human reproduction and development. The adverse health effects of acute poisoning include a severe form of acne (chloracne), swelling of the upper eyelids, discoloration of the nails and skin and numbness in the arms and legs. The condition is characterized by weakness, muscle spasms, chronic bronchitis and problems related to the nervous system. Some evidence links prolonged high-level exposure to an increased incidence of cancer, particularly of the liver and kidney. Human studies demonstrating associations between exposure to PCB mixtures and adverse immunological, reproductive, dermatological and carcinogenic effects are limited because the exact exposure data are unknown, and inconsistency in the results of studies of long-term human exposures prevents accurate risk estimations. Toxicological *in vivo* and *in vitro* studies of the influence of PCBs did not produce any solid evidence that PCBs display genotoxic potential.

The quantification of PCBs in biological samples usually entails extraction from the sample matrix, cleanup procedure and determination by gas chromatography with a suitable detector. Gas chromatography coupled with an electron capture detector is most frequently used for sensitive determination of PCBs; however, confirmation by mass spectrometry is recommended when measurement of multiple individual congeners is required [72, 73].

POLYCYCLIC AROMATIC HYDROCARBONS

Polycyclic aromatic hydrocarbons (PAHs) are a group of approximately 10,000 ubiquitous organic compounds consisting of three or more fused benzene rings and containing only carbon and hydrogen atoms. At room temperature, polycyclic aromatic hydrocarbons are solids with low volatility. They are soluble in many organic solvents but are relatively insoluble in water. Polycyclic aromatic hydrocarbons are formed when complex organic substances are exposed to high temperatures or pressures. Even though the compounds can be formed naturally by forest fires and volcanoes, most polycyclic aromatic hydrocarbons in ambient air result from the incomplete burning of coal, wood, petroleum, oil and the emission from motor vehicles. In general, burning of materials at low temperatures, as in wood fires or cigarettes, contribute to more PAHs formed. The most common polycyclic aromatic hydrocarbons include benzo[a]pyrene, benzo[e]pyrene, benz[a]anthracene, chrysene, pyrene, benzo[ghi]fluoranthene, benzo[ghi]perylene, coronene, dibenzo[a,h]anthracene, indeno[1,2,3-cd]pyrene etc. (Figure 9.4.). Benzo[a]pyrene is among the strongest known carcinogenic substances.

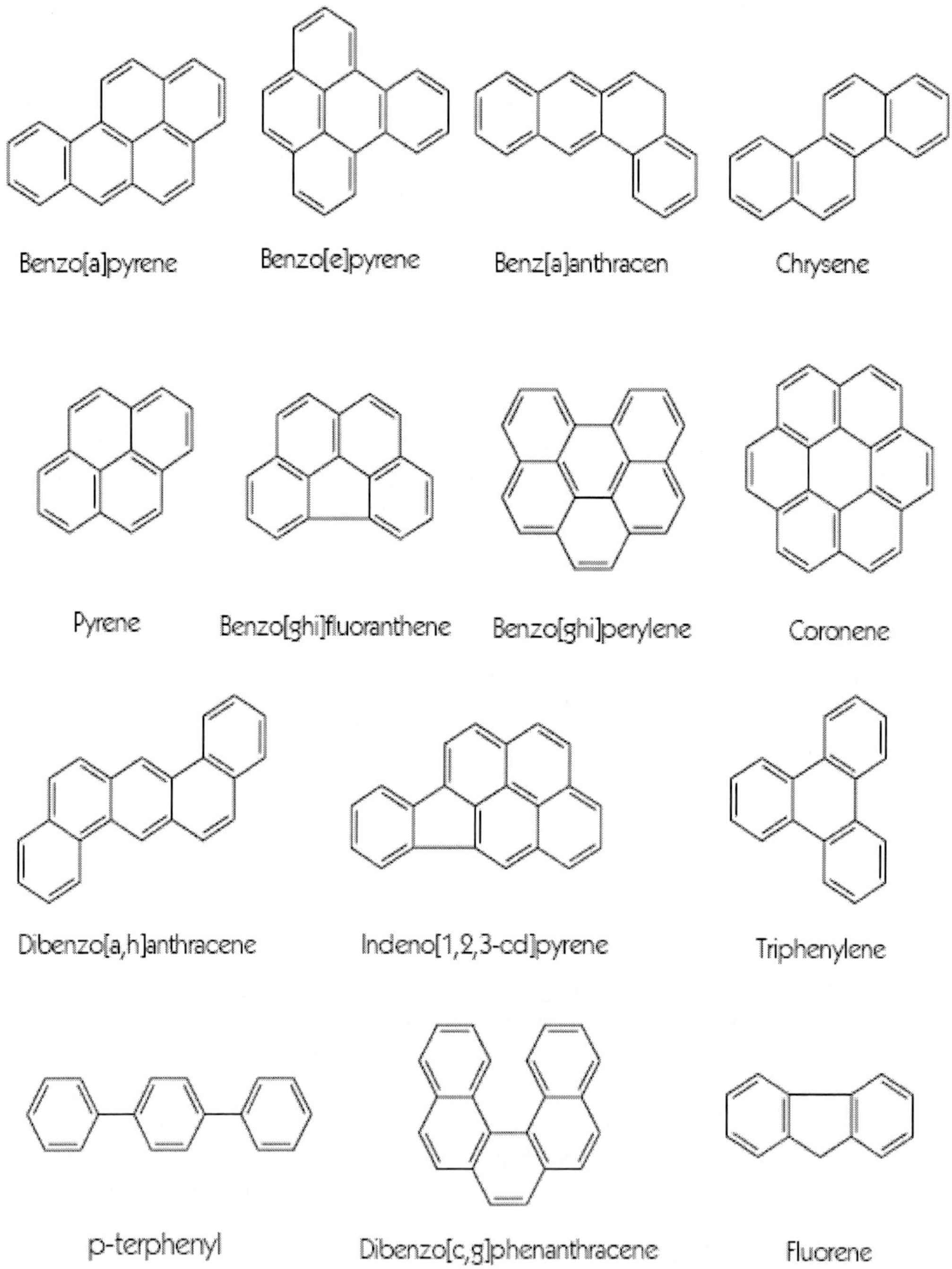

Figure 9.4. The structure of some polycyclic aromatic hydrocarbons.

Percival Pott, a British surgeon, was the first to report a connection between occupational exposure to tar and carcinogenesis. In 1775, he described an unusually high incidence of scrotal cancer among chimney sweeps in London. Years later, in the 1930s, benzo[a]pyrene was isolated from coal tar and was identified and recognized to be carcinogenic when applied to the skin of the test animals. Since that discovery, hundreds of polycyclic aromatic hydrocarbons have been described and many of them are classified as carcinogenic.

Water and soil contain measurable amounts of airborne polycyclic aromatic hydrocarbons [74] . Water contamination occurs as a result of industrial effluents and accidental spills. Most polycyclic aromatic hydrocarbons can be photo-oxidized in the environment and degraded to simpler substances.

Due to pyrolytic (300 °C) decomposition mainly of fats, but also of other food macronutrients, polycyclic aromatic hydrocarbons are present in food and their concentrations vary depending on many factors. Charring meat or barbecuing food over a charcoal fire greatly increases the concentration of formed PAHs. However, the contamination is significantly lower as the distance from the charcoal increases. Cooked and smoked meats and fish may contain up to 2.0 mg/kg of polycyclic aromatic hydrocarbons. Roasted peanuts and coffee, refined vegetable oil and many other foodstuffs contain polycyclic aromatic hydrocarbons in amounts from several to several hundred ppb. According to some evidence, certain plants, such as wheat, rye and lentils, may synthesize polycyclic aromatic hydrocarbons. One of the greatest sources of exposure to polycyclic aromatic hydrocarbons is tobacco smoke. One cigarette yields 10 to 50 ng of benzo[a]pyrene, 18 ng of chrysene, 40 ng of dibenz[a,h]anthracene and 12 to 140 ng of benz[a]anthracene.

Polycyclic aromatic hydrocarbons have low acute toxicity. As highly lipophilic after absorption the compounds enter the lymph system. They are metabolized primarily in the liver and kidneys. In addition to liver and kidneys activity, metabolism of polycyclic aromatic hydrocarbons also occurs in adrenal and thyroid glands, testes, lungs, skin and the small intestine. Polycyclic aromatic hydrocarbons are transformed initially to epoxides, which are converted to dihydrodiol derivatives and phenols. Glucuronide and sulfate conjugates of these metabolites are excreted in bile and glutathione conjugates are further metabolized in the kidney to mercapturic acids and are excreted in urine. Billiary and urinary excretion of polycyclic aromatic hydrocarbons is relatively efficient because of the high activity of biotransformation enzymes involved in their transformation to more polar metabolites. Due to their lipophilic nature, polycyclic aromatic hydrocarbons accumulate in the breast milk and adipose tissue.

Although they are shown not to be acutely very toxic, polycyclic aromatic hydrocarbons are potent carcinogens. The major site of PAH-induced carcinogenesis is the skin, but long-term exposure also increases the risk of lung, bladder and gastrointestinal cancer. Inhaled PAH in cigarette smoke has been implicated in cancer genesis in the respiratory system.

Polycyclic aromatic hydrocarbons are known inductors of biotransformation enzymes, inducing primarily cytochrome P450, which is responsible for their bioactivation, and increasing the sensitivity to subsequent exposures. The toxic effects of PAHs occur due to their metabolites, epoxides, which interact with DNA, RNA and other macromolecules. The rate of epoxide formation and the activity of detoxifying enzymes in various tissues are important determinants of tissue-specific toxicity. PAHs such as benzo[a]pyrene bind to DNA by means of activated form, 7,8-dihydrodiol 9,10-epoxide, which shows affinity to 2-amino position in guanylic acid. After the mechanism of genotoxicity was explained for benzo[a]pyrene, numerous other carcinogens were found to interact with DNA and to provoke the toxic effects by similar mechanisms. Inducibility of aryl hydrocarbon hydroxylase has been shown to be an important genetic factor influencing the susceptibility to lung and laryngeal cancer from exposure to PAHs.

In vitro mutagenicity studies show that interactions between different polycyclic aromatic hydrocarbons in simultaneous exposures produce both synergistic and antagonistic effects. The results of the test indicate that polycyclic aromatic hydrocarbons might have teratogenic effects in humans.

The most frequently applied analytical technique for the determination of polycyclic aromatic hydrocarbons in various samples is gas chromatography coupled with mass detector to ensure the sufficient sensitivity and reliable identification of numerous possible compounds that may be present in the sample [75, 76]. The application of liquid chromatography has also been reported [77, 78]. A great number of antibodies have been produced for organic pollutants, such as polycyclic aromatic hydrocarbons, polychlorinated byphenyls and dioxins, enabling reliable, simple and time-saving screening [79, 80]. The application of immunoassay may be suitable as a simple and sensitive routine test for monitoring PAHs content, and it is considered to be the current state of the art in this type of analysis.

References

[1] Szymanska, JA; Sapota, A; Frydrych, B. The disposition and metabolism of tetrabromobisphenol-A after a single i.p. dose in the rat. *Chemosphere* 45, 2001, 693-700.

[2] WHO. Tetrabromobisphenol A and Derivatives. *Envir. Health Crit.*. 172. Geneva:World Health Organization; 1995.

[3] Mariussen, E; Fonnum, F. The effect of pentabromodiphenyl ether, hexabromocyclododecane and tetrabromobisphenol-A on dopamine uptake into rat brain synaptosomes. *Organohal.. Comp.* 57, 2002, 395-399.

[4] Meerts, IA; Assink, Y; Cenihn, P; Weijers, B; van den Berg, H; Koeman, J. Distribution of the flame retardant tetrabromobisphenol A in pregnant and fetal rats and effect on thyroid hormone homeostasis. *Organohal. Comp.* 40, 1999, 375-379.

[5] Lyman, WJ. *Handbook of Chemical Property Estimation Methods*. Washington, DC: American Chemical Society; 1990.

[6] Reistad, T; Fonnum, F; Mariussen, E. Neurotoxicity of the pentabrominated diphenyl ether mixture, DE-71, and hexabromocyclododecane (HBCD) in rat cerebellar granule cells in vitro. *Arch. Tox.* 80(11), 2006, 785-796.

[7] de Wit CA. An overview of brominated flame retardants in the environment. *Chemosphere* 46, 2002, 583-624.

[8] Morck, A; Hakk, H; Orn, U; Klasson-Wehler, E. Decabromodiphenyl ether in the rat--absorption, distribution, metabolism and excretion. *Drug Metab. Dis.* 31, 2003, 900-908.

[9] Meironyte, D; Noren, K; Bergman, A. Analysis of polybrominated diphenyl ethers in Swedish human milk. A time-related trend study, 1972-1997. *J. Tox. Environ. Health* 58, 1999, 329-341.

[10] Viberg, H; Fredriksson, A; Eriksson, P. Neonatal exposure to the brominated flame retardant 2,2',4,4',5-pentabromodiphenyl ether causes altered susceptibility in the cholinergic transmitter system in the adult mouse. *Toxicol. Sci.* 67, 2002,104-107.

[11] Hallgren, S; Sinjari, T; Hakansson, H; Darnerud PO. Effects of polybrominated diphenyl ethers (PBDEs) and polychlorinated biphenyls (PCBs) on thyroid hormone and vitamin A levels in rats and mice. *Arch. Tox.* 75, 2001, 200-208.

[12] Zhou, T; Taylor, MM; DeVito, MJ; Crofton, KM. Developmental exposure to brominated diphenyl ethers results in thyroid hormone disruption. *Tox. Sci.* 66, 2002, 105-116.

[13] Kester, MH; Bulduk, S; van Toor, H; Tibboel, D; Meinl, W; Glatt, H. Potent inhibition of estrogen sulfotransferase by hydroxylated metabolites of polyhalogenated aromatic hydrocarbons reveals alternative mechanism for estrogenic activity of endocrine disrupters. *J. Clin. Endocrinol. Metab.* 87, 2002, 1142-1150.

[14] Guerra, P; de la Torre, A; Angeles, M; Martínez, R; Eljarrat, E; Barceló, D. Identification and trace level determination of brominated flame retardants by liquid chromatography/quadrupole linear ion trap mass spectrometry. *Rapid. Comm. Mass spectrom.* 22(7), 2008, 916 – 924.

[15] Morris, S; Bersuder, P; Colin, R; Allchin, CR; Zegers, B; Jan, P; Boon, JP; Pim, EG, de Boer, LJ. Determination of the brominated flame retardant, hexabromocyclodocane, in sediments and biota by liquid chromatography-electrospray ionisation mass spectrometry. *Trends Anal. Chem.* 25(4), 2006, 343-349.

[16] Hyötyläinen, T; Hartonen, K. Determination of brominated flame retardants in environmental samples. *Trends Anal. Chem.* 21(I), 2002, 13-30.

[17] Hardie, DWF. In: Standen, A. *Kirk-Othmer Encyclopedia of Chemical Technology*, second ed., Vol. 5. Edition: New York, USA; Interscience, 1964, pp. 231-240.

[18] Campbell, I; McConnell, G. Chlorinated paraffines and the environment. Environmental occurrence. *Environ. Sci. Technol.* 14, 1980, 1209–1214.

[19] Svanberg, O; Bengtsson, BE; Linden, E. Paraffins - A case of accumulation and toxicity to fish. *Ambio* 7(2), 1978, 64-65.

[20] Fisk, T; Westmore, AT; Muir, DCG. Environmental chemistry and toxicology of polychlorinated *n*-alkanes. *Rev. Environ. Contam. Tox.* 158, 1998, 53–128.

[21] Bucher, JR; Alison, RH; Montgomery, CA; Huff, J; Haseman, JK; Farnell, D; Thompson, R; Prejean, JD. Comparative toxicity and carcinogenicity of two chlorinated paraffines in F 344/n rats and B6C3F1 mice. *Fund. Appl. Tox.*. 9, 1987, 454–468.

[22] Roy, TA; Dalbey, WE; Mackere, CR.C12-chlorinated paraffins: Workplace exposure and risk assesment. *The Toxicologist* 42, 1998, 225-229.

[23] Burýšková, BL; Bláha, L; Vršková, D; Šimková, K; Maršálek B. Sublethal Toxic Effects and Induction of Gutathione S-transferase by Short-Chain Chlorinated Paraffins (SCCPs) and C-12 alkane (dodecane) in *Xenopus laevis Frog Embryos. Acta Vet. Brno* 75, 2006, 115-122.

[24] Hollies, JI; Pinnington, DF; Handley, AJ; Baldwin, MK; Bennett, D.The determination of chlorinated long-chain paraffins in water,sediment and biological samples. *Anal. Chim. Acta,* 111, 1979. 201-213.

[25] Zitko, V. Chromatography of chlorinated paraffins on alumina and silica columns. *J. Chromatogr.* 81, 1973, 152-155.

[26] Schmid, PP; Müller, MD. Trace level detection of chlorinated paraffins in biological and environmental samples, using gas chromatography/mass spectrometry with negative-ion chemical ionization. *J. Assoc. Off Anal. Chem.* 68, 1985, 427-430.

[27] Tommy, GT; Stern, GA; Muir, DCG; Fisk, AT; Cymbalisty, CD; Westmore, JB. Quantifying C_{10}–C_{13} polychloroalkanes in environmental samples by high-resolution gas chromatography/electron capture negative ion high-resolution mass spectrometry. *Anal. Chem.* 69, 1997, 2762–2771.

[28] Gjos, N; Gustavsen, K. Determination of chlorinated paraffins by negative ion chemical ionization mass spectrometry. *Anal. Chem.* 54, 1982, 1316-1318.

[29] Coelhan, M. The determination of short chain polychlorinated paraffines in fish samples by short-column GC–MS using negative chemical ionisation. *Anal. Chem.* 71, 1999, 4498–4505.

[30] Mark, E; Hahn, M. Dioxin Toxicology and the Aryl Hydrocarbon Receptor: Insights from Fish and Other Non-traditional Models. *Marine Biotech.* 3(1), 2001, 224-238.

[31] Eljarrat, E; Barceló, D. Special Feature: Perspective Congener-specific determination of dioxins and related compounds by gas chromatography coupled to LRMS, HRMS, MS/MS and TOFMS. *J. Mass Spec.* 37(11), 2002, 1105 – 1117.

[32] Hanberg, A; Waern, F; Asplund, L; Haglund, E; Safe, S. Swedish dioxin survey: Determination of 2,3,7,8-TCDD toxic equivalent factors for some polychlorinated biphenyls and naphthalenes using biological tests. *Chemosphere* 20(7-9), 1990, 23-28.

[33] Overmeire, IV; Clark, GC; Brown, DJ; Chu, MD; Cooke, WM; Denison, MS; Baeyens, W; Srebrnik, S; Goeyens, L. Trace contamination with dioxin-like chemicals: evaluation of bioassay-based TEQ determination for hazard assessment and regulatory responses. *Envir. Sci.Policy* 4(6), 2001, 345-357.

[34] Davies, AG. *Organotin Chemistry*, second Edition, Wiley-VCH Verlag GmbH and Co. KGaA; 2004.

[35] lanigen, D; Weinberg, EL. The use of estertin stabilizers in pvc. In: Zucherman, JL. *Organotin compounds: new chemistry and applications.* Washington DC: American Chemical Society; 1976.

[36] Schramel, P; Samsahl, K; Pavlu, J. Some determinations of Hg, As, Se, Sb, Sn, and Br in water plants, sediments and fishes in bavarian rivers. *Int. J. Environ. Stud.* 5, 1973, 37-40.

[37] Mazaev, VT; Golovanov, OV; Igumov, AS; Tsay, VN. Problem of the transformation of organotin compounds in a water medium. *Gig. I Sanit,* 3, 1976, 17-20.

[38] Freitag, KD; Bock, R. Degradation of triphenyltin chloride on sugar beet plants. *Pest. Sci.* 5, 1974, 731-739.

[39] Bowen, HJM. *Trace elements in biochemistry*. New York: Academic Press; 1966.

[40] Hall, CA; Ludwig, PD. Evaluation of the potential use for several organotin compounds against the sheep blowfly (*lucilia* spp.). *Vet. Res.* 90, 1972, 29-32.

[41] Ascher, KRS; Nissim, S. Organotin compounds and their potential use in insect control. *World Rev. Pest. Control,* 3, 1964, 188-211.

[42] Torack, R; Gordon, J; Prokop, J. Pathobiology of acute tri-ethyltin intoxication. *Int. Rev. Neurobiol.* 12, 1969, 45-86.

[43] Barnes, JM; Magee, PN. The biliary and hepatic lesion produced experimentally by dibutyltin salts. *J. Pathol. Bacteriol.* 75, 1958, 267-279.

[44] Kreyberg, S; Torvik, A; Bjorneboe, A; Wiik-Larsen, W; Jacobsen, D. Trimethyltin poisoning:report of a case with postmortem examination. *Clin. Neuropat.* 11,1992, 256-259.

[45] Snoey, NJ; Van Iersel, AAJ; Penninks, AH; Seinen, W. Toxicity of organotin compounds: Comparative in vivo studies with a series of trialkyltin compounds and triphenyl chloride in male rats. *Tox. Appl. Pharm.* 81, 1985, 274-286.

[46] Barnes, JM; Stoner, HB. The toxicology of tin compounds. *Pharm. Rev.* 11, 1959, 211-231.

[47] Stoner, HB; Barnes, JM; Duff, JI. Studies on the toxicity of alkyl tin compounds. *Br. J. Pharm.* 10, 1955, 16-25.

[48] Barnes, JM; Magos, L. The toxicology of organometallic compounds. *Organomet.. Chem. Rev.* 3, 1968, 137-150.

[49] Aldridge, WN; Street, BW. Oxidative phosphorylation: The relation between the specific binding of trimethyltin and triethyltin to mitochondria and their effects on various mitochondrial functions. *Biochem. J.* 124, 1971, 221-234.

[50] Studer, RK; Siegel, BA; Morgan, J; Potchen, EJ. Dexamethasone therapy of triethyltin induced cerebral edema. *Exp. Neurol.* 38, 1973, 429-437.

[51] Cai, Y; Rapsomanikis, Sand Andreae, MO. Determination of butyltin compounds in river sediment samples by gas chromatography–atomic absorption spectrometry following in situ derivatization with sodium tetraethylborate. *J. Anal. At. Spectrom.* 8, 1993, 119 – 125.

[52] Ritsema, R. Speciation of organotin and organoarsenic in water samples. *Microchim. Acta* 109(1-4), 2005, 43-48.

[53] Wasik, A; Ciesielski, T. Determination of organotin compounds in biological samples using accelerated solvent extraction, sodium tetraethylborate ethylation, and multicapillary gas chromatography-flame photometric detection. *Anal. Bioanal. Chem.* 378(5), 2004, 1357-1363.

[54] Tomomi, I; Kiwao, K; Daisuke, J; Yoshifumi, H; Manabu, S. Determination of organotin compounds in water and sediment samples by isotope dilution GC/MS. *Bunseki Kagaku* 48(6), 1999, 555-561.

[55] Le Gac, M; Lespes, G; Potin-Gautier, M. Rapid determination of organotin compounds by headspace solid-phase microextraction. *J. Chrom. A* 999(1), 2003, 123-134.

[56] Olle, N. Determination of occupational exposure to organotin compounds after multivariate optimization of a liquid chromatography flame atomic absorption spectrometry. *Spectrom. Acta B* 48(8), 2006, 977-983.

[57] Giesy, JP; Kannan, K. Global distribution of perfluorooctane sulfonate in wildlife. *Envir. Sci. Techn.* 35, 2001, 1339–1342.

[58] Hu, W; Jones, PD; Upham, BL; Trosko, JE; Lau, C; Giesy, JP. Inhibition of gap junctional intercellular communication by perfluorinated compounds in rat liver and dolphin kidney epithelial cell lines *in vitro* and Sprague-Dawley rats *in vivo*. *Tox. Sci.* 68, 2002, 429–436.

[59] Widdows, J; Donkin, P. Mussels and environmental contaminants: Bioaccumulation and physiological aspects. In: Gosling, E. *The Mussel Mytilus: Ecology, Physiology, Genetics and Culture*. Amsterdam; Elsevier; 1992.

[60] Seacat, AM; Thomford, PJ; Hansen, KJ; Olsen, GW; Case, MT; Butenhoff, JL. Subchronic toxicity studies on perfluorooctanesulfonate potassium salt in cynomolgus monkeys. *Tox. Sci.* 68, 2002, 249–264.

[61] Luebker, DJ; Hansen, KJ; Bass, NM; Butenhoff, JL; Seacat, AM. Interactions of fluorochemicals with rat liver fatty acid-binding protein. *Toxicology* 176, 2002,175–185.

[62] Hu, W; Jones, PD; DeCoen, W; King, L; Fraker, P; Newsted, J. Alterations in cell membrane properties caused by perfluorinated compounds. *Comp. Biochem. Physiol. Tox. Pharm.* 135, 2003, 77–88.

[63] Lau, C; Thibodeaux, JR; Hanson, RG; Rogers, JM; Grey, BE; Stanton, ME. Exposure to perfluorooctane sulfonate during pregnancy in rat and mouse. II. Postnatal evaluation. *Tox. Sci.* 74, 2003, 382–392.

[64] Inoue, K; Okada, F; Ito, R; Kawaguchi, M; Okanouchi, N; Nakazawa. Determination of perfluorooctane sulfonate, perfluorooctanoate and perfluorooctane sulfonylamide in human plasma by column-switching liquid chromatography–electrospray mass spectrometry coupled with solid-phase extraction. *J. Chrom. B* 810(1), 2004, 49-56.

[65] Powley, CR; Michael, J; Michalczyk, Mary, A. Kaiser, MA; Buxton LW. Determination of perfluorooctanoic acid (PFOA) extractable from the surface of commercial cookware under simulated cooking conditions by LC/MS/MS. *Analyst* 130, 2005, 1299-1302.

[66] Fiedler, H; Hoff, H; Tolls, J; Mertens, C; Gruber, A; Hutzinger, O. Environmental Fate of Organochlorines in the Aquatic Environment. *Organohal.. Comp.* 15, 1994, 56-62.

[67] Burbach, KM; Poland, A; Bradfield, CA. Cloning of the Ah-receptor cDNA reveals a distinctive ligand-activated transcription factor. *Proc. Natl. Acad. Sci. USA*. 89(17), 1992, 8185-8189.

[68] Adachi, J; Mori, Y; Matsui, S. Indirubin and indigo are potent aryl hydrocarbon receptor ligands present in human urine. *J. Biol. Chem.* 276(34), 2001, 31475–31478.

[69] Denison, MS; Nagy, SR. Activation of the aryl hydrocarbon receptor by structurally diverse exogenous and endogenous chemicals. *Annu. Rev. Pharm.. Tox.. 43, 2003, 309–334.*

[70] Kester, MHA; Bulduk, S; Tibboel, D. Potent inhibition of estrogen sulfotransferase by hydroxylated PCB metabolites: A novel pathway explaining the estrogenic activity of PCBs. *Endocrinology* 141(5), 2000, 1897-1900.

[71] Arnold, DL; Bryce, F; McGuire, PF. Toxicological consequences of Aroclor 1254 ingestion by female rhesus (*Macaca mulatta*) monkeys. Part 2. Reproduction and infant findings. *Food Chem. Tox.* 33, 1995, 457–474.

[72] Ruddy, BA; Qadah, DT; Aldstadt, JH; Bootsma, HA. Improving analytical confidence in the determination of PCBs in complex matrices by a sequential GC-MS/MS approach. *J. Envir. Anal.Chem.* 88(5), 2008, 337 – 351.

[73] Moret, I; Gambaro, A; Piazza, R; Ferrari, S; Manodori, L. Determination of polychlorobiphenyl congeners (PCBs) in the surface water of the Venice lagoon. *Mar. Poll. Bull.* 50(2), 2005, 167-174.

[74] Kravić, SŽ, Marjanović, NJ, Pucarević, MM, Suturović, ZJ, Švarc-Gajić, JV. Determination of polycyclic aromatic hydrocarbons in soil by gas chromatography – mass spectrometry. *Acta Period. Technol.* (APTEFF) 36, 2005, 99-109.

[75] Sánchez-Brunete C, Esther M, José LT. Rapid method for the determination of polycyclic aromatic hydrocarbons in agricultural soils by sonication-assisted extraction in small columns. *J. Sep. Sci.* 29(14), 2006, 2166 – 2172.

[76] Marjanović, NJ ,Kravić, SZ, Suturović, ZJ, Švarc-Gajić, JV. Determination of sensitivity limit in quantitative analysis of polyciclic aromatic hydrocarbons by GC-MS. *Acta Period. Technol. (APTEFF)* 35, 2004, 111 - 119.
[77] Hiroshi, M; Miho, I; Shusaku, Y; Hidekazu, M; Jean-François, A. Determination of Polycyclic Aromatic Hydrocarbons in Sediment by Liquid Chromatography-Atmospheric Pressure Photoionization-Mass Spectrometry. *Anal. Sci*, 20(2), 2004, 375-381.
[78] Kishikawa, N; Wada, M; Kuroda, N; Akiyama, S; Nakashima, K. Determination of polycyclic aromatic hydrocarbons in milk samples by high-performance liquid chromatography with fluorescence detection. *J. Chrom. B*, 789 (2), 2003, 257-264.
[79] Schoket, B, Doty, WA, Vincze, I, Strickland, PT, Ferri, GM, Assennato, G; Poirier, MC. Increased sensitivity for determination of polycyclic aromatic hydrocarbon-DNA adducts in human DNA samples by dissociation-enhanced lanthanide fluoroimmunoassay (DELFIA). *Cancer Epid. Biom. Prev.* 2(4), 2007, 349-353.
[80] Knopp, Seifert M, Väänänen V and Niessner R. Determination of polycyclic aromatic hydrocarbons in contaminated water and soil samples by immunological and chromatographic methods. *Envir. Sci. Tech.*34(10), 2000, 2035-2041.

Chapter X

FUNGAL AND BACTERIAL TOXINS

MYCOTOXINS

Mycotoxins are pharmacologically active secondary metabolites produced by microfungi. They represent mold metabolites, which belong to various chemical classes and are characterized by mammalian toxicity. Many of them elicit carcinogenicity, inhibition of protein synthesis, immunosuppression, dermal irritation and other metabolic perturbations. Some mycotoxins and their derivatives are used in the production of drugs such as antibiotics or growth promotants, while others have been implicated as biochemical warfare agents. Mycotoxins usually enter the body via ingestion of contaminated food, but inhalation of toxigenic spores and direct dermal contact are also possible routes of exposure.

The symptoms of a mycotoxicosis, a dietary, respiratory, dermal or other exposure to toxic fungal metabolites, depends on the type of mycotoxin, its amount and duration of the exposure to it. The age of the exposed individual, as well as health and gender, significantly influence the manifestations of mycotoxicosis. Other parameters that influence the manifestations of poisoning include possible synergistic effects with dietary factors and interactions with other toxic substances.

While all mycotoxins are of fungal origin, not all toxic compounds produced by fungi are mycotoxins. Mycotoxins are toxic to vertebrates and other animal groups in low concentrations, while fungal metabolites such as ethanol, that are toxic only in high concentrations, are not considered mycotoxins. Fungal products that are toxic mainly to bacteria, such as penicillin, are usually called antibiotics.

Due to the diverse chemical structures of mycotoxins, their classification is very difficult and they are grouped on the basis of several criteria. Depending on the organ they affect, mycotoxins are classified as hepatotoxins, nephrotoxins, neurotoxins and immunotoxins. Classification schemes can further encompass grouping in classes of teratogens, mutagens, carcinogens and allergens, or according to their chemical structure as lactones, coumarins, etc. Mycologists group mycotoxins on the basis on fungal genera that produce them on *Aspergillus* toxins, *Penicillium* toxins etc.

Aflatoxins

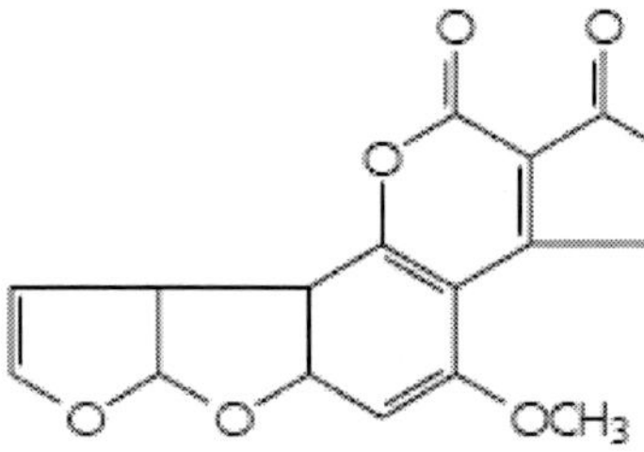

Figure 10.1. Aflatoxin B_1.

Aflatoxins were isolated and characterized after eluciding turkey X disease, which caused the death of more than 100,000 turkey poults after consumption of mold-contaminated peanut meal. Aflatoxins are difuranocoumarin derivatives produced by many strains of *Aspergillus flavus* and *Aspergillus parasiticus. Aspergillus bombycis, Aspergillus ochraceoroseus, Aspergillus nomius* and *Aspergillus pseudotamari* are also aflatoxin-producing strains, but they are encountered less frequently in food contamination [1]. The four major aflatoxins are B_1, B_2, G_1 and G_2, identified on the basis of their fluorescence under UV light (blue or green) and relative chromatographic mobility in thin-layer chromatography [2]. Aflatoxin B_1 (Figure 10.1.) is the most potent natural carcinogen known and is usually the major aflatoxin produced by toxigenic strains.

Some *Aspergillus flavus* strains may produce more than 10^6 µg/kg of aflatoxins [3]. Many substrates, such as cereals, figs, oilseeds, nuts and tobacco, support the growth and production of aflatoxin by aflatoxigenic molds. Milk products may also serve as an indirect source of aflatoxin. After cows consume aflatoxin-contaminated feeds, metabolic biotransformation of aflatoxin B_1 produces hydroxylated form called aflatoxin M_1, which is excreted in their milk [4].

Aflatoxins are associated with both toxicity and carcinogenicity in human and animal populations. Acute aflatoxicosis results in death, whereas chronic exposure to aflatoxins results in cancer genesis and immune suppression. The liver is the primary target organ and the organ where carcinogenesis takes place. The mechanisms of carcinogenesis include the formation of 8,9-epoxides catalized by cytochrome P450 enzymes, which are capable of binding to both DNA and proteins [5]. Aflatoxin B_1-DNA adducts can result in GC to TA transversions. The presence of hepatitis B virus infection, an important risk factor for liver cancer, highly contributes to the development of neoplasms and complicates research on sole aflatoxin influence. Within a given species, the magnitude of the response is influenced by age, sex, weight, diet, exposure to infectious agents and presence of other mycotoxins and pharmacologically active substances.

Further biotransformation of aflatoxins occur in phase II reactions. A reactive glutathione S-transferase system from both cytosol and microsomes, catalyzes the conjugation of activated aflatoxins with reduced glutathione and promotes the excretion of aflatoxin conjugates. Variations in the level of glutathione transferase system, as well as variations in cytochrome P450 activity, are thought to contribute to variations in interspecies susceptibility.

Due to differences in aflatoxin susceptibility in test animals, it is difficult to predict possible effects of aflatoxins on humans; however, it has been estimated that acute poisoning in humans occurs after ingestion of approximately 10 to 20 mg of aflatoxin B1.

Biomonitoring of exposure to aflatoxin may be performed by analyzing the presence of aflatoxin metabolites in blood, milk and urine, as well as by detecting the blood protein adducts [6]. An aflatoxin B_1-N^7-guanine adduct represents the most reliable urinary biomarker for aflatoxin exposure but reflects only recent events.

Citrinin

Citrinin (Figure 10.2.) was first isolated from *Penicillium citrinum* before World War II and later it was identified in over a dozen species of *Penicillium*, including certain strains of *Penicillium camemberti* used in cheese production, and several species of *Aspergillus* (*Aspergillus terreus*, *Aspergillus niveus, Aspergillus oryzae*). *Aspergillus oryzae* is used to produce sake, miso and soy sauce [7]. Substrates which support citrinin biosynthesis include wheat, oats, rye, corn, barley, rice, as well as certain vegetarian foods colored with industrially produced red *Monascus* pigments isolated from *Monascus purpureus* [8, 9].

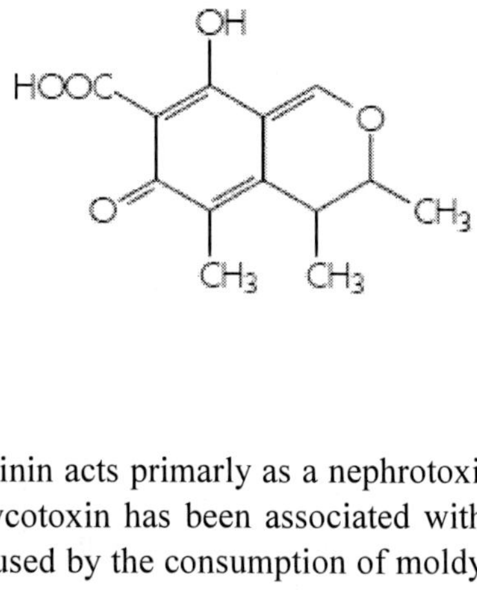

Figure 10.2. Citrinin.

In all animal species, citrinin acts primarly as a nephrotoxin, and its acute toxicity varies for different species. The mycotoxin has been associated with yellow rice disease in Japan and with numerous deaths caused by the consumption of moldy rice imported from Southeast Asia [10].

Yellow rice disease involves cardiac problems acompanied with vomiting, ascending paralysis, convulsions and respiratory arrest. Even though the symptoms of cardiac malfunction resembled beriberi disease, the overall clinical features of yellow rice disease, as well as its epidemic character, distinguished it from vitamine B1 defficiency and beriberi disease. Today, improved techniques of harvesting and storage of rice prevent further outbreaks and citrinin-induced poisonings.

The mechanism of citrinin toxicity is not yet elucidated, but some evidence suggest that citrinin may act by supressing RNA synthesis in the cells, and it is speculated that in this action, citrinin acts synergistically with ochratoxin.

Ergot Alkaloids

Human ergotism, also called St. Anthony's Fire, was a frequent health condition in Europe in the Middle Ages and was caused by the consumption of grain, usually rye, contaminated with ergot alcaloids. Between the 6th and 18th centuries, 132 epidemics of ergotism were reported in Europe [11].

Ergot alkaloids are indole compounds whose structure is similar to lysergic acid (Figure 10.3.) and which are produced by *Claviceps* species, common plant pathogens. Centuries before these epidemics, it was observed that grazing on grass infected with ergot alkaloids caused the abortion in pregnant farm animals. The condition is provoked by several ergot alkaloids with the ability to induce smooth muscle contractions. These observations led to adopting the fungus in a folk medicine to induce abortion and also to accelerate uterine contractions of women in labor [12].

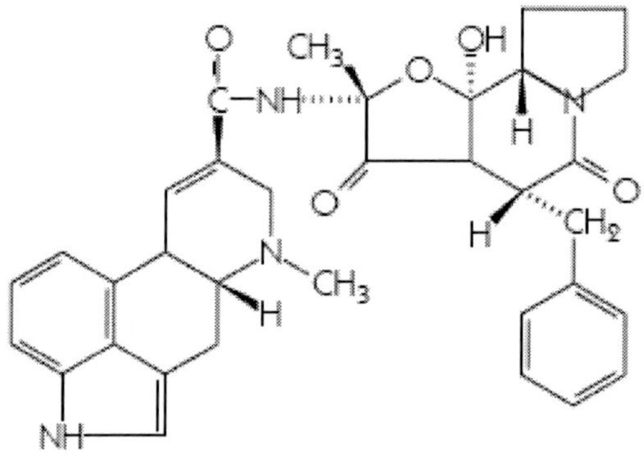

Figure. 10.3. Ergotamine.

In poisoning with substances produced by *Claviceps* species, two forms of ergotism are recognized, gangrenous and convulsive, depending on the dominating group of produced compounds. Synthetized substances produce clinical signs that are generally the result of vasoconstriction and psychoactive activity. The gangrenous form of ergotism arises predominantly due to affected blood supply to the extremities, while in convulsive ergotism, the central nervous system is perturbed [13]. Produced ergot alkaloids affect alpha-adrenergic, dopamine and serotonin receptors, inducing the constriction of smooth muscle fibres and the walls of small blood vessels. *Claviceps purpurea* besides producing ergot alkaloids, also produces numerous other biologically active substances, such as acetylcholine, histamine and tyramine.

Clinical symptoms of poisoning with ergot active substances include reduced food intake, convulsions, tremor and incoordination. The condition is characterized by a rise in the pulse rate and gastrointestinal problems, such as hypersalivation, diarrhea and abdominal cramps. In intoxicated persons, the pupils at first contract but in later stages they dilate. Affected blood vessels provoke a feeling of numbness and coldness in extremities followed with the development of gangrene, in severe cases. The severity of ergot poisoning arises from the fact that in serious cases neurological symptoms may persist even after the source of infection has been removed, and by the fact that the damage to blood vessels caused by their intense constriction is likely to be irreversible.

Physiologically active substances produced by *Claviceps purpurea* species were used in the past in folk medicine, but some of these compounds still attract medical attention. More recently, pure ergotamine was used in the treatment of migraine headaches, and it is known that the antimigraine effects of the compound are mediated by neuronal and vascular serotonin receptors. Other ergot derivatives are used as prolactin inhibitors, in the treatment of Parkinsonism and for cerebrovascular insufficiency [12].

Fumonisins

Fumonisins are mycotoxins produced by some *Fusarium* species, notably *Fusarium verticillioides, Fusarium proliferatum* and *Fusarium nygamai,* as well as by *Alternaria alternata* [14]. *Fusarium verticillioides* is present in virtually all corn samples. Most fungus strains do not produce the toxin, and the presence of molds does not necessarily mean that contamination with fumonisin has occured.

The most abundantly produced member of the fumosine family is fumonisin B_1 (Figure 10.4.)

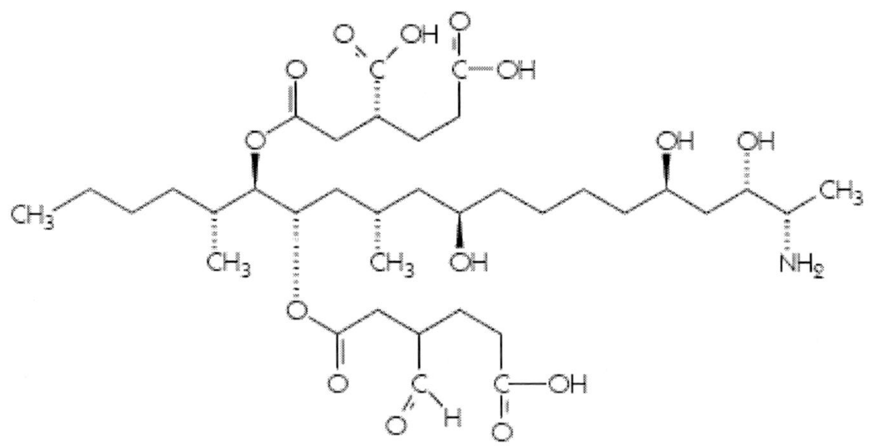

Figure 10.4. Fumonisin B_1.

Mammalian toxicity caused by fumonisins is attributed to interferrence with the sphingolipid metabolism [15]. It has also been demonstrated that this class of compounds is hepatotoxic and carcinogenic, and some research results indicate a probable link between exposure to fumosins and esophageal cancer [16]. The International Agency for Research on Cancer has evaluated the cancer risk of fumonisins to humans and classified them as group 2B substances (probably carcinogenic) [17].

Unlike other mycotoxins that are soluble in organic solvents, fumonisins are hydrophilic, and they may be extracted from food samples with aqueous methanol or aqueous acetonitrile.

Fumosins are the only known hydrophylic class of mycotoxins, and it is speculated that many other hydrophilic toxic products of fungal metabolism might exist and are not yet discovered because the scientific community is focused primarily on the lipophilic fraction. High-performance liquid chromatography with fluorescent detection is the most widely used analytical technique applied for the determination of fumosins in contaminated foodstuffs [18].

Ochratoxin

Members of the ochratoxin family are metabolites of many different species of Aspergillus, including Aspergillus alliaceus, Aspergillus auricomus, Aspergillus carbonarius, Aspergillus glaucus, Aspergillus melleus and Aspergillus niger [19]. Penicillium verrucosum, a common contaminant of barley, is the only confirmed ochratoxin producer in this genus [20]. Because Aspergillus niger is used widely in the production of enzymes and citric acid for human consumption, it is important to ensure that industrial strains do not produce mycotoxins. Mycotoxins from ochratoxin family are also frequently found in pork intended for human consumption.

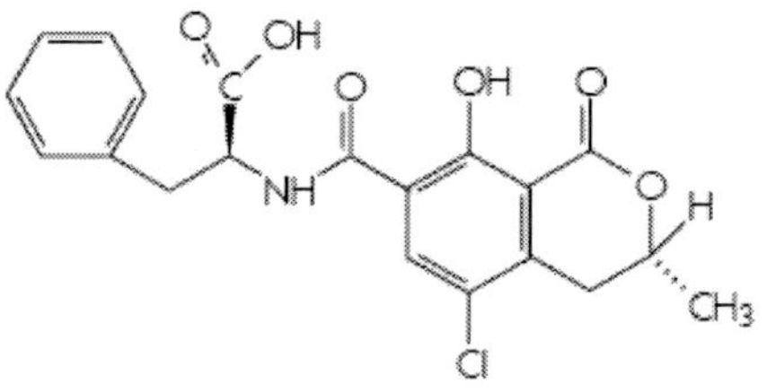

Figure 10.5. Ochratoxin A.

As in the case of other mycotoxins, the substrate on which the molds grow, as well as the moisture level, temperature and the presence of competitive microflora, influence the level of the produced toxin. Ochratoxin A (Figue 10.5.) has been found in barley, oats, rye, wheat, coffee beans and other plant products, with barley having a particularly high likelihood of contamination. There is also concern that ochratoxin may be present in certain wines, especially those made from grapes contaminated with *Aspergillus carbonarius*.

The primary target organ of adverse health effects of ochratoxins is the kidney. Ochratoxin A has been shown to be nephrotoxic in all animal species studied and most likely in humans, too. Some speculation links ochratoxin exposure with the disease called Balkan endemic nephropathy [22]. The diseases is observed in rural populations of Bulgaria, Romania and the territory of the former Yugoslavia. Ochratoxin A has been detected in the blood samples of people living in endemic nephropatic regions. The disease is characterized by the significant reduction in kidney size and impaired kidney function. Histological evidence of pathophysiology include tubular degeneration, intersticial fibrosis and hyalinization of glomeruli. In addition to being nephrotoxic, ochratoxin A is shown in animal studies also to be a liver toxin, an immune suppressant, a potent teratogen and a carcinogen

[22]. The International Agency for Research on Cancer rates ochratoxin as a possible human carcinogen. Ochratoxin A disturbs cellular physiology in multiple ways, but it seems that the primary effects are associated with the enzymes involved in phenylalanine metabolism. The mechanism of toxicity involves the inhibition of the enzyme involved in the synthesis of the phenylalanine-tRNA complex, inhibition of mitochondrial ATP production [23] and stimulation of lipid peroxidation [24].

The half-life of ochratoxin A is shown to be longer in humans than in other mammals. The compound is partially excreted in milk.

Patulin

Patulin, 4-hydroxy-4H-furo[3,2c]pyran-2(6H)-one (Figure 10.6.) is produced by many different molds, but was first isolated as an antimicrobial active principle during the 1940s from *Penicillium patulum*. During the 1950s and 1960s it became apparent that, in addition to its antibacterial, antiviral and antiprotozoal activity, patulin is also toxic to both plants and animals, precluding its clinical use as an antibiotic.

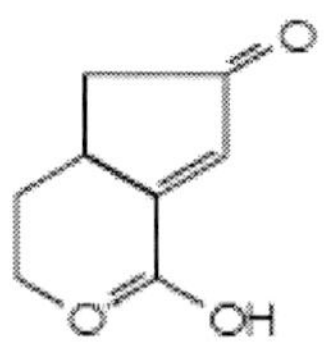

Figure 10.6. Patulin.

Penicillium expansum, the blue mold that causes soft rot of apples, pears, cherries and other fruits, is recognized as one of the most common offenders in patulin contamination. Patulin is regularly found in unfermented apple juice in levels usually below 50 μg/litre; however, apple juice can occasionally be heavily contaminated with patulin, and necessitating the routine analysis of apple juice samples. In addition to being toxic to animals, mutagenic and teratogenic, some evidence exist that in higher concentrations patulin might be carcinogenic to humans. Speculation concerning probable carcinogenicity derives from conducted animals studies; however, the incidence of patulin-induced tumorogenesis is not sufficient to make reliable conclusions. The International Agency for Research on Cancer has evaluated the toxicity data and classifies patulin as a group 3 carcinogen, i.e., a compound for which there is not enough data to support its classification. Conducted animal studies also show that oral administration of patulin induces intestinal injuries, including epithelial cell degeneration, inflammation, ulceration and hemorrhages.

Trichothecenes

Trichothecenes constitute a family of more than sixty sesquiterpenoid metabolites produced by a number of fungal genera, including *Fusarium, Myrothecium, Phomopsis,*

Stachybotrys, *Trichoderma*, *Trichothecium* and others [3]. *Fusarium* is the major genus implicated in the production of nonmacrocylic trichothecenes, and many members of this genus are significant plant pathogens. Trichothecenes are commonly found as food and feed contaminants, and intake of these mycotoxins can result in hemorrhage in the gastrointestinal tract and vomiting. In direct skin contact they may cause dermatitis [25].

Trichothecenes are classified as macrocylic and nonmacrocyclic, whereas nonmacrocylic trichothecenes in turn can be subclassified as type A (Figure 10.7a.) and type B (Figure 10.7b.). Common representatives of class A nonmacrocylic trichothecenes are T-2 toxin, neosolaniol and diacetoxyscirpenol, while fusarenon-x, nivalenol and deoxynivalenol belong to class B of nonmacrocylic trichothecenes.

Trichothecenes are extremely potent inhibitors of eukaryotic protein synthesis. Different trichothecenes interfere with initiation, elongation or termination stages of the polypeptide chain synthesis.

Deoxynivalenol is one of the most common mycotoxins of trichothecene class and is often found in grains - barley, wheat corn and rye, as well as in sunflower seeds and mixed feeds. When ingested in high doses by agricultural animals, it causes nausea, vomiting and diarrhea. At lower doses, animals exhibit weight loss and food refusal [26]. In animal studies T-2 and diacetoxyscirpenol have been shown to be the most potent among class A nonmacrocylic trichothecenes. They express immunosuppressive effects that result in decreased resistance to infectious microbes [27]. They cause a wide range of gastrointestinal, dermatological and neurologic symptoms. It is hypothesized that T-2 and diacetoxyscirpenol are associated with a human disease called alimentary toxic aleukia provoked by necrosis in lymphoid and hematopoietic tissue. The symptoms of this disease include inflammation of the skin, vomiting, bloody feces, weakness and damage to hematopoietic tissues. Acute poisoning is accompanied by necrosis in the oral cavity, bleeding from the nose, mouth, and vagina, as well as central nervous system disorders. Chronic intoxication with these toxins results in bone marrow aplasia, diminished hemostasis and severe lymphatic tissue alteration. Clinical signs of chronic exposure may include dyspnea, dehydration and loss of weight. It is possible that in the past alimentary toxic aleukia was often misdiagnosed with diphtheria or scurvy due to some common symptoms [28].

Macrocyclic trichothecenes are produced intensively by *Myrothecium, Stachybotrys* and *Trichothecium* species. Glutinosin, a mixture of macrocyclic trichothecenes verrucarin A and B, was originally identified as an antimicrobial agent [29].

Recently, trichothecenes produced by *Stachybotrys atra* (*Stachybotrys chartarum*) have received much attention. Produced compounds include satratoxins (Figure 10.8.), roridins, verrucarins and atranones.

Stachybotrys grows well on all sorts of wet building materials with high cellulose content, for example, ceiling tiles, wood fiber boards, and even on dust-lined air conditioning ducts. According to some studies the presence of *Stachybotrys* has been associated with pulmonary bleeding in infants [30].

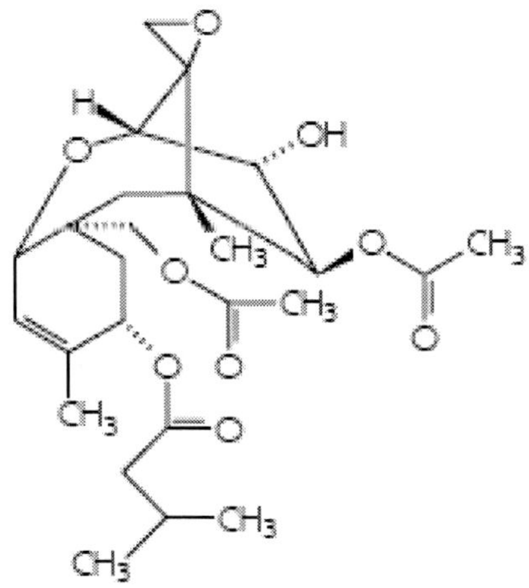

a.

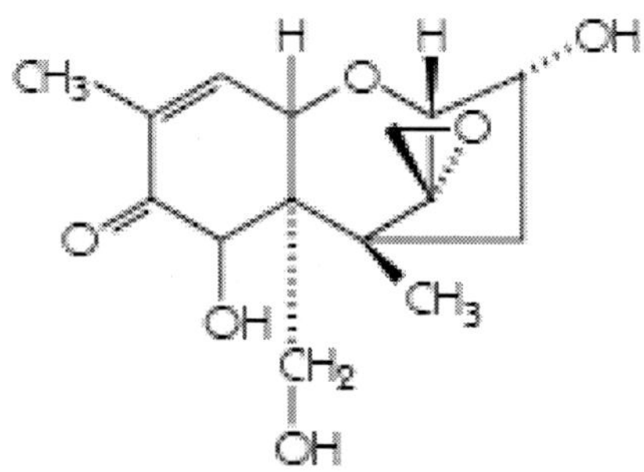

b.

Figure 10.7. Nonmacrocylic trichothecenes (a) Class A: T-2 toxin; (b) Class B: Deoxynivalenol.

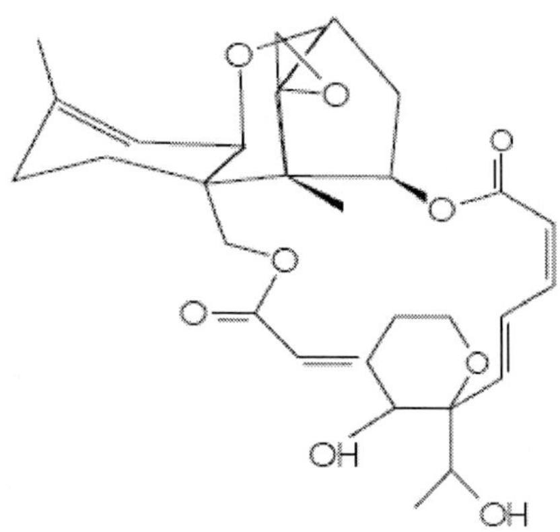

Figure 10.8. Satratoxin H.

Zearalenone

Zearalenone (6-[10-hydroxy-6-oxo-*trans*-1-undecenyl]-B-resorcyclic acid lactone) is a secondary metabolite produced by *Fusarium graminearum, Fusarium culmorum, Fusarium equiseti* and *Fusarium crookwellense*. The compound is frequently found in bread and grains, and the production of the compound is considered to occur in significant amounts after harvesting. Under proper environmental conditions, zearalenone is readily produced in storage on corn and small grains, such as barley, oats, wheat, rice and others. Alternating low and moderate temperatures in storage promote production of this toxin. The toxin is heat-stable, and it is not destroyed by baking or roasting.

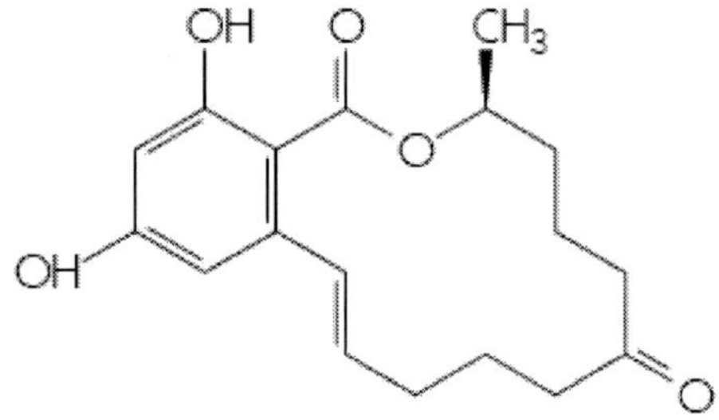

Figure 10.9. Zearalenone.

The zearalenone molecule (Figure 10.9.) is a biologically active compound due to its resemblance to 17β-estradiol, the principal estrogen hormone, and it is able to bind to estrogen receptors in mammalian target cells. The biological potency of the zearalenone family of fungal metabolites is high, but actual acute toxicity is low. The LD_{50} in female rats is greater than 10,000 mg/kg, in female guinea pigs it is 5,000 mg/kg, while as little as 1 μg/kg may create a detectable uterogenic response in female swine.

Dietary concentrations of zearalenone as low as 1.0 ppm have been shown to provoke hyperestrogenic syndrome in pigs, while higher concentrations in prolonged exposure can lead to disrupted conception, abortion and other reproductive problems [31].

Zearalenone may be used for the treatment of postmenopausal symptoms in women [32], and both zearalenol (a reduced form of zearalenone) and zearalenone have been patented as oral contraceptives [33].

Other Mycotoxins

Penicillium roqueforti and *Penicillium camemberti*, species used to manufacture mold-ripened cheeses, produce a number of toxic metabolites, including penicillin acid, roquefortine, isoflumigaclavines A and B, PR toxin and cyclopiazonic acid [34]. *Penicillium crustosum* produces penitrem A, a compound implicated in human tremor, vomiting and bloody diarrhea [35]. *Penicillium cyclopium* and many other Penicillium species, as well as several species of *Aspergillus*, including *Aspergillus flavus*, synthetize a cyclopiazonic acid that acts like specific inhibitor of calcium-dependent ATPase and induces alterations in ion transport across cell membranes [36]. Certain species of Aspergillus, Penicillium and

Claviceps induce tremors due to neurological response produced by mycotoxins like penitrems, janthitrems, lolitrems, aflatrem, paxilline, paspaline, paspalicine, paspalinine and paspalitrem A and B [37].

The chemical diversity of mycotoxins, and substrates in which they may occur, pose serious challenges for analytical chemists. Each group of compounds and each substrate have different chemical and physical properties, so the methods for the separation of the toxins from substrates must be developed for each sample individualy. Mycotoxins are often produced in trace concentrations, so sensitivity of the detection systems is also essential for the applied method. Traditional methods for mycotoxin determination rely on various chromatographic techniques for quantitating different compounds after appropriate cleaning of the sample extract. More recently, immunogenic assays were developed which avoid tedious solvent clean-up procedures and enable high selectivity.

Bacterial Toxins

At chemical level, there are two main types of bacterial toxins, lipopolysaccharides, which are associated with the cell wall of Gram-negative bacteria and belong to endotoxins, and proteins, which are released from bacterial cells and may act at sites remote from the site of bacterial growth. These diffusible toxins are referred to as exotoxins.

Only small amounts of endotoxins may be released into surrounding fluids from the living bacteria, because they constitute a part of bacterial cell wall. Greater amounts are released upon bacterial death and cell walls disintegration. The release of endotoxin may also occur from lysed cells due to an effective host's immune response or the activities of certain antibiotics. Endotoxins generally act in the vicinity or presence of bacterial growth . They are less potent compared to exotoxins, and larger amounts of endotoxins are needed to induce disease symptoms. Also, they are heat resistant and cannot be converted into toxoids. A toxoid is a bacterial toxin whose toxicity is suppressed either by chemical (formalin, phenol) or heat treatment, while other properties, typically immunogenicity, are maintained. Toxoids are used in vaccines as they induce an immune response to the original toxin or increase the response to similar antigen.

Exotoxins, on the other hand, are mainly proteins that are secreted by a bacterial cell into surrounding fluids, and they are produced by both Gram-negative and Gram-positive bacteria. The bacterial toxins of this class are extrememly powerful biological poisons, as in the case of tetanus, where it has been estimated that 1 mg of purified tetanus toxin is sufficient to cause the death of millions of mice. Bacterial protein toxins are the most powerful human poisons known. Some protein toxins have a very specific cytotoxic activity, and they attack specific types of cells. For example, tetanus and botulinum toxins attack only neurons. Others (produced by staphylococci, streptococci, clostridia, etc.) have broad cytotoxic activity and cause nonspecific death of various types of cells, damaging the tissues and eventually resulting in necrosis. Bacterial exotoxins are chemically unstable and can be readily inactivated by heat. However, even when inactivated they retain their antigenic properties. This property is used for the production of toxoids. Some bacterial exotoxins may provoke the impairment of immunologic functions by acting directly on T cells. One large family of bacterial protein toxins produced by staphylococci and streptococci express biological activities by potent stimulation of the immune system and by provoking pyrogenicity. Immunostimulatory properties arising from toxin binding to T-cell receptors results in a proliferation of peripheral T cells. The result of T-cell proliferation is a massive release of cytokines from the lymphocytes (e.g., interleukin-2, tumor necrosis factor ß, gamma interferon) and monocytes (e.g. interleukin 1, interleukin 6, tumor necrosis factor a) [38, 39]. These cytokines act as mediators of characteristic symptoms of intoxination, such as hypotension, high fever and rash. Staphylococcal exotoxins also act in this way; however, it has not been proved that vomiting and diarrhea are caused by the same mechanism.

There are two major mechanisms of toxin entry into target cells. In direct entry, the toxin binds to a specific receptor on the target cell and induces the formation of a pore in the membrane through which it is transferred into the cell cytoplasm. In an alternative mechanism, the native toxin binds to the target cell and is taken into the cell by the process of receptor-mediated endocytosis.

Various bacterial toxins have different mechanisms of toxic action. Some of the most important extracellular toxins and the modes of their toxic action are summarized in Table 10.1.

There are great number of bacterial infections that can be manifested at localized level or expressed as a systematic response. Many of them, if not treated, can provoke lethal outcome. The infection may occur in numerous ways, such as by inhalation of spores or by dermal absorption, especially if the barrier properties of the skin are damaged by a skin injury or wound. Most frequently, bacterial infections occur due to the consumption of contaminated food. Over the last few decades, people have made significant changes in their lifestyle and habits, and the number of food-borne poisonings has decreased. Furthermore, improvements in the regulation of food quality and control contributes to a dramatic reduction in the number of bacterial infections. Nevetheless, the risk of ubiquitous pathogenic bacteria is constanly present in our surroundings, and some of the most common pathogenic bacteria are briefly mentioned below.

Campylobacter jejuni is commonly associated with poultry and naturally colonizes the gastrointestinal tract of many bird species and animals. Infection with this Gram-negative species has been associated with dysentery-like gastroenteritis, as well as with other types of infections, including attack of the central nervous system in humans. Gastrointestinal symptoms cause diarhea, which usually contains blood and leukocytes. Other symptoms include fever, abdominal pain, nausea, headache and muscle pain. The illness usually occurs 2-5 days after ingesting contaminated food or water, and lasts 7-10 days.

Botulism is a rare but serious paralytic illness caused by a nerve toxin produced by bacterium *Clostridium botulinum*. Foodborne botulism is caused by the consumption of foods that contain botulism toxin. Wound botulism is caused by a toxin produced from a wound infected with *Clostridium botulinum*. Infant botulism is caused by consuming spores of the botulinum bacteria, which then grow in the intestines and release the toxin. All forms of botulism can be fatal. The classic symptoms of botulism include double vision, blurred vision, drooping eyelids, slurred speech, difficulty swallowing, dry mouth, and muscle weakness. Infants with botulism appear lethargic, feed poorly, and have a weak cry and poor muscle tone. If untreated, symptoms may progress to cause paralysis of the arms, legs, trunk and respiratory muscles. In foodborne botulism, symptoms generally begin 18 to 36 hours after ingesting contaminated food, but they can occur as early as 6 hours or as late as 10 days.

Clostridium perfringens, which produces a huge array of exotoxins (including α, β, ε and ι) causes wound and surgical infections that lead to gangrene. Clostridial hemolysins and extracellular enzymes such as proteases, lipases, collagenase and hyaluronidase, contribute to the invasive nature of ongoing processes in the body. Enterotoxin produced by *Clostridium perfringens* is a common source of food poisoning. Bacteria is usually encountered in improperly sterilized (canned) foods in which endospores have germinated.

Tetanus is a serious disease that affects the body's muscles and nerves. It typically arises from a skin wound that becomes contaminated by a bacterium *Clostridium tetani*, which is often found in soil. The bacteria produces a neurotoxin, tetanospasmin, which causes muscle spasms. The toxin can travel throughout the body via the bloodstream and lymph system. As it distributes more widely, the toxin interferes with the normal activity of the nerves throughout the body, leading to generalized muscle spasms. Without treatment, tetanus can be fatal. Muscle spasms in the jaw can be accompanied by difficulty swallowing and stiffness or pain in the muscles of neck, shoulders and back. These spasms can spread to the muscles of abdomen, upper arms and thighs. Today tetanus immunization is part of the routine DTP (diphtheria, tetanus, acellular pertussis) vaccination.

Diphtheria exotoxin is a single polypeptide chain with a molecular weight of 60,000 daltons, produced by *Corynebacterium diphtheriae*. The toxin is synthesized in high yield only after exhaustion of the exogenous iron supply to the bacteria.

Table 10.1. Activities of extracellular bacterial toxins

Bacteria	Toxin	Mechanism of toxicity
Bacillus anthracis	Anthrax toxin	Zn^{2+} dependent protease induces cytokine release and is cytotoxic to cells by an unknown mechanism. An adenylate cyclase enzyme increases the levels in intracellular cyclic AMP in phagocytes. Formation of ion-permeable pores in cell membrane leads to edema and decreased phagocytic responses
Bordetella pertussis	Adenylate cyclase toxin	Acts locally to increase levels of cyclic AMP in phagocytes. Formation of ion-permeable pores in cell membranes
Bordetella pertussis	Pertussis toxin	ADP ribosylation of G proteins blocks inhibition of adenylate cyclase in susceptible cells
Clostridium botulinum	Botulinum toxin	Zn^{2+} dependent protease that inhibits neurotransmission at neuromuscular synapses resulting in flaccid paralysis
Clostridium difficile	Toxin A/Toxin B	Modifies small GTP-binding proteins that are regulators of the actin cytoskeleton. Deamination of the glutamine residue to a glutamic acid produces a protein unable to hydrolyze bound GTP. Pathological result is cell necrosis and bloody diarrhea associated with colitis

Table 10.1. (Continued)

Clostridium perfringens	Perfringens enterotoxin	Stimulates adenylate cyclase leading to increased cAMP in epithelial cells. Result is diarrhea
Clostridium tetani	Tetanus toxin	Zn^{2+} dependent protease that inhibits neurotransmission at inhibitory synapses resulting in spastic paralysis
Corynebacterium diphtheriae	Diphtheria toxin	Inhibition of protein synthesis in target cells
Escherichia coli	*E. coli* LT toxin	Similar to cholera toxin
Escherichia coli	*E. coli* ST toxins	Binding of the heat-stable toxins to a guanylate cyclase receptor results in adverse effects on electrolyte flux. Promotes secretion of water and electrolytes from intestinal epithelium leading to diarrhea
Pseudomonas aeruginosa	Exotoxin A	Inhibits protein synthesis similar to diphtheria toxin
Shigella dysenteriae *E. coli* O157:H7	Shiga toxin	Enzymatically cleaves eucaryotic rRNA resulting in inhibition of protein synthesis in susceptible cells. Results in diarrhea, hemorrhagic colitis and hemolytic uremic syndrome
Staphylococcus aureus	Alpha toxin	Protein subunits assemble into an oligomeric structure that forms an ion channel (pore) in the cell plasma membrane
Staphylococcus aureus	Exfoliatin toxin	Cleavage within epidermal cells (intraepidermal separation)
Staphylococcus aureus	Staphylococcus enterotoxins	Causes massive activation of the immune system, including lymphocytes and macrophages, as well as emesis
Staphylococcus aureus	Toxic shock syndrome toxin	Acts on the vascular system causing inflammation, fever and shock
Streptococcus pyogenes	Erythrogenic toxin	Inflammation, fever and shock can cause localized erythematous reactions (scarlet fever)
Vibrio cholerae	Cholera enterotoxin	ADP ribosylation of G proteins stimulates adenlyate cyclase and increases cAMP in cells of the gastrointestinal tract, causing secretion of water and electrolytes and leading to diarrhea

The genus *Corynebacterium* consists of a diverse group of bacteria, including animal and plant pathogens, as well as saprophytes. Diphtheria toxin enters its target cells by either direct entry or receptor-mediated endocytosis, inducing an upper respiratory tract illness characterized by sore throat and low fever. Diphtheria toxin causes death in eucaryotic cells and tissues by the inhibition of protein synthesis. Since the initiation of immunization against diphtheria, disease caused by infection with *Corynebacterium diphtheriae* has virtually disappeared.

Listeria monocytogenes is a Gram-positive rod-shaped bacterium. The two main clinical manifestations of infection with this bacteria are sepsis and meningitis. *Listeria monocytogenes* is presumably ingested with raw, contaminated food. Influenza-like symptoms, including persistent fever, gastrointestinal symptoms such as nausea, vomiting, and diarrhea, may precede serious forms of listeriosis poisoning or may be the only symptoms expressed. The onset time to gastrointestinal symptoms is unknown but probably exceeds 12 hours.

Salmonella species are frequent cause of foodborne illness, especially from poultry, raw eggs, unprocessed milk, meat and water. The bacteria may also be carried by pets like turtles and birds. *Salmonella* primarly attacks the stomach and intestines but in more serious cases, the bacteria may enter the lymph system and the blood. The onset symptoms of poisoning arise 12-72h after infection. Cases of less serious infection are usually manifested only by diarrhea, which may last several days. Most mild types of salmonella infection subside in four to seven days without requiring any treatment. More severe infections may cause excessive diarrhea, stomach cramps and general health problems. Symptoms usually last 4-7 days without leaving long-term effects; however, in serious poisonings eye irritation, painful urination as well as painful joints, may persist up to a year.

Shiga toxin and Shiga-like toxins belong to the group of protein toxins that bind to the cell surface, and after entry into the cytosol, inhibit protein synthesis. *Shigella dysenteriae*, some strains of *Escherichia coli*, as well as other bacteria can secrete such toxins, which may cause serious health complications. Shiga toxin primarly acts on the lining of the blood vessels and on the vascular endothelium. When the protein is inside the cell, it interacts with the ribosomes and inactivates them, leading to distruption in protein synthesis and causing cell death. The first response to poisoning with Shiga toxin is bloody diarrhea, abdominal pain, vomiting and bloody urine, which are typical symptoms of hemolytic uremic syndrome arising in shiga infections. A specific target for the toxin appears to be the vascular endothelium of the glomerulus. Destruction of these structures leads to kidney failure and the development of often deadly hemolytic uremic syndrome. The toxin is effective in damaging small blood vessels, such those as found in the digestive tract, kidneys, and lungs, but it does not affect the large vessels such as the arteries or major veins. Besides attacking specifically kidneys, food poisoning with shiga toxin may also affect the lungs and nervous system.

Of the genus *Vibrio,* most clinically significant to humans is the toxin *Vibrio cholerae,* the agent of cholera, which can be ingested with uncooked and undercooked food or by the fecal-oral route. In its extreme manifestation, cholera is one of the most rapid fatal illnesses known. Commonly, the disease progresses from the first liquid stool to shock in 4-12 hours, with death following in 18 hours to several days. The cholera enterotoxin activates the adenylate cyclase enzyme in the intestinal cells, converting them into pumps which extract the water and the electrolytes from the blood and tissues and pump it into the lumen of the intestines. Loss of potassium ions from this action may result in cardiac complications and circulatory failure. Vibrios are one of the most common organisms in surface waters. *Vibrio parahaemolyticus* is an invasive organism affecting primarily the colon, while *Vibrio cholerae* affects the small intestine through secretion of an enterotoxin. *Vibrio vulnificus* is an emerging pathogen in humans. This organism causes wound infections and gastroenteritis. A vibrio-like organism, *Helicobacter pylori*, causes duodenal and gastric ulcers and gastric cancer.

Yersinia enterocolitica belongs to a family of rod-shaped bacteria. Other species of bacteria in this family include *Yersinia pseudotuberculosis*, which causes illness similar to *Yersinia enterocolitica*, and *Yersinia pestis*, which causes plague. Only a few strains of *Yersinia enterocolitica* cause illness in humans. The strains of *Yersinia enterocolitica* that cause human illness are present in domestic animals, such as pigs, rodents, rabbits, sheep, cattle, horses, dogs and cats. Infection is most often acquired by eating contaminated food, especially raw or undercooked pork products. Drinking contaminated unpasteurized milk or untreated water can also transmit the infection. Symptoms typically develop 4 to 7 days after exposure and may last 1 to 3 weeks or longer. Strong abdominal pain and fever may be the predominant symptoms, and may be confused with symptoms of appendicitis. In a small number of cases, complications such as skin rash, joint pains or the spread of bacteria to the bloodstream may occur.

REFERENCES

[1] Klich, MA; Pitt JI. Differentiation of *Aspergillus flavus* from *Aspergillus parasiticus* and other closely related species. *Trans. Br. Mycol. Soc.* 1988 91, 99-108.

[2] Goldblatt, L. Aflatoxin, scientific background, control, and implications. New York: Academic Press; 1969.

[3] Cole RJ; Cox RH. *Handbook of toxic fungal metabolites*. New York: Academic Press; 1981.

[4] Van Egmond, HP. Aflatoxin M_1: occurrence, toxicity, regulation. In: Van Egmond HP. *Mycotoxins in dairy products*. London: Elsevier Applied Science; 1989.

[5] Eaton, DL; Gallagher, EP. Mechanisms of aflatoxin carcinogenesis. *Annu. Rev. Pharm. Tox.*. 34, 1994, 135-172.

[6] Sabbioni, G; Sepai O. Determination of human exposure to aflatoxins. In: Sinha, KK; Bhatnagar, K. *Mycotoxins in agriculture and food safety*. New York: Marcel Dekker, Inc.; 1994.

[7] Manabe, M. Fermented foods and mycotoxins. *Mycotoxins* 51, 2001, 25-28.

[8] Blanc, PJ; Loret, MO; Goma, G. Production of citrinin by various species of *Monascus*. *Biotechnol. Lett.* 17, 1995, 291-294.

[9] Saito, M; Enomoto, M; Tatsuno, T. Yellowed rice toxins: luteroskyrin and related compounds, chlorine-containing compounds and citrinin. In: Ciegler, A; Kadis, S; Ajl, SJ. *Microbial toxins,* vol. VI*: fungal toxins*. New York: Academic Press; 1971.

[10] Chu, FS. Current immunochemical methods for mycotoxin analysis. In: Vanderlaan, M; Stanker, LH; Watkins, HR; Roberts, DW. *Immunoassays for trace chemical analysis: monitoring toxic chemicals in humans, food, and the environment.* Washington, D.C: American Chemical Society; 1991.

[11] Haller, JS. Ergotism. In: Kiple, KF. *The Cambridge world history of human disease.* Cambridge, United Kingdom: Cambridge University Press; 1993.

[12] Riddle, JM. Eve's herbs. *A history of contraception and abortion in the west.* Cambridge: Harvard University Press; 1997.

[13] Bennett, JW; Bentley, R. Pride and prejudice: the story of ergot. *Persp. Biol. Med.* 42, 1999, 333-355.

[14] Marasas, WFO; Miller, JD; Riley, RT; Visconti, A. Fumonisins—occurrence, toxicology, metabolism and risk assessment. In: Summerell, BA; Leslie; JF; Backhouse, D; Bryden, WL; Burgess, LW. *Fusarium.* Paul E. Nelson Memorial Symposium, St. Paul, Minn: APS Press; 2001.

[15] Wang, E; Norred, WP; Bacon, CW; Riley, RT; Merrill, AH. Inhibition of sphingolipid biosynthesis by fumonsins. *J. Biol. Chem.* 266, 1991, 14486-14490.

[16] Sydenham, EW; Shephard, GS; Thiel, PG; Marasas, WFO; Stockenstrom, S. Fumonsin contamination of commercial corn-based human foodstuffs. *J. Agric. Food Chem.* 39, 1991, 2014-2018.

[17] Rheeder, JP; Marasas, WF; Vismer, HF. Production of fumonisin analogs by *Fusarium* species. *Appl. Environ. Microb.* 68, 2002, 2102-2105.

[18] Plattner, RD; Weisleder, D; Poling, SM. Analytical determination of fumonisins and other metabolites produced by *Fusarium moniliforme* and related species on corn. *Adv. Exp. Med. Biol.* 392, 1996, 57-64.

[19] Abarca, ML; Bragulat, MR; Sastella, G; Cabanes, FJ. Ochratoxin A production by strains of *Aspergillus niger* var. niger. *Appl. Environ. Microb.* 60, 1994, 2650-2652.

[20] Pitt, JI. Penicillium viridicatum, Penicillium verrucosum, and production of ochratoxin A. Appl. Environ. Microb. 53, 1987, 266-269.

[21] Hult, K; Piestina, R; Habazin-Novak, V; Radic, B; Ceovic, S. Ochratoxin A in human blood and Balkan endemic nephropathy. *Arch. Tox.* 51, 1982, 313-317.

[22] Kuiper-Goodman, T; Scott, PM. Risk assessment of the mycotoxin ochratoxin A. *Biomed. Environ. Sci.* 2, 1989, 179-248.

[23] Meisner, H; Meisner, P. Ochratoxin A, an inhibitor of renal phosphoenolpyruvate carboxylase. *Arch. Biochem. Biophys.* 208, 1981, 146-151.

[24] Rahimtula, AD; Bereziat, JC; Bussacchini-Griot, V; Bartsch, H. Lipid peroxidation as a possible cause of ochratoxin A toxicity. *Biochem. Pharm.* 37, 1988, 4469-4475.

[25] Beasley, VR. *Trichothecene mycotoxicosis: pathophysiologic effects*, Vol. I:. Boca Raton, Fla: CRC Press; 1989.

[26] Rotter, BA; Prelusky, DB; Pestka, JJ. Toxicology of deoxynivalenol (vomitoxin). *J. Tox. Environ. Health* 48, 1996, 1-34.

[27] Peraica, M; Radic, B; Lucic, A; Pavlovic, M. Toxic effects of mycotoxins in humans. *Bull. WHO* 77, 1999, 754-766.

[28] Matossian, MK. *Poisons of the past: molds, epidemics, and history*. New Haven: Conn, Yale University Press; 1989.

[29] Grove, JF. The constituents of glutinosin. *J. Chem. Soc.* 4, 1968, 810-815.

[30] Centers for Disease Control and Prevention. Acute pulmonary hemorrhage/hemosiderosis among infants. *Morb. Mortal. Wkly Rep.* 43, 1994, 881-883.

[31] Kurtz, HJ; Mirocha, J. Zearalenone (F2) induced estrogenic syndrome in swine. In: Wyllie, TD; Morehouse, LG. *Mycotoxic fungi, mycotoxins, mycotoxicoses.* Vol. 2: New York: N.Y., Marcel Dekker; 1978.

[32] Urry, WH; Wehrmeister, HL; Hodge, EB; Hidy PH. The structure of zearalenone. *Tetrahedron. Lett.* 27, 1966, 3109-3114.

[33] Hidy, PH; Baldwin, RS; Greasham, RL; Keith, CL; McMullen, JR. Zearalenone and some derivatives: production and biological activities. *Adv. Appl. Microb.* 22, 1977, 59-82.

[34] Scott, PM. Toxins of *Penicillium* species used in cheese manufacture. *J. Food Prot.* 44, 1981, 702-710.

[35] Hocking, AD; Holds, K; Tobin, NF. Intoxication by tremorgenic mycotoxin (penitrem A) in a dog. *Aust. Vet. J.* 65, 1988, 82-85.

[36] Riley, RT; Goeger, DE. Cyclopiazonic acid: speculations on its function in fungi. In: Bhatnagar D; Lillehoj, EB; Arora, DK. *Handbook of applied mycology, Mycotoxins in ecological system,* Vol. 5:. New York: Marcel Dekker, Inc.; 1992.

[37] Steyn, PS;Vleggaar, R. Tremorgenic mycotoxins. In: Herz, W; Griseback, H; Kirby GW; Tamm, Ch. Progress in the chemistry of organic natural products, Vol. 48:. Vienna, Austria: Springer-Verlag; 1985.

[38] Hallas, G. The production of pyrogenic exotoxins by group A streptococci. *J. Hyg.* 95, 1985, 47-57.

[39] Hackett, SP; Schlievert, PM; Stevens, DL. Cytokine production by human mononuclear cells in response to streptococcal exotoxins. *Clin. Res.* 39, 1991, 189-195.

Chapter XI

PRIONS AND PRION RELATED DISEASES

Prion diseases are family of rare, progressive neurodegenerative disorders that affect both humans and animals. This unusual disease drew public attention for the first time in 1986 during the epidemic of mad cow disease. For decades scientists tried to identify the causative agent of the disease but without success. Scientists had already demonstrated that the cause was none of the common infectious agents, such as bacteria, viruses or parasites. Adding to their confusion their discovery that the agent was resistent to UV irradiation, which normally destroys the DNA of any organism, It had been speculated that the disease was caused by a slow virus, which can remain latent in a host for years. Then, years later, hundreds of chemists confirmed that a simple protein, the proteinetious infectious particle (prion), was the infectious agent.

An infectious prion is an abnormal, transmissible agent that is able to induce abnormal aggregation of normal cellular proteins in the brain, leading to brain damage and to other characteristics signs and symptoms of the disease [1]. Prion diseases are usually rapidly progressive, always fatal, and are distinguished by long incubation periods and characteristic spongiform changes in the brain associated with neuronal loss. Normal cellular prion proteins, PrP^c (PrP – Prion related Proteins or Protease resistant Proteins) are transmembrane glycoproteins found at the surface of certain cells, such as neural and stem cells. They are attached to the surface of neurons by a glycophosphatidylinositol anchor. It is presumed that these cells have a role in transferring copper ions into the brain, and are important for the activity of brain cuproenzymes. Two-thirds of a protein molecule at the carboxyl terminal have the structure of alpha helix and the remaining part at the amino end contains a copper-binding part [2]. PrP^c demonstrate the activity of superoxide dismutase and may play a role in cellular resistance to oxidative stress. Similar proteins, acting as prion proteins, have been found in yeast and fungus, but the diseases related to them have not yet been identified. Prion diseases are caused by abnormal protein, denoted as PrP^{Sc} (Sc – Scarpie), which has the same amino acid sequence as the normal protein, but with a dominant beta conformation in its secondary structure (Figure 11.1). The PrP^{Sc} protein is insoluble in solvents and is highly resistant to digestion by proteases.

The infection occurs when the abnormal PrP^{Sc} protein comes in contact with normal PrP^C, converting it to a pathogenic form that has the ability to aggregate. Metal ions enchance this interaction and may contribute to disease progression. A particular abnormal PrP^{Sc} can convert only PrP^C molecules of the same or a similar primary structure and it acts as a template for converting PrP^C to PrP^{Sc} [3]. As a result of this interaction, small aggregates are formed which further propagate, forming large insoluble deposits in the nervous system, that can be observed in brain autopsies in patients infected with prion diseases. Amyloid deposits similar to those in prion diseses can be found in patients with Alzheimer`s and Parkinson`s diseases, indicating that these pathological conditons also may be related to prions. Most cells, including neurons in the brain, contain proteasomes, which are responsible for degrading the aggregated proteins. Formation of prion aggregates, in addition to representing the mechanical perturbation, blocks the ability of proteasomes to process other proteins that are the normal candidates for destruction and disturbes normal cellular metabolism leading to cell death.

According to another theory, the so-called "X" hypothesis, a so-far unidentified "X" protein is responsible for PrP^C/PrP^{Sc} conversion by forming the complex with both forms of the molecules. Variability in "X" protein is likely to be a cause of low interspecies transmission of the disease, reducing the probability for the disease to be transferred across species.

The incubation period of prion diseases depends on the animal species due to variations in the prion gene (PrP PRNP) that encodes prion protein. Mutations in the prion gene enchance the spontaneous conversion PrP^c in PrP^{Sc}, making such persons more susceptible to prion diseases. In interspecies transmission, the incubation period may be very long, whereas with every subsequent transmission within species, the incubation period decreases and finally stabilizes. Measurement of the incubation period can be performed by using an intracerebral injection of prion-containing brain tissue homogenizate in mice and monitoring the disease progression and manifestation in usually 12 topological brain spots. Oral infection by prions through diet also is possible, because mucosa- associated limphoid tissue adopts prion particles via phagocytosis, transferring them to other limphoid tissues, where they multiply. Since all limphoid tissues, such as the lymph nodes, tonsils and spleen, are highly innervated, further transfer of prions to spinal cord is probable.

Naturally high content of normal PrP^c proteins usually leads to more probable disease development. Even though all prion diseases are fatal, a few vaccines have been developed which tend to stabilize PrP^c conformation by binding to it. Inspired by the same principle, research on the development of vaccines for Alzheimer`s disease are currently being conducted.

Compounds that interfere with PrP^c/PrP^{Sc} interaction are also predicted to be useful in the treatment of prion-related diseases. Many substances, such as congo red, express an affinity to the amyloid conformation of proteins and may also be used in therapy, as well as substances influencing endocytosis, exocytosis and protein degradation. In the diagnosis of prion diseases, bioaffinity tests, such as the Enzyme-Linked Immunosorbent Assay (ELISA) , can be used. In analyzing blood plasma, the sandwich ELISA enables the detection limit of 50 pg/ml of prion proteins, but requires previous discrimination between normal and protease resistant forms.

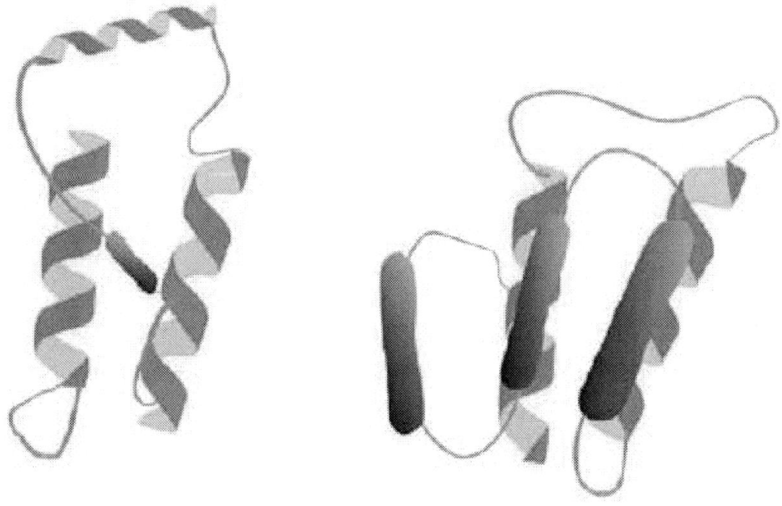

Figure 11.1. The dominant alpha helix conformation in PrPc prion (left) and dominant beta blocks in PrP^{Sc} prion (right).

HUMAN PRION DISEASES

Creutzfeldt-Jakob disease was recognized in the early 1920s. This disease is a neurodegenerative alteration of the brain that evokes no apparent immune or inflammatory response and results in dementia, blindness and motor disorders. The condition is characterized by frequent muscle spasms and uncontrollable trembling. Death usually follows within six months of the onset of symptoms and there are no documented recoveries from the disease.

The most common form of classic *Creutzfeldt-Jakob disease* is believed to occur sporadically and is caused by spontaneous transformation of normal prion proteins into abnormal prions. The causative agent of Creutzfeldt-Jakob disease has been found in brain, spinal cord, cerebrospinal fluid and in lesser amounts in lymph tissues, liver, kidneys, lungs and blood. The risk of Creutzfeldt-Jakob disease increases with the age, and in persons over 50 years of age, the annual rate is approximately 3.4 cases per million. About 10% - 15% of the cases of Creutzfeldt-Jakob disease are believed to occur due to inherited mutations in prion protein genes. Besides being sporadic, numerous cases of infection occurring through organ transplantation, implantation of the infected electrodes in the brain or through the use of contaminated surgical instruments have been reported. The disease also may be transmitted by injecting the growth hormone derived from infected people. Surgical instruments may remain infectious for two years after their sterilization [4]. It is speculated that the disease also may be transmitted by blood transfusion and mothers milk. It is believed that the occurence of the Creutzfeldt-Jakob disease is more frequent than reported, due to misdiagnoses as Alzheimer's disease, because two diseases express some common symptoms.

Variant Creutzfeldt-Jakob disease appeared as a consequence of the epidemic of bovine spongiform encephalopathy that struck Great Britain in 1985 and was first described in 1996.

Identified prion sequences suggest that the illness occurs as a result of the consumption of contaminated beef [5]. In contrast to traditional forms of *Creutzfeldt-Jakob* disease, the variant of the disease, which affects younger people of the average age of 29 years, has a relatively longer duration and is strongly linked to exposure. In the early stage of the illness, patients usually experience psychiatric symptoms, which most commonly take form of depression or, less often, a schizophrenia-like psychosis. Unusual sensory symptoms, such as "stickiness" of the skin, also have been experienced in infected persons. Neurological signs, including unsteadiness, difficulty walking and involuntary movements, develop as the illness progresses and, by the time of death, patients become completely immobile and mute.

Gerstmann-Straussler-Scheinker syndrome is an unusual variant of a Creutzfeldt-Jakob disease that was first diagnosed in 1928 by Gerstmann and was described, along with seven additional cases in 1936, by all three researchers. The illness appears very similar to Creutzfeldt-Jakob disease but differs in several important ways. Gerstmann-Straussler-Scheinker syndrome is very rare with an occurrence of 1 in 10 million of population and has an earlier onset of symptoms at approximately the 4th decade of life and a longer duration of symptoms, typically 5 years. The symptoms are dementia, cerebral ataxia and signs of disturbed function of the cerebral cortex, with death being the final outcome. Gerstmann-Straussler-Scheinker syndrome appears to be an inherited form of Creutzfeldt-Jakob disease affecting only certain families, of which several hundred have been identified throughout the world.

Kuru is another prion disease that causes rapid deterioration of mental functions and loss of muscle coordination. This disease used to occur in tribes of the New Guinea highlands where it was related to ritual cannibalism. The cannibalistic ritual involved consumption of the brain tissue of dead relatives as a sign of respect. Kuru was probably initiated by consumption of prion-contaminated tissue from a person affected by Creutzfeldt-Jakob disease. The illness was more common among women and children because they were more frequently honoured to consume the brain of the dead relative. Many of these rituals have since been abandoned, and kuru has been virtually eradicated. Symptoms of kuru included loss of muscle coordination and difficulty in walking. Abnormal involuntary movements, such as repetitive, slow writhing or rapid jerking of the limbs and body, were observed. Emotions switched suddenly from sadness to happiness, with sudden outbursts of laughter. People with kuru became demented and unable to speak. Death usually occured 3 to 24 months after the first appearance of symptoms.

Fatal familial insomnia is an autosomal heritable disease characterized by severe untreatable insomnia and cognitive disorder. In affected persons, endocrine and motor systems are highly impaired [6]. Cognitive disorder, especially in first stages of the illness, is manifested as an attention disturbance, with short-term memory deficits, which evolve towards a state of confusion, dementia and finally death [7]. Symptoms of the disease usually start to develop at the age of 50, and typical disease duration is between 7 and 36 months. The disease is associated with a point mutation in codon 178 of the prion protein gene [8]. Neuropathological findings demonstrate selective atrophy of the thalamus in affected persons [9]. The signs of spongiosis are also found in the cortex. Affected areas show neuronal loss and astrocytic proliferation; however, deposits of PrP^{Sc} are not always observed as in the cases of other prion diseases.

Animal Prion Diseases

The most common form of prion disease in animals is *Scrapie*, a fatal, degenerative disease that affects the nervous systems of sheep and goats. The disease has been known for over 200 years and was the first case of neurodegenerative spongiform encephalopathy to be studied [10]. Afflicted animals loose coordination and become irritable. In some cases, an intense itching developes forcing the animals to scrape off their wool or hair, and due to this the disease was denoted as "Scrapie." The disease can be transmitted from animal to animal or by feed contaminated with infected nerve tissue. The scrapie prion has been shown to survive in the soil for years and can also be transmitted to other animals via placental tissue that remains on grassland.

An epidemic of bovine spongiform encephalopathy (BSE or mad cow disease) strucen Great Britain in 1985. The spread of the disease was initiated by the use of feed containing the brain tissue of sheep infected with scrapie. After the cause of the mad cow disease was identified, the use of feed constituents of animal origin was banned. Animals affected with bovine spongiform encephalopathy might display changes in temperament, such as nervousness or aggression. Infected animals also demonstrate an abnormal posture and incoordination and have the difficulty rising or standing. Despite continued good appetite, infected animals experience loss of body weight and decreased milk production with the onset symptoms.

Several other similar prion diseases affecting other animals are known and include transmissible mink encephalopathy, chronic wasting disease in mules, deer and elk, and feline spongiform encephalopathy [11].

References

[1] Prusiner, SB; Collinge, J; Powell, J; B. Anderton, B. *Prion Diseases Of Humans And Animals*. London: Ellis Horwood; 1992.

[2] Prusiner, SB. Molecular Biology Of Prion Diseases. *Science* 252, 1991, 1515-1522.

[3] Cohen, FE; Pan, KM; Huang, Z; Baldwin, M; Fletterick, RJ; Prusiner, SJ. Structural Clues To Prion Replication. *Science* 264, 1994, 530-531.

[4] Belay, E; Schonberger, L. Variant Creutzfeldt-Jakob Disease and Bovine Spongiform Encephalopathy. *Clin. Lab. Med.* 22, 2002, 849-862.

[5] Belay, E. Transmissible Spongiform Encephalopathies in Humans. *Annu. Rev. Microb.* 53, 1999, 283-314.

[6] Lugaresi, E; Medori, R; Montagna, P; Baruzzi, A; Cortell, P; Lugaresi, A; Tinuper, P; Zucconi, M; Gambetti, P. Fatal familial insomnia and dysautonomia with selective degeneration of thalamic nuclei. *N. Engl. J. Med.* 315, 1986, 997-1003.

[7] Gallassi, R; Morreale, A; Montagna, P; Cortelli, P; Avoni, P; Castellani, R; Gambetti, P; Lugaresi, A. Fatal familial insomnia: behavioral and cognitive features. *Neurology* 46, 1996, 935-939.

[8] Medori, R; Tritschler, HJ; LeBlanc, A; Villare, F; Manetto, V; Chen, HY, Xue R. Fatal familial insomnia, a prion disease with a mutation at codon 178 of the prion protein gene. *N. Engl. J. Med.* 326, 1992, 444–449.
[9] Guilleminault, C. *Fatal Familial Insomnia: Inherited Prion Diseases, Sleep, And The Thalamus.* Raven Press; 1994.
[10] Parry, HP. *Scrapie Disease In Sheep* . Academic Press; 1983.
[11] Belay, ED; Schonberger, LB. The Public Health Impact of Prion Diseases. *Annu. Rev. Public Health* 26, 2005, 191-212.

Chapter XII

GENETICALLY MODIFIED ORGANISMS (GMO)

A genetically modified organism (GMO) is an organism whose genetic structure has been altered by incorporating a gene from another plant, animal or bacterial species, aiming at the expression of a desirable feature. Combining genes from different organisms is known as recombinant DNA technology, and the resulting organism is denoted as genetically modified or transgenic. Genetic modification implies technologies that alter the genetic material of animals, plants or bacteria, while biotechnology is a more general term and refers to using living organisms or their products, such as enzymes, in the production of food and beverages, like wine, cheese, beer, as well as for the production of medicines and vaccines. Biotechnology also includes bioremediation processes that exploit modified or native microorganisms or plants in the treatment of polluted environments. Some trees have been succesfully genetically engineered to remediate heavy metal contamination of the soil and some strains of *Escherichia coli*, after adaptation in laboratory, can be used to treat soil contaminated with petroleum products. Many other plants and microorganisms are applied for bioremediation and may be succesfully used to reduce pollution in the environment.

One of the factors limiting recombinant DNA technology was the determination of the exact location of the genes responsible for certain features, but that limitation has been successfully resolved by the development of programs for use in genome sequencing.

Genetic engineering began to develop intensively after genetic engineering was successfully used to incorporate the fluorescent pigment from jelly fish into mice. The first food product obtained by recombinant DNA technology and evaluated by FDA was the so-called Flavr Savr tomato [1]. Upon ripening, tomatoes become soft due to activity of enzymes that catalize the disintegration of their cell walls. In the Flavr Savr tomato, a gene that regulates the activity of those enzymes was modified to significantly reduce the damage to the cell walls, and the Flavr Savr tomato acquired a feature that conveniently preserved firmness and quality sustainability during transportation.

The technologies used in genetic modification offer dramatic potential for improving the quality of plants intended for food production, such as enhanced taste, reduced maturation time and increased nutrient content. Other possible benefits from transgenetic plants include enchanced stress tolerance and improved resistance to diseases, pests and herbicides. In animals, recombinant DNA technology aims to increase resistance, productivity, hardiness and feed efficiency. These tailored benefits promise better yields of meat, eggs and milk and overall improved animal health, which reduces the need for the application of chemicals, which would eventually reach humans. Benefits on the environmental level include conservation of soil, water and energy. Bioprocessing enables better waste management, and application of bioherbicides and bioinsecticides dramatically reduces environmental contamination.

Even though it is a subject of much controversy and dispute, the production of food derived from genetically modified organisms clearly comes with the benefits of reduced costs, as well as consumer benefits in terms of the nutritional implications. Recombinant DNA technology offers the opportunity to produce plants using fewer pesticides and less mechanical effort, and to grow plants that are richer in nutrients. "Golden Rice," for example, which contains beta carotene, a source of vitamin A, and iron, was tailored in order to decrease childhood blindness and maternal anemia in developing countries where the main dietary staple is rice. Other examples of succesful genetic modification include the production of potatoes and rice containing the soybean glycinin gene, responsible for the biosynthesis of glycinine, a rich storage protein, then herbicide- and insect-resistant soybeans, as well as modified corn, cotton, canola and alfalfa. Cotton and tomatoes were modified using a gene from *Bacillus thuringiensis* subsp. that provides increased protection against pests [2]. *Bacillus thuringiensis* is a naturally occurring bacterium that produces proteins lethal to insect larvae. This toxin affects many species of insect larvae and the benefit of using this modification lies in the fact that it is impossible to selectively killl only larvae by any other mean [3]. A variety of plants, such as tobacco and potatoes, were re-designed to be able to survive weather extremes by incorporating an antifreeze gene from a cold water fish. Plants engineered to withstand long periods of drought, or high salinity content, have made formerly inhospitable land useable for agriculture. Future perspectives in the production of genetically modified organisms lies in the efforts to produce bananas engineered to produce antibodies intended for human vaccines against infectious diseases such as hepatitis; to design the fish that mature more quickly and cows that are resistant to bovine spongiform encephalopathy. Production of fruit and nut trees that produce their yields more quickly and plants that produce new plastic materials might also be attractive for future consideration.

The most obvious health concern regarding the use of food produced from genetically modified organisms is the risk of allergic reactions. More than 90% of real food allergies occur in response to specific proteins, usually from milk, eggs, wheat, fish, nuts, soybeans or shellfish. The risk for allergic reactions lies in the possibility that proteins from modified foods may be indirectly introduced into a food that does not provoke a known allergic reaction and where the risk is not anticipated and thus avoided. A proposal to incorporate a gene from Brazil nuts into soybeans was rejected, because of the risk of introducing a common allergen in the Brazil nut. Brazil nuts contain an albumin consisting of 18% of methionine. The proposal for modification was to introduce the gene encoding the synthesis of that protein into soybean plants. In the modified soybean, the methionine content would be significantly increased, eliminating the need to use expensive methionine supplements in soy-derived livestock feed. For approval of certain genetically modified foods, the FDA requires each presenter of a product to show scientific evidence that allergenic substances have not been incorporated into the product. If the presenter cannot provide this evidence, the FDA requires a label on the product to alert the consumer of its possible allergic potential [4].

A major problem in conducting a valid study on the health impact of genetically modified organisms is the impossibility to identifying a large enough control group of people who do not consume GMOs, since the majority of food products available on the market contain at least some components derived from modified organisms. Furthermore, the evaluation of the safety of derived foods is much more difficult than for individual chemicals, drugs or food additives, because foods are very complex and their composition varies. Testing for safety is further complicated by the fact that the raw ingredients of most food products originate from many different sources. For these and other reasons, there is a lack of consistency in publications describing clinical studies on the investigation of the influence of genetically modified products on the human health; and existing animal studies are rare, inadequate and controversal.

When a new, genetically modified product is proposed for commercial distribution, the presenter must provide accurate information about the incorporated gene in order to receive approval. The FDA must evaluate whether the newly incorporated protein is similar to other common proteins found in the food. If the new protein is significantly different, it is treated as a food additive and will require pre-market approval by the FDA.

The use of genetically modified organisms is a subject of much controversy. Most of controversy focuses on human and environmental safety, labeling, intellectual property rights, ethics, food security and poverty issues. Another concern is that the insertion of specific genes into the genome may result in unintended effects. These effects may be reduced by selection, but the ways in which the new genes will affect the functioning of the modified species are not completely predictable and may lead to the development of unknown toxic or allergic manifestations.

Another concern relates to the mutation of gastrointestinal bacteria, which may uptake the GMO plasmids because the ingested DNA material is not fully destroyed in the gastrointestinal tract.

The possible human health impacts of introduced allergens, transfer of antibiotic resistance and other unknown effects are barriers to unconditional acceptance of genetically modified products. Other important issues of concern include potential environmental impacts through the unintended transfer of transgenes, their unidentified effects on other organisms and the potential loss of biodiversity.

In many countries, the labeling of genetically modified foods is still an unsolved problem. Many different dimensions, such as financial, ethical, legal and practical, are interwoven in this area. Agricultural industries believe that labeling should be voluntary, while consumer interest groups demand mandatory labeling. The FDA feels that genetically modified foods are substantially equivalent to non-genetically modified foods, and therefore not subject to more stringent labeling.

It is certain that genetically modified foods have the potential to solve many of the world's hunger and malnutrition problems, and to help to protect and preserve the environment by increasing yields and reducing reliance upon chemical pesticides. Yet, many challenges also exist in the areas of safety testing, regulation, international policy and food labeling. This technology, which offers such enormous potential for benefit, should not be ignored;but the creation and introduction of new products, as well as the continued use of products already developed, requires great precaution and should proceed in small steps.

GMO Testing

When the type of genetic modification in question is known, a genetic modification can be confirmed by a specific and sensitive biochemical method, known as a Polymerase Chain Reaction (PCR) technique., The PCR was invented by Kary Mullis in 1983, and for his work he received the Nobel Prize in Chemistry. In order to obtain necessary information, the PCR technique must be applied in concert with either gel or capillary electrophoresis. If these two techniques are applied, the presence and in certain PCR techniques also the quantity, of well established modifications may be determined. PCR technique is also an important tool in the production of genetically modified organisms, because it enables the amplification of a specific gene. After the gene is amplified, it can be inserted into a vector such as a plasmid, and introduced into another organism. This biochemical technique is also widely used in biochemistry for detecting hereditary diseases, diagnosis of infectious diseases, paternity testing and for the identification of the genetic fingerprint. The detection of hereditary diseases is performed by amplifying the gene in question using the appropriate primers and subsequent sequencing in order to detect a mutation. Viral diseases can be detected using PCR through the amplification of viral RNA. PCR analysis provides reliable results right upon infection, while actual symptoms may occur from several days to several months later. A variation of PCR technique also may be used to determine evolutionary relationships between organisms.

In order to avoid contaminations from the DNA of bacteria, viruses or staff, the reaction mixture for the PCR process is prepared in laminar flow cabinets with UV lamps. PCR technique typically amplifies only short DNA fragments, usually up to 10 kilo base pairs (10 000 nucleotide pairs), but certain methods can copy fragments up to 25 kilo base pairs [5]. In the amplification process, the gene content increases exponentionally. The DNA fragment to be amplified is determined by the selection of *primers*. Primers are short, artificial DNA strands, often not more than 50 and usually only 18 to 25 base pairs long, that are complementary to the beginning or the end of the DNA fragment to be amplified. They adhere to the DNA template at these starting and ending points, defining the fragment which will be amplified. The tagged fragment is then enzymatically amplified in a thermal cycler with the help of polymerase activity. DNA polymerases occur naturally in living organisms. Their function in cells is to duplicate the DNA during cell division. Polymerase acts by binding to single DNA strand and creating the complementary strand. DNA polymerase isolated from thermophilic bacteria is stable at high temperatures and is not inactivated during the denaturation step, a preliminary step of the PCR technique. One of the first thermostable DNA polymerases was obtained from *Thermus aquaticus (Taq)*. Use of polymerases such as Pwo (*Pyrococcus woesi*i) or Pfu (*Pyrococcus furiosus*) significantly reduces the number of mutations that occur in the copied DNA sequence during PCR testing. However, the Pwo and Pfu enzymes polymerize the DNA at much slower rate than the Taq enzyme. However, combinations of Taq and Pfu are now available and provide both fast polymerization and high accuracy in duplication of DNA.

The process of enzymatic amplification usually consists of series of twenty to thirty-five cycles. Each cycle consists of three steps. In the first step. the double-stranded DNA is denaturated by heating to 94-96°C in order to separate the strands. Also, certain polymerases are activated at this stage. After the DNA strands are separated, the temperature is lowered so the primers can attach themselves to the single DNA strands. This step is called annealing. The temperature of this stage depends on the primers and is usually 5°C below their melting temperature. The temperature during the annealing step is critical, as primers may not bind to the template DNA, or bind at random, if the temperature is not right. If the primers are too short, they may anneal at several positions on a DNA template, and this results in non-specific copies. On the other hand, the length of a primer is limited by maximum melting temperature of a specific primer, as melting temperature increases with the length of the primer. A melting temperatures that is too high (above 80°C) can cause the loss of polymerase activity. The optimum length of a primer is generally from 15 to 40 nucleotides with a melting temperature between 55°C and 65°C [6]. If the same gene is to be amplified from different organisms, degenerate primers, i.e. the mixture of similar, but not identical primers, are used. Use of degenerate primers can greatly reduce the specificity of PCR amplification. Non-specific gene amplification may occur due to partial primer binding, leaving the 5' end unattached. Manipulation of the annealing temperature and the addition of a magnesium ion, which stabilizes DNA and RNA interactions, improves the specificity. In the last step of the technique, the DNA polymerase copies the DNA strands. The polymerase starts its function at the annealed primer and adds the nucleotides along the DNA strand. This step is called elongation. The elongation temperature depends on the DNA polymerase. Obtained gene fragment can be identified by gel electrophoresis.

The efficiency of PCR amplification is reliable for DNA fragments of two to three thousand base pairs in lengths. It is possible to amplify larger pieces of up to 50,000 base pairs with a slower heating cycle and special polymerases which are fused to DNA-binding protein [7].

In *nested* PCR, two sets of primers are used in two successive PCR runs. The second set is intended to amplify a secondary target, which is the product of the first run. The technique requires more detailed knowledge of the involved sequences. *Inverse PCR* is used when only one internal sequence is known. This technique is especially useful in identifying insertation mutations and involves the activity of endonuclease and self ligation, resulting in known sequences at either end of the unknown sequence. *Reverse Transcription PCR (RT PCR)* is the method used to amplify, isolate or identify a known sequence of the RNA from a cell or tissue and is import because it may be used to convert RNA to cDNA. *Assembly PCR* is an artificial synthesis of long gene products by performing PCR with oligonucleotides with short overlapping segments. Overlapping segments selectively determine the final PCR product. *Asymmetric PCR* is used to preferentially amplify one strand of the DNA. In this technique, PCR is carried out as usual, but with substantial excess of primers for the chosen strand. Due to the slow, arithmetic amplification used in this technique, extra cycles of PCR are required. *Quantitative PCR (Q PCR)* is used to rapidly measure the quantity of the PCR product and is an indirect method for determining starting amounts of DNA, cDNA or RNA. This technique is commonly used to determine the quantity of a specific gene in the sample. The amount of a certain gene fragment is measured by using fluorescent dyes and probes. *Touchdown PCR* is a variant of PCR that reduces nonspecific primer annealing by more gradually lowering the annealing temperature between cycles. *Hot-start PCR* is a technique that reduces non-specific primer annealing which might occur during the preparation of the reaction components. This technique may be performed manually by simply heating the reaction components briefly at the melting temperature of the primers before adding the polymerase. For this technique specialized enzyme systems have been developed in order to inhibit the polymerase's activity at the ambient temperature, either by binding of an antibody or by the presence of covalently bound inhibitors that dissociate only after a high-temperature activation step. *Multiplex-PCR* uses multiple primer sets within a single PCR reaction to produce different DNA sequences. By targeting multiple genes simultaneously, a single test can provide additional information. In order to be successful, annealing temperatures for each of the primer sets must be optimized to work correctly within a single reaction, and produced fragments must be well separated by electrophoresis step.

REFERENCES

[1] Redenbaugh, K; Hiatt, W; Martineau, B; Kramer, M; Sheehy, R; Sanders, R; Houck, C; Emlay, D. *Safety assessment of genetically engineered fruits and vegetables*. CRC Press, Inc.; 1992.

[2] Healey, J. *Genetically Modified Foods.* Spinney Press, 2000.

[3] Huang, J.; Hu, R; Rozelle, S; Pray, C. Insect-Resistant GM Rice in Farmers' Fields: Assessing Productivity and Health Effects in China. *Science* 308(5722), 2005, 688 – 690.

[4] Goodburn, K. *EU Food Law: A Practical Guide.* Woodhead Publishing; 2001.

[5] Mullis, K. *Dancing Naked in the Mind Field.* New York: Pantheon Books; 1998.

[6] Rabinow, P. *Making PCR: A Story of Biotechnology.* Chicago: University of Chicago Press; 1996.

Chapter XIII

VENOMS

Toxinology is the branch of toxicology that studies the occurance, toxicity and poisoning treatment related to venoms. A broader definition would explain toxinology as a scientific discipline dealing with microbial, plant and animal venoms, poisons and toxins [1]. This scientific discipline encompasses the chemistry and the mode of toxic action, the biology of venom-producing organisms, as well as the structure and function of the venom apparatus.

Venoms are group of naturally occurring substances produced by living organisms and characterized by very rapid toxic action when introduced into mammalian organisms. This class of poisonous substances evolved in nature to provide a defensive mechanism that protects an animal or plant by very rapid action. It is assumed that the first venoms appeared about 60 millions years ago and since then the molecular structure of venoms has been changed and evolved simultaneously with the evolution of animals as they adjusted to changes in their surrounding.

Even though venoms were not scientifically investigated thoroughly until the end of the 19th century, they were commonly applied in homeopathic remedies. Later, modern medicine accepted and used them as components of medications and as therapeutic or diagnostic agents. Today the application of venoms in medicine occurs mostly in blood disorders and as analgetic agents in cases of severe, inbearable pain. More recently the toxycodynamics of venom demonstrated that they may be used possible models for designing new drugs.

Venoms may be classified according to the animal that produces them, according to the mechanism of their action and on the basis of their structure. To evaluate the toxic potency of a certain venom, is very important to consider the quantity of produced venom, the effectivness of its toxic mechanism and the existence of an effective antidote.

NEUROTOXIC EFFECTS OF VENOMS

The toxicity of venoms can arise by diferent mechanisms. Adverse effects on the nervous system have been shown to be common for great number of venoms originating from different animal species, including snakes, frogs, fish and snails, as well as for many bacterial toxins and alkaloids. In nature, neurotoxins play a key role in prey immobilization by provoking paralysis, disorientation or depressed respiration. Venoms often contain multiple neurotoxins, which act synergistically to impair the nervous system.

Table 13.1. Different modes of neurotoxic action for venoms

Mode of action	Species	Venom
Sodium channel blockers	Spiders	Robustoxin
		Versutatatoxin
		μ-neurotoxin
Potassium channel blockers	Scorpions	Charibdotoxin
		Noxiustoxin
	Snakes	Dendrotoxins
	Poison arrow frogs	Alkaloidal toxins
Calcium channel blockers	Spiders	ω-neurotoxin
		Polyamines
	Cone snails	ω-conotoxins
Sodium channel prolongers	Cone snails	King-kong peptides
		δ-conotoxin
	Pufferfish	Tetrodotoxin
	Poison arrow frogs	Tetrodotoxin
	Blue-ringed octopus	Tetrodotoxin
Acetylcholine vesicle mobilization	Spiders	Latrotoxin
Nicotinic receptor antagonist	Cone snails	α-conotoxin
	Snakes	α-conotoxin
Nicotinic receptor agonist	Poison arrow frogs	Epibatadine

Neurotoxins act by disturbing the transmission of nerve impulses through synaptic clefts. The cytoplasm of the nerve cells is negatively charged, while its external, extracellular portion is positively charged. Transmission of nerve impulses results from the depolarization of cell membranes and the release of neurotransmitters from their vesicles. Ion channels present on the nerve cells have an important role in conducting nerve impulses, and many neurotoxins act by targeting ion channels. Sodium ion channels are present both presynaptically and postsynaptically, while calcium channels are present only presynaptically. The role of sodium channels in conducting the nerve impulses is to influx sodium ions into the nerve cell. Sodium influx depolarizes the nerve cells and provokes the release of neurotransmitters. Sodium channels are known to be tetrodotoxin- and saxitoxin-sensitive. Even though calcium channels are not directly involved in the conducting nerve impulses, they prolong the depolarization of the cells via the inward movement of calcium ions.

Neurotoxic agents usually interfere with the release of acetylcholine and provoke paralysis. They also act by increasing the amount of exocytitically released acetylcholine, causing muscle cramps. The toxins of *Clostridium* bacteria are representative of neurotoxins that interfere with the release of acetylcholine and are similar to botulinum toxin, which causes paralysis through specifically targeting cholinergic motor nerve endings. Neurotoxins of tetanus bacteria, on the other hand, cause spastic paralysis by selectively targeting spinal neurons and increasing the release of acetylcholine.

In the synapses, acetylcholine may bind to nicotinic or muscarinic receptors. Nicotinic acetylcholine receptors occur primarily at neuromuscular junctions, while muscarinic acetylcholine receptors occur primarily in the central nervous system. The two receptors also differ functionally: nicotinic receptors are ligand-gated ion channels, while muscarinic receptors belong to the class of G-protein-coupled receptors. Binding the two molecules of acetylcholine to nicotinic acetylcholine receptor causes conformational changes of the ion channel, resulting in its opening and permeability to sodium and calcium ions. Ion influx further causes cell depolarization and muscular contraction. Neurotoxins targeting nicotinic receptors reversibly bind to the receptors increasing the cations inflow. In venoms that attack muscarinic receptors, the venoms activate the intracellular enzymes and initiate an intracellular cascade leading to an increase in calcium ions [2]. The difference between these two types of venom is that venoms bound reversibly to nicotinic receptors prevent the binding of acetylcholine, while venoms bound irreversibly to muscarinic receptors continually stimulate the receptor [3]. Different modes of neurotoxic action of venoms produced by wide variety of animals are illustrated in Table 13.1.

Snake and Lizard Venoms

Venomous snakes are found throughout the world and pose a significant mortality and morbidity risks. Most snake venoms act neurotoxically; however, cardiotoxic effects, inhibiting or promoting coagulation, and nephrotoxic effects are not uncommon. Some venoms do not act systematically, but instead, act necrotoxically to cause local tissue damage.

Important factors in discussing the toxicity of snake venoms include both commonly adopted measures of acute toxicity, such as LD_{50}, and the amounts of produced and injected toxic substance, which may vary significantly (30 fold) in different snake species. Another consideration is the most appropriate route of administration to experimental animals in dosing procedure, which is subcutaneous as that most closely simulates the snake bite. Snake venoms taken orally are readily inactivated by enzymes of the digestive system and do not provoke toxic effects. Data on the acute toxicity of all known snake venoms is summarized in Appendix VI.

The two principal mechanisms of toxicity in snake venom are related to to neurotoxic activity and to the disturbance of normal coagulation (clotting) processes .

The clotting process is a very complex event and requires the pre-activation of many co-factors in order for the coagulation cascade to proceed. Most snake venoms influence the activity of coagulation co-factors. Some snake venoms, for example that of *Daboia russelli*, activate the coagulation factors V, X, IX, as well as protein C, and may be used as diagnostic agents for the determination of clotting factor X deficiency observed in individuals with impaired coagulation ability. Due to its ability to induce coagulation, the venom of *Daboia russelli* in the past was used to treat hemophilia.

An important group of snake venoms, with respect to their coagulant activity, are prothrombin activators, which initiate the conversion of prothrombin to thrombin. Prothrombin activators can be divided into four groups. Group I activators convert prothrombin to meizothrombin, which is insensitive to coagulation cofactors. Group II and III prothrombin activators cleave the peptide bonds in prothrombin and initiate the conversion of prothrombin to thrombin. Group IV activators are proteases which transform prothrombin into a non-active precursor form of thrombin [4]. The diagnostic usefulness of prothrombin activators lies in their ability to act independently on any co-factor in the coagulation process, and they can be used to measure the content of prothrombin in the body. Group II prothrombin activators may also be applied in diagnostic application to determine the blood level content of factor V, which is necessary for the conversion of prothrombin to thrombin [5]. Since factor V is a necessary co-factor in the conversion of prothrombin to thrombin, the amount of generated thrombin by adding group II prothrombin activator is directly proportional to factor V levels and can be used for to accurately determine factor V levels in the blood.

Some thrombin-like venoms influence levels of plasma fibrin by fibrinogen conversion event in the presence of heparin. This allows the monitoring levels of plasma fibrinogen even if the patient is undergoing heparin therapy. The venom originating from Central American pit viper (*Bothrops moojeni*) shows this effect and may be used not only as a diagnostic agent but also as a part of the therapy.

(Author André Karwath; License http://creativecommons.org/licenses/by-sa/2.5).

Figure 13.1. The venomous bearded dragon (*Pogona vitticeps*).

Besides numerous species of poisonous snakes, other reptiles include a few venomous lizard species, such as Mexican Beaded Lizard (*Heloderma horridum*), Gila Monster (*Heloderma suspectum*), Monitor Lizards (family Varanidae) and Iguanas (family Iguanidae). Studies of the effects of Lace Monitor Lizard (*Varanus varius*) venom show that the active substance in its venom has a dramatic effect on the victim's blood pressure and clotting, and that it provokes internal bleeding, revealing that its toxicity arises by the disrupton of coagulation processes. It is believed that only nine types of venoms are produced by venomous lizards. The saliva of the Commodore Lizard (*Varanus komodoensis*), besides containing venom, poses a health risk because it contains more than 50 different bacteria.

What distinguishes venomous systems in snakes from those of lizards is that venom producing glands are located in the upper jaw in snakes and in the lower jaw in lizards. The lizard-like bearded dragon (*Pogona vitticeps*) (Figure 13.1.) has venom systems in both upper and lower jaws. This lizard secretes small quantities of neurotoxic venom into the saliva.

FROG VENOMS

Deadly frog toxins were well known in the past, and were used by Amazonian tribal peoples to tip their arrows in order to improve the efficiency of hunting.

(Used with permission of prof. Joseph T. Collins).

Figure 13.2. Arrow-poison frog (genus *Dendrobates*).

True poison arrow frogs (also called arrow poison frogs) are those of the Dendrobatid family (Figure 13.2.), which includes the genera *Dendrobates, Epipedobates, Minyobates* and *Phyllobates*. However, some poison arrow frogs have developed a very rare adaptive feature whereby they mimic the color shemes of other poison frogs. The production of venom in frogs is highly dependent on the animal's diet and as toxic substances, they can vary significantly in type. Some frogs produce toxic alkaloids after ingesting insects and the alkoloids are mobilized into glands that are situated under their skin. Venoms are released from the glands when the frog is in threat. The potency of some frog venoms, such as batrachotoxin, produced by *Phyllobates terribilis*, is 250 times greater than strychnine. It is not uncommone for frog offspring withdrawn from their natural environment not to produce th toxins.

Most frog toxins are alkaloid in structure; however, many other classes of compounds also have been isolated from the skin of frogs. The diversity of poison arrow frog alkaloids ranges from small and simple decahydroquinoline to large and extremely complex batrachotoxin. The most important frog venoms are batrachotoxin, decahydroquinolines, histrionicotoxins, indolizidines, pumiliotoxins, pyrrolizidines, quinolizidines and tetrodotoxin (Figure 13.3). Pumiliotoxins are widespread alkaloids found in a number of species, and they act by increasing the rate of sodium channel opening and closing. Batrachotoxin is a lipid-soluble neurotoxin that binds to sodium channels. It is among the most powerful animal venoms known. While sodium channels are commonly targeted, frog toxins also may affect potassium channels. *Dendrobates histrionicus* (the Colombian arrow poison frog) produces two alkaloids unique to this species: gephyrotoxin and histrionicotoxin. Gephyrotoxin is a tricyclic alkaloid that, like histrionicotoxin, blocks the passage of potassium ions. Epibatidine, a toxin isolated from *Epipedobates tricolor* (Figure 13.3.), has 120 times greater analgetic potency compared to nicotine, and it is of tremendous interest as a potential drug. The isolation of toxins from frog skin for research and medical purposes is not always succesful; and sometimes frogs lose their ability to produce the toxins in captivity.

The salamander (Figure 13.4.) is an amphibian that is about 20 cm long, but some giant species found in Japan and China may reach up to 1.8 m in length and 30 kg in weight. Some species of salamander possess the ability to produce and secrete venoms through their skin glands like other amphibians. These animals secret a defensive poisonous liquid which contains salamandrin and other steroidal alkaloids. Salamandrin is a strong neurotoxin and the poisoning with salamandrin is usually accompanied by convulsions.

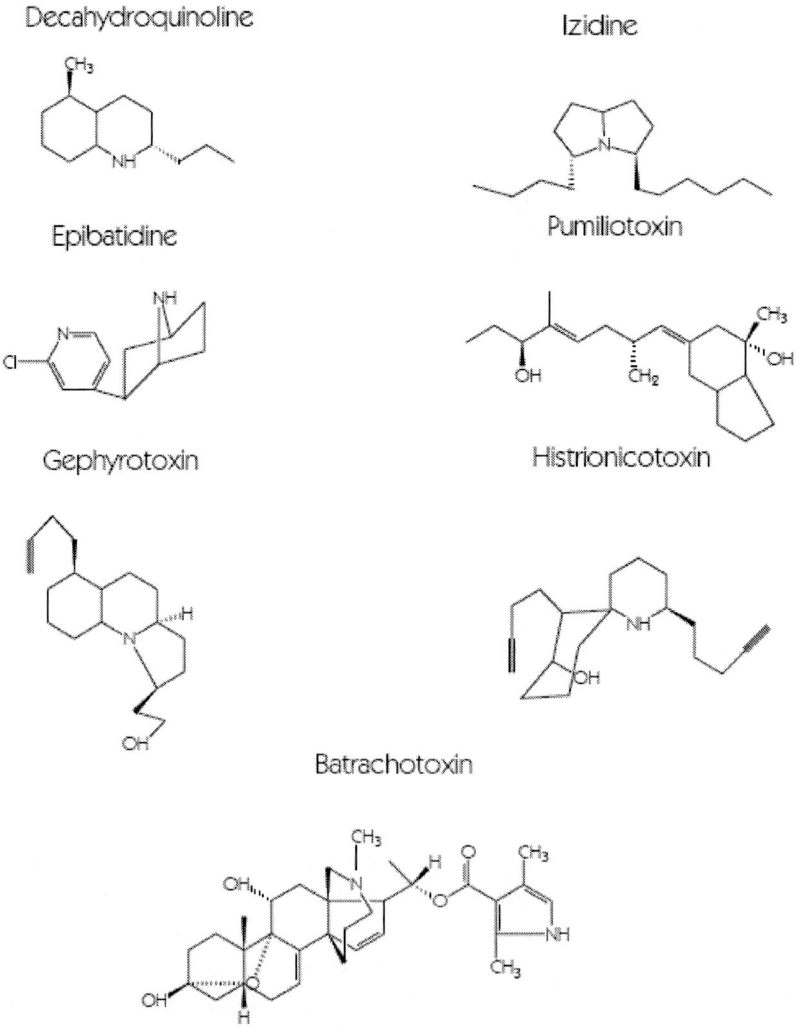

Figure 13.3. The structure of common poison arrow frog venoms.

Figure 13.4. Fire salamander (*Salamandra salamandra*).

VENOMS OF MARINE ORGANISMS

Numerous marine organisms are characterized not only by the wide variety of their colors, shapes and species, but also by the high number of venomous animals that produce great number of chemically diverse toxins. Some species of corals, jellyfish, snails, octopus and fish are known for their dangerous and sometimes life-threatening venoms. Other marine organisms may also pose more or less serious risks of envenomation. The starfish, for example, is covered with thin venomous skin that may cause swelling, nausea and vomiting in direct contact with a wound. In a variety of possible toxic substances, the complexity of the venom structure is independent on the developmental level of the organism. In addition to having a great variety of chemical structures, venoms of different marine organisms differ in the mechanisms of their toxic activity and the manifestations of envenomation.

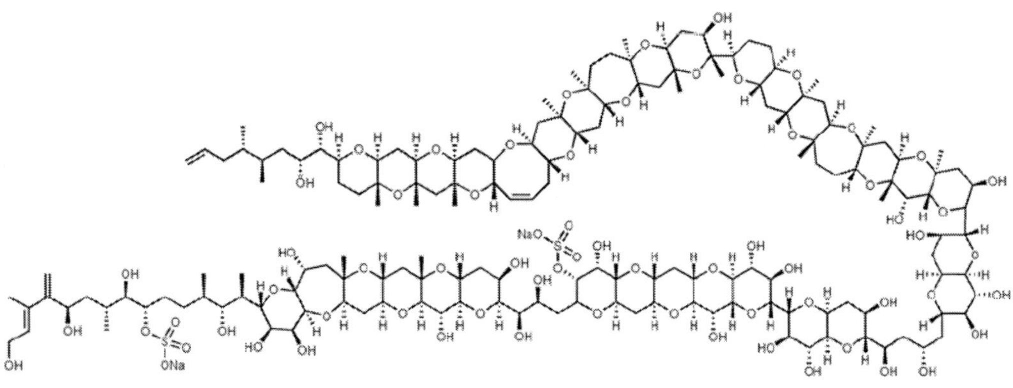

a.

Figure 13.5. (Continued).

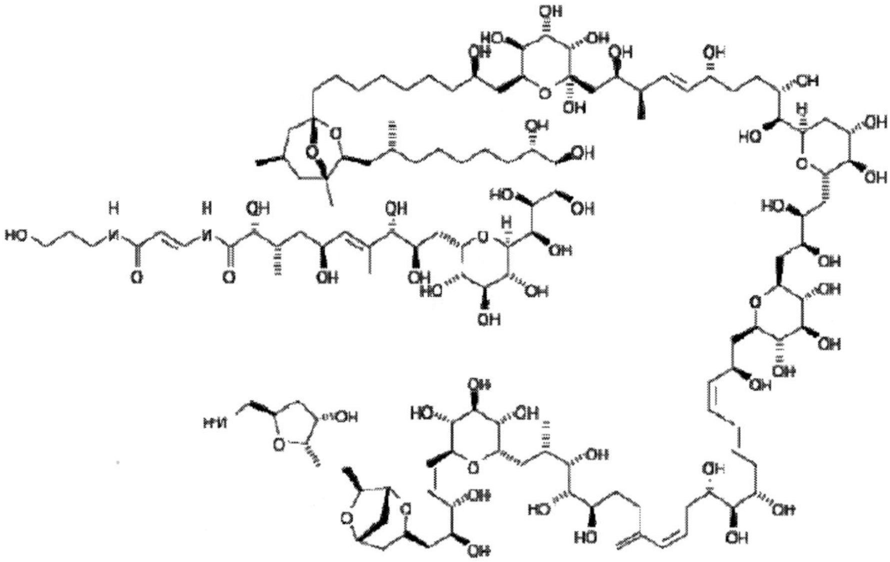

b.

(Courtesy of authors (a) Azulene and (b) Auroranorth).

Figure 13.5. The chemical structure of non-peptide toxins (a) maitotoxin from algae (*Gambierdicus toxicus*) and (b) palytoxin produced by soft corals (Order Zoantharia).

Some unicellular algae (e.g., *Gambierdicus toxicus*) produce extremely potent toxin, called maitotoxin, which is not toxic for fish feeding on algae, but represents potential indirect risk for man because it may be transmitted in fish consumed as food. High acute toxicity of maitotoxin is demonstrated by its lethal does in mice being 0.13 µg/kg in intraperitoneal injection. The mechanism of maitotoxin toxicity includes first, the activation of non-selective calcium channels, leading to increased levels of cytosolic Ca^{2+} ions. Then, an uncontrolled influx of calcium ions ultimately provokes cell lysis. Together with palytoxin, maitotoxin represents structurally the most complex non-protein toxic substance produced by a marine organism (Figure 13.5.).

Palytoxin, a substance with a very complex structure, was first isolated from soft corals. For a long time, all efforts to synthetize palytoxin were unsuccessful due to the complexity of its structure, but in 1994 a research group with Yoshito Kishi managed to synthetize the substance, which is considered by many to be one of the greatest synthetic accomplishments. Palytoxin acts by binding to sodium-potassium pump proteins, allowing passive transport of both sodium and potassium ions. Typical symptoms of palytoxin poisoning are chest pain, breathing difficulties, tachycardia, unstable blood pressure and hemolysis. The symptoms develop rapidly and end with lethal outcome. Vasodilatators, such as papaverine and isosorbide nitrate, may be used as antidotes in palytoxin poisoning. Antidotes are effective only if they are injected directly into the heart immediately following exposure.

Cone snail (Figure 13.6.) venoms are composed of small peptides, 12-30 aminoacids long, which are highly constrained due to their dense disulphide bonds. Venoms produced by cone snails are known as conotoxins and are among the most potent and diverse neurotoxins known. Three main classes of paralytic conotoxins have been identified: α-conotoxins, µ-conotoxins and ω-conotoxins. The toxicity of α-conotoxins arises from binding to nicotinic acetylcholine receptors, causing their inhibition. μ-conotoxins act by direct inactivation of muscle action by binding to post-synaptic sodium channels. ω-conotoxins are dangerous because they prevent the release of acetylcholine in the synapses by blocking the calcium entry.

(Courtesy of author Kerry Matz; National Institute of General Medical Services).

Figure 13.6. The cone snail (*Conus geographicus*).

Cone snail venoms are produced in a long tubular duct that is often several times the length of the snail itself. The composition of the venom, as well as pharmacological activity and caused symptoms, are different with each injection. Symptoms vary from weakness to loss of coordination. In most stings, the sting area swells, after which blisters and scars usually develop. The feeling of numbness and local pain arises first and the victim's vision, speech and hearing become disturbed in later stages. In cases of severe envenomation, paralysis of the respiratory muscles may lead to death. There is no antivenom/antidote developed for cone snail stings, but tetanus prophylaxis is useful in preventing infection in the wound.

Jellyfish are equipped with a specialized venom apparatus intended for defense and feeding. A capsule inside the venom apparatus contains a stinging structure. The stinging structure varies according to species, but generally consists of a hollow coiled thread with barbs lining its length. Venomous capsules are activated by contact with an object. Pressure within the capsule forces the stinging thread to rapidly uncoil and to inject the paralyzing toxin. Stings can paralyze or kill only small marine organisms, such as fish and small crustaceans. The severity of these stings in humans depends on the jellyfish species, the penetrating power of the venom apparatus, the thickness of the exposed targeted skin and the individual's sensitivity. *Chironex fleckeri*, the box jellyfish, is the most lethal jellyfish and its toxin causes rapid depression of the cardiac and respiratory systems. The chemical structure of the venom is still unidentified, but it is speculated that they are large proteins, as is the case with the most venoms produced by jellyfishes. Envenomation by Irukandji jellyfish (*Carukia barnesi*), the small box jellyfish, causes a specific syndrome which includes delayed pain from severe muscle cramping. Irukandji syndrome encompasses an array of systemic symptoms, including severe headache, chest and abdominal pain, sweating, nausea and vomiting. Anxiety, hypertension and pulmonary edema are also commonly displayed. Symptoms generally abate in 4 to 30 hours, but complete resolution may take up to two weeks. Many of symptoms occur due to impairment of the sodium channels function. In an overall picture of the syndrome, catecholamines probably participate to some extent, but the real cause of heart failure remains undefined. In Irukandji envenomation, antihistamines may be beneficial for pain relief; however, most cases require intravenous analgesia with morphine or other strong analgetics. *Chiropsalmus* sp. and numerous Carybdeids produce less potent venoms in comparison to box jellyfish, and the symptoms of envenomation are much less severe. Moon jelly fish (*Aurelia marginalis*) are known to be slightly venomous. Contact with them can produce symptoms ranging from immediate prickly sensations to mild burning. The sting of the *Lion's Mane* jellyfish (*Cyanea capillata*) provokes pain that is usually restricted to the contact area. Its symptoms are similar to those of the moon jelly, but are usually more intense. Pain is often described as burning rather than stinging. The man-of-war or bluebottle jellyfish (*Physalia physalis*) can inflict extremely painful stings whose symptoms include severe shooting pain and intense joint and muscle pain. The pain may be accompanied first with headaches, fever, nausea and vomiting, and may proceed with hysteria, chills and faintness. The sting of a bluebottle jellyfish is especially dangerous to children, elderly people, asthmatics and people with allergies, and it can cause respiratory failure.

Sea cucumbers possess a specialized defense system, called Cuvierian organs, which produce and store poisonous substances. The system is mobilized when the animal is mechanically stimulated, resulting in the discharge of the system`s tubules. An active substance, holothurin, was isolated from the Cuvierian defense organ, and it was observed that besides demonstrating the toxic effects the substance also expresses the beneficial ones. The substance exhibits an antimicrobial activity and specific biochemical potency that may be used in tumor prevention. The holothurin toxin was succesfully used to treat skin fungi and to produce a general stimulating effect. The compound was also used for the production of drugs intended to regulate cardiac activity and to improve the metabolism. The toxicity of holothurin arises from its disturbance of nerve impulse transmittion. If the content of Cuvierian tubules were put directly into the eye, it is possible that its active substances might cause permanent blindness. In addition to holothurin, other useful glycosides have been isolated from the venom apparatus of sea cucumbers.

The blue-ringed octopus is a very small, venomous octopus, reaching up to 100 g in weight and 20 cm across spread tentacles. It is normally yellowish brown, but when disturbed, the octopus changes its appearance and reveals its bright blue rings. The venom of the blue-ringed octopus is excreted in its saliva, and is produced in a two glands about same the size as its brain. The venom has two components. One of them is probably most effective on crabs and relatively harmless to humans. The other is very potent—tetrodotoxin, which is identical to the toxin produced by the puffer fish. Tetrodotoxin probably serves the octopus as a defense against predatory fish and acts by blocking sodium channels and provoking motor paralysis and respiratory failure. The most common cause of death after envenomation by blue-ringed octopus is respiratory failure.

In addition to the venomous terrestrial snakes described above, some other sea snakes, which are in fact close relatives of the cobra, can also produce venoms. Their venom is injected into the victim in small concentrations, and imobilize the prey. After a painless bite, symptoms ressembling stiffness, muscle aches and spasms of the jaw soon develop. Powerful neurotoxins in the venom of sea snakes cause blurred vision, drowsiness and respiratory paralysis. The bite may be treated with sea snake antidote or polyvalent antidote.

Fish Venoms

Many venomous marine fishes belong to the colorful family *Scorpaenidae* and are used in aquarium exhibits. Within the family, the genera *Pterois*, *Parapterois*, *Brachypterois*, *Ebosia* and *Dendrochirus* have many representatives that may pose more or less serious health threats and are commonly known as Lionfish, Dragon Fish, Turkey Fish and by other names. Lionfish (Figure 13.7.) have a venomous spine that is deadly to its prey, but usually not to humans. Envenomation is accompanied with severe pain, possible strong headaches and vomiting. In order to denaturate the venom the afflicted area should be soaked with hot water.

The stone fish (*Synanceia verrucosa)* (Figure 13.8.) is one of the the most dangerous venomous fishes. Its venom consists of a mixture of proteins, including hemolytic stonustoxin, neurotoxic trachynilysin and cardioactive cardioleputin. If the venom penetrates into deeper layers of the skin, local tissue necrosis may be observed at the injury site. Victims suffer localized nerve damage leading to atrophy. Shock brought on by the poisoning may develop into paralysis. Immediate first aid treatment requires immobilization of the venom at the penetration site by firm constrictive bandaging and, since all of the venom's constituents are proteins, heat should be applied at the injury site to denaturate the toxins. Some speculation exists concerning potential beneficial effects of stone fish venom on arthritic symptoms, such as improved mobility and reduction in joint pain. However the exact compound responsible for such benefits has not yet been identified.

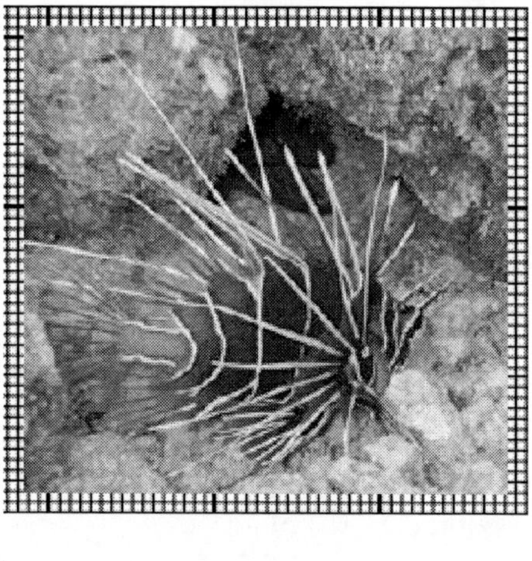

(Courtesy of author Jan Derk).

Figure 13.7. Clearfin Lionfish (*Pterois Radiata*).

(available at http://commons.wikimedia.org/wiki/Image:Stone_Fish_at_AQWA_SMC2006.jpg; License ttp://creativecommons.org/licenses/by/2.5).

Figure 13.8. The stone fish (*Synanceia verrucosa*).

Figure 13.9. The boxfish.

Figure 13.10. The pufferfish.

Boxfishes (Figure 13.9.) are closely related to pufferfishes (Figure 13.10.) and filefishes, and are notable for the hexagonal shape of their skeletons. They are also known as cofferfishes, cowfishes or trunkfishes, and are commonly used in aquarium exhibits even though they are very poisonous. In their mucus, boxfishes produce a very potent hemolytic toxin called ostracitoxin, which is secreted when the fish is threatened. The venom is potent enough to kill other marine organisms, including other boxfishes. The secreted toxin is unique among known fish poisons and is somewhat similar to the venom of sea cucumber. Ostracitoxin is not a protein poison and therefore it cannot be denaturated or inactivated by heating.

"Fugu" is the Japanese name for pufferfishes. Nearly all fugu fishes are poisonous and after the venom is produced, it can be localized in various parts of the fish, such as skin, liver, ovaries or elsewhere. The toxic component of fugu venom is a deadly nerve poison, tetrodotoxin. Tetrodotoxin produced in fish organisms and other animals, such as salamander or frog, is structurally identical and its production is attributed to the activity of bacteria that live symbiotically in the bodies of the animals.

There are about 35 species in the stingray family, and some of them are armed with a saw-edged envenomed spine. The sting of these fishes injects venom that consists of a mixture of various proteins. Active venom compounds cause profuse bleeding and intense pain at the site of injury. Usually large swelling develops around the sting area. The pain and unpleasant sensation may last for months but is most severe in first 30–60 minutes after the sting. Other manifestations of envenomation by stingrays include nausea, fatigue, headaches and fever. Fatal stings of stingrays are very rare; however, lethal outcomes after envenomation with stingray venom have been reported.

Spider and Scorpion Venoms

Most spiders produce venom, but the quantity of the synthetized active substances and their toxic potency vary from species to species. Spiders inject venom into their victims through hollow fangs. Spider venoms are intended to paralyze the victim and to aid in their digestion. Most spiders are not dangerous to humans because their fangs are too fragile and too short to penetrate into human skin; nevertheless. Human envenomation by certain spider species may cause serious consequences.

Figure 13.11. Black widow female spider (*Latrodectus mactans*).

Latrodectus family includes about six species of spiders living around the world and all of them are venomous; however, their bites are rarely fatal to humans. The female

Latrodectus (Figure 13.11.), also called the red-back spider or black widow, is about four times larger than the male and much more dangerous. After mating, the female spider kills and eats the male. A black widow spider bite gives the appearance of a target, with a pale area surrounded by a red ring. If the bite is not treated immediately, systemic symptoms of poisoning may develop. In first 2 hours, severe muscle pain and cramps may be experienced. Strong pain in the muscles is caused by a potent neurotoxin. Cramps are usually felt first in the back, shoulders, abdomen and thighs, and are accompanied by difficult breathing and increased blood pressure. Other symptoms include weakness, sweating, headache, anxiety, itching, nausea and vomiting. Young children, the elderly and those with high blood pressure are at the highest risk of developing serious symptoms from a black widow spider bite. Poisoning with black widow venom should be treated with a specific antivenom, and full recovery usually occurs within 2-3 month. Without treatment, lethal outcome is probable and will usually occur within 1-2 days.

The venom of the violin-shaped brown recluse spider (*Loxosceles reclusa*), commonly found under stones or in dark corners inside buildings, may cause severe mechanical damage in the form of necrotic lesions due to the destruction of blood vessel walls; however, fatal bites of the brown recluse spider are rare. This spider's bite usually causes pain or burning sensation in first 10 minutes, accompanied by itching, and leaving a painful ulcer. A generalized red, itchy rash usually appears in the first 24-48 hours. Other symptoms include fever, chills, nausea, vomiting, muscle aches and hemolysis. Treatment of envenomation by brown recluse spider should include washing the wound and applying an antibiotic in order to prevent infection.

Other species of spiders, like the running spider, jumping spider, wolf spider, tarantula, sac spider, orbweaver spider and the hobo spider, are venom-producing species and their bites may be an unpleasant experience. With the bites of most of these spiders, pain and burning sensations at the bite site are felt within first 10 minutes, and the bite sites will have a common "target" appearance. The center of the wound is usually a blister surrounded by a reddened area. A pale or blanched area may surround the discolored, reddened area. The blister may rupture, leaving an open ulcer. In severe cases the ulcer may become deep, infected and necrotic. 24-72 hours after envenomation, it is not uncommon to develop a red, itchy rash over the torso, arms and legs. Human bite victims may experience pain in the muscles and joints, fever, chills, swollen lymph nodes, headaches, nausea and vomiting. Symptoms of a jumping spider bite usually last about 1-4 days. The bite of large, hairy spiders, known as wolf spiders, in humans causes swollen lymph glands and blackening of the skin area at the bite site as well as swelling and pain, which may last up to ten days. Tarantulas are also large, hairy spiders. Handling a tarantula may result in irritation to the skin. Most tarantula bites do not produce any significant poisoning symptoms. However, the bites may be quite painful because of the large size of the spider. The hobo spider (*Tegenaria agretis*) often causes a bite that leaves an open, slow-healing wound. and bites from this spider are often mistakenly thought to be those of the brown recluse spider bites.

Scorpions use their sting for a variety of purposes, such as to capture prey, in defense and during mating. All species of scorpions are poisonous to their prey, which are usually insects. A very small number of more than 1,050 known scorpion species can be dangerous to humans. The scorpion's venom comprises of a variety of compounds, most of which have not yet been identified. The venom from a single scorpion may include several neurotoxins, histamine, serotonin, enzymes, enzyme inhibitors, and other unidentified compounds. Each neurotoxin is believed to target a specific kind of animal. After the scorpion stings, immediate local reaction with intense pain is displayed, followed by numbness, tenderness, and tingling at the site of the sting. In the next stages, the pain radiates towards the body, and anxiety and increased body temperature are experienced. Swelling of the face, tongue and throat, pain or tightness in the chest or back also occur and are attributed in part to histamine. Some scorpion species are capable of injecting relatively large quantities of highly toxic venoms. In these cases, fever, excessive salivation, nausea and vomiting, may occur along with confusion, convulsions and coma. Increased or decreased heart rate and pulmonary edema may also develop. A lethal outcome may be expected due to heart or respiratory failure. The lethality of scorpion envenomation depends on many factors, but is mostly determined by the acute toxicity of the specific scorpion venom and quantity injected.

INSECT VENOMS

The medical effects of ant poisons have been known to humans since ancient times and have been used for various purposes. Two thousand years ago, ants were used in the treatment of skin diseases and African tribes treated neurological diseases by burying patients in a formicary (anthill). Local doctors in New Guinea, after performing a light surgical operation, used bulldog-ants for wound treatment instead of sewing it. Ant venoms possess potent antimicrobial activity against bacteria like streptococci and staphylococci, as well as against many pathogenic fungi. The natural purpose of antiseptic chemicals in ant venoms is to protect underground ant nests, eggs, larvae and stored food supplies from infection. Some ant species produce poisonous fluid containing formic acid and alkaloids which ants can spray a distance of up to 10 cm. Due to toxic components of ant venom, envenomation may be very serious and even fatal in individual cases. Symptoms of envenomation include rash, facial edema and wheezing. In sensitive or allergic persons, serious systemic reactions such as shock, apnea and cardiac arrest may be life threatening.

Bombardier beetles are historically linked to Napoleon`s crossing of the Pyrenees. Soldiers were injured by bombardier-beetle attacks, which left the permanent spots on their bodies. The bombadier beetle has two glands that secrete poisinous substances. The active compound in the venom is apitoxin, which is a colorless, clear, bitter liquid with a pleasant smell. The physiological effects of apitoxin were known in the past and used for the treatment of rheumatism. The produced venom is stored in specialized chambers. When the beetle is in danger, the liquid venom is expelled from the insect's anus up to a distance of 30 cm. Such an explosive release occurs because the temperature in the venom chamber reaches 100° C due to specific chemical reactions. The symptoms of poisoning with bombadier-beetle venom are highly dependent on the location of the skin that was exposed to the venom. Envenomation may cause simply a sharp localized pain, swelling and reddening, but entry of the poison into blood vessels of the face or the neck provokes more severe symptoms, including vomiting, abdominal pain and tachicardia.

In the past, *Scolopendra* (centipede) poison was highly valued in Chinese folk medicine as a cure for rheumatism, kidney stones and some skin diseases and for the treatment of severe scarring. The poison of scolopendra (Figure 13.12.) is a complex mixture, containing many different substances, including acethycholine, histamine, serotonin,and other substances. Their poisoned bite is always accompanied by a local sharp pain, severe swelling, weakness and fever. Female scolopendra are more dangerous and poisonous than males. The animal secretes venom and causes painful biting only when it senses danger. The victim's skin may become red and swolen due to inflammation, even where the scolopendra has only crawled on the surface. .

The Spanish fly (Figure 13.13.) is a beetle with an unpleasant smell. When it feels threatened, it secretes a liquid that causes reddening and rash on the skin. The substance cantharidin, identified as an active component of this venom, was isolated from the beetle's secretion. The lethal dose of cantharidin for humans is about 5 μg/kg. High doses of cantharidin cause excessive salivation, stomach, kidneys and urino-genital system inflammation, as well as headache, vomiting and bloody diarrhea. After ingestion, distribution throughout the body and biotransformation, cantharidin is excreted in the urine. During urination the excreted metabolites cause irritation of the urogenital tract and leads to itching and swelling of the genitals. In the past, the swelling of genital area was mistaken for sexual arousal and gave rise to the erroneous belief that beetle venom had aphrodisiacal qualities. In reality, the swelling of genitals is a result of serious inflammation and can be very painful and unpleasant. The kidneys also suffer severe inflammation and may be permanently damaged. Orally taken, the highly toxic cantharidin can result in severe gastrointestinal disturbance, convulsions and coma. In cases of serious poisoning, coma may end in death within 24 hours.

The active compound of Spanish fly secretion has had many uses. Cantharidin tincture was used to stimulate the hair growth and to heal rheumatism, pneumonia, swelling and gout. The substance also was used to stimulate sexual desire. Misuse and inadequately determined doses often led to severe poisoning. The Marquis de Sade, known for his unconventional sexual habits and behaviour, is remembered also for the mass poisoning of his guests, who were offered sweet containing powdered Spanish fly. For this, de Sade finally was sent to prison for life.

Figure 13.12. Peruvian giant yellowleg centipede (*Scolopendra gigantea*).

(Author Franco Christophe; available at http://en.wikipedia.org/wiki/Image:Lytta-vesicatoria.jpg; License http://creativecommons.org/licenses/by-sa/2.5).

Figure 13.13. The spanish fly.

A honey bee that is away from the hive rarely stings, but honey bees may collectively sting if they perceive that their hive is threatened. By injecting the venom into the victim, they also release pheromones that enable communication with other bees and alert them to attack en masse. The active portion of bee venom is a complex mixture of proteins that causes local inflammation and acts as an anticoagulant. One of the strongest active components of honey bee venom, apitoxin, is a bitter, colorless liquid and is also a component of venoms produced by other insects. Apitoxin is similar to certain snake venoms and can be deactivated with ethanol. A honey bee can inject 0.1 mg of venom via its stinger. and it is estimated that 1% of the population is allergic to bee stings.

In a wasp sting, small amounts of neurotoxin are transmitted into the victim's skin. The manifestations of a wasp sting are similar to those of a bee sting and are usually mild. Potential life-threatening risk arises only due to possible allergic reactions. In cases where a sting causes anaphylactic shock, corticosteroids should be administrated intravenously and adrenaline subcutaneously.

Termites do not pose a serious risk for humans. Their secretion contains benzochinon and various triens, which in reaction with atmospheric oxygen, thicken and harden. This poisonous liquid is poured over the victim by the contraction of the abdominal muscles, and hinders the victim's escape. If an entire termite colony is in danger, all of the termites will release their liquid at the same time by pheromone communication.

MAMMALIAN VENOMS

The platypus (*Ornnithorhynchus anatinus*) (Figure 13.14.) is a small, semiaquatic mammal that lives in freshwater basins along Eastern Australia and Tasmania. Mature males have a pair of venomous spurs on the inside of their back legs and use them in sexual combat. Their venom is composed of various enzymes that have a paralytic effect on the affected body area. The venom is produced in endothelial cells, and it is believed that it helps regulate the animal's blood flow by vasodilatation effects. About 19 different peptides have been identified in platypus venom. Contact with this secreted venom causes intense pain and local tissue damage in humans. Envenomation is accompanied with swelling due to release of histamine from the victim's mastocytes. The active compounds in this venom have been shown to influence cellular processes by altering the membrane potential and targeting primarily calcium channels. After being poisoned, the victims are usually unable to move their limbs for days. There is no specific treatment for platypus envenomation; however, analgetics and tetanus prophylaxis are recommended [6].

(Author Tom Mc Hugh, available at http://en.wikipedia.org/wiki/Image:Ornithorhynchidae-00.jpg; License http://creativecommons.org/licenses/by-sa/2.5).

Figure 13.14. The platypus (*Ornithorhynchus anatinus*).

(Author Gilles Gonthier; License http://creativecommons.org/licenses/by/2.0/).

Figure 13.15. Blarina (*Blarina brevicauda*), a small venomous mammal from the family Soricidae.

Blarina (*Blarina brevicauda*) are poisonous mammals that belong to family Soricidae (Figure 13.15.) and are related to mole and their bite can cause unpleasant symptoms. These animals have an underground, ground and amphibian way of life and feed on insects, spiders, worms, snails and the like. Poison produced by the animals contains a neurotoxin that is stored in their front teeth and injected into the victim in a bite. The venom serves to immobilize victims such as snakes and birds, which could harm the blarina. The poison causes paralysis and death to small mammals, but in humans it produces only weak local inflammation on the site of the bite. Another active component of the venom is protease, which causes vasodilatation, contraction of smooth muscles, increased vascular permeability and pain [7, 8].

Figure 13.16. Slow loris (*Nycticebus coucang*).

The slow loris (Figure 13.16.) is a venomous primate and a member of the family Lorisidae, which include lories, galagos and pottos. These animals produce toxins and excrete them in the glands located on the skin on the inside of the elbows. The toxin is used to cover their babies in order to protect them from predators. The active compound of slow loris venom is believed to be approximately a 18 kDa large protein, consisting of 70-90 amino acids.

Members of Mephitidae family, which include skunks and stink badgers, have chemical defense mechanisms that enable the ejection of venom when the animal is threatened. The noxious fluid is produced by glands located near the animal's anus. In dermal contact, the venom can cause painful skin irritation, and if it comes into the contact with eyes, it may cause temporary blindness. [9].

Numerous non-mammalian vertebrates have evolved lethal venoms to aid either in securing the prey or as a protection from predators., However, venom-producing apparatus in mammals is quite rare. In addition to the poisonous mammals briefly described above, a few small animals belonging to Solenodontidae family should also be mentioned. The Cuban solenodon (*Solenodon cubanus*) and the Haitian solenodon (*Solenodon paradoxus*) have dangerous bites. Their venom is delivered from modified salivary glands via hollow grooves. Even though the animals are able to cause serious injury in other small animals, solenodon bites are not fatal for humans.

Antidote Therapy

Serums for treatment of venom poisonings are produced by injecting the venom into the bloodstream of an animal, most often a horse, sheep or rabbit, and isolating the produced immunoglobulins.Rabbits, as small animals, are not very convenient for large-scale production of the immunoglobulins. The concentration of the injected venom is gradually increased in order to avoid a lethal outcome in the animal. In envenomated animals, the normal immunological response to a foreign compound triggers the production of the antibodies. The principal components of the venom antidote are IgG antibodies., These are antibodies that are separated from the other serum components and in which the antigen binding fragments (or Fab (Fragment antigen binding) fragments), which have reacted with the injected venom, have been released by digestion. For neurotoxic venoms and venoms disturbing the coagulation processes, antidotes produced in such manner are less effective. Also, antivenom serums produced in this way may pose a serious risk to humans by provoking systematic allergic reactions, i.e., anaphylaxis, delayed allergic response and the risk of transmitting infectious diseases. Future perspectives in impoving the quality of produced serums include the broadening of a serum`s specificity to more than one venom and increasing the affinity to venoms. There are some indications that the production of high-affinity human IgG antivenoms will soon be made possible by *in vitro* techniques that rely on human cell cultures. The production of stable, synthetic antivenoms that can be taken orally and are not inactivated after ingestion is also predicted.

Identification of a specific toxic compound in the blood of envenomated person is performed by applying high-pressure liquid chromatography in conjunction with mass spectrometer (HPLC MS). Selective detection is required due to the wide chemical diversity of possible venoms. Detection kits for urine and blood analysis have been developed for some poisons to enable the rapid indentification of the encountered venom and efficient selection of an adequate antivenom. Administration of antivenom should be avoided unless evident signs of envenomation or positive laboratory findings, such as coagulation defect, reveal a high risk. In snake bites, fang marks alone are not an indication that antivenom should be applied. If the only manifestation is moderate local pain, antivenom therapy is not indicated at that stage.

Most antivenoms should be administrated intravenously. If the patient has previously demonstrated an allergic response to antivenoms, the application of steroids is also recommended. The antivenom dose for different patients may vary considerably. While for some patients application of antivenoms is not required, others may need multiple doses of antivenom. A common practice is also to apply a premedication, such as antihistamines, hydrocortisone or adrenaline, before injecting the antivenom. Adults should receive subcutaneously 0.25 mg of adrenaline, while children should receive 0.005 - 0.07 mg/kg of adrenaline as premedication treatment. Orally administrated antihistamines may aid in premedication treatment, but might also provoke sedative effects and cause hypotension.

Medical Use of Venoms

Honey bees produce specific active substances that were once used in medical treatments to cure various diseases. Their venom contains at least 18 active substances that have been shown to be beneficial for patients suffering arthritis and other inflammatory or degenerative diseases [10]. In some cultures, ant venom was used for the same purpose. Despite the fact that bee venom contains compounds that are demonstrated to provide relief in some medical conditions, bee envenomation remains an unpleasant experience. Swelling around the sting area arises due to secreted histamine. The role of histamine is to soften the tissue and facilitate the flow of other substances into the victim. Neurotransmitters such as dopamine and serotonin also have been identified in bee venom. The principal active component of bee venom is a polypeptide melittin consisting of 26 amino acids. Melittin is a powerful stimulator of phospholipase A2 and one of the most potent antiinflammatory agents known. It is 100 times more potent than hydrocortisol. In addition to melittin, many other components of a honey bee venom demonstrate antiinflammatory or analgetic activity. Apamin consists of a polypeptide chain built of 18 amino acids. It is believed that apamin enhances the transmittion of impulses in the central nervous system by selective blocking of Ca^{2+}-activated K^+ channels.

Bee venom therapy is still used to reduce both pain and swelling in patients who suffer from arthritis and other systemic inflammations. Research on the effects of bee venom also has revealed that the venom may be used for esthetic purposes, because the active compounds soften scar tissues and after repeated application significantly reduce the scars and improve their appearance. Furthermore, bee venom is currently under research for the use in the treatment of multiple sclerosis [11].

Some neurologists believe that scorpion venoms may be used to fight glioma, a form of brain cancer for which there is as yet no cure.. Chlorotoxin, a peptide consisting of 36 amino acids that was isolated from the venom of the giant Israeli scorpion, demonstrates promising chemotherapeutic effects due to its selective influence on cancerous cells: This compound specifically binds to glial tumor cells, but not to normal cells. Tarantula venom is scheduled for research in the treatment of heart attacks and pathological conditions like brain tumors.

Many snake venoms act by disturbing normal coagulation processes. Blood coagulation may be initiated or prevented, depending on the venom type. After the elucidation of toxic mechanisms in snake venoms, various medical applications for them were developed, for both diagnostic and therapeutic purposes. Proteins found in snake venoms, which normally cause heavy bleeding and death, when applied in small, well-controlled doses, actually reduce blood clotting and may be effectively used to prevent and treat heart disease and strokes.

The successful use of scorpion venom in tumor treatment has encouraged the scientists to search for other venom applications, and to investigate venoms derived from different animals. All over the world, research is currently being conducted on venoms for use in some very promising and attractive medical applications.

REFERENCES

[1] Mebs D. *Venomous and Poisonous Animals.* CRC Press: Boca Raton; 2002.

[2] Wickramaratna, JC; Fry, BG; Loiacono, RE; Aguilar, MI; Alewood, PF; Hodgson, WC, Isolation and Characterisation at Cholinergic Nicotinic Receptors of a Neurotoxin from Venom of the *Acanthophis* Sp. Seram Death Adder, *Biochem. Pharm.*, 68, 2004, 383-394.

[3] Fry, BG; Lumsden, NG; Wuster, W; Wickramaratna, JC; Hodgson, WC; Kini, RM. Isolation of a neurotoxin (alpha-colubritoxin) from a nonvenomous colubrid: evidence for early origin of venom in snakes. *J. Mol. Evol.* 57, 2003, 446-452.

[4] Fry, BG; Winkel, DK; Wickramaratna, JC; Hodgson, W; Wuster, W. Effectiveness of Snake Antivenom: Species and Regional Venom Variation and Its Clinical Impact. *J. Tox.* 22(1), 2003, 23-24.

[5] Fry, BG. Structure functione properties of venom components from Australian elapsids. *Toxicon.* 37, 1999, 11-32.

[6] Kourie, JI. Characterization of a C-type natriuretic peptide (CNP-39)-formed cation-selective channel from platypus (*Ornithorhynchus anatinus*) venom. *J. Physiol.* 518(2), 1999, 359-369.

[7] Kita, M; Nakamura, Y; Okumura, Y; Ohdachi, SD; Oba, Y; Yoshikuni, M;Kido, K; Uemura, D. Blarina toxin, a mammalian lethal venom from the short-tailed shrew *Blarina brevicauda*: Isolation and characterization. *Proc. Natl. Acad. Sci. USA* 101(20), 2004, 7542–7547.

[8] Kruska, D. Mustelidae. In: Grzimek, B. *Grzimek's Encyclopedia of Mammals*, Vol. 3, 1st Edition. New York: McGraw-Hill; 1990, pp. 388-449.

[9] Simics, M. Bee Venom Therapy - Making an Informed Decision. *Apitherapy Education Service - Apitronic Services*, Richmond, BC, Canada; 2000, pp. 10.

[10] Santilli, J; Rockwell, WJ; Wallerstedt, DB; Bellanti, JA. The Use of Bee Venom Therapy or Chronic Progressive Multiple Sclerosis - A Case Report. *J. AAS* 5(4), 1999, 6-7.

APPENDIX I. THE LIST OF CHEMICAL TERATOGENS AND POTENTIAL CHEMICAL TERATOGENS

Abovis
Acebutolol
Acebutolol hydrochloride
Acemetacin
Acepreval
Acetaldehyde
Acetamide
5-Acetamide-1,3,4-thiadiazole-2-sulfonamide
Acetazolamide sodium
Acetic acid methylnitrosaminomethyl ester
Acetohydroxamic acid
Acetonitrile
3-(alpha-Acetonyl-para-nitrobenzyl)-4-hydroxy-coumarin
para-Acetophenetidide
17-Acetoxy-19-nor-17-alpha-pregn-4-en-20-yn-3-one
Acetoxyphenylmercury
Acetoxytriphenylstannane
1-alpha-Acetylmethadol hydrochloride
Acetylsalicylic acid
Acetyltryptophan
Acid red 92
4,-(9-Acridinylamino) methanesulphon-meta-anisidide
Acrylic acid
Acrylonitrile
Actihaemyl
Actinomycin
Actinomycin C
Actinomycin D
Acyclovir
Acyclovir sodium salt
Adalat
1-Adamantanamine hydrochloride
Adapin
Adenine
Adenosine-3'-(alpha-amino-p-methoxyhydrocinnamamido)-3'-deoxy-N,N-dimethyl
Adipic acid bis (2-ethylhexyl) ester
Adipic acid dibutyl ester
Adipic acid di(2-hexyloxyethyl) ester
Adobiol
Adona trihydrate
1-Adrenaline chloride
Adrenocorticotrophic hormone
Adriamycin
Aflatoxin
Aflatoxin B1
Afridol blue
Agent orange
Alclometasone dipropionate
Alcohol sulphate
Aldactazide
Aldecin
Aldimorph
Aldrin
alpha-Alkenesulfonic acid
Alkyl dimethylbenzyl ammonium chloride
3-(Alkylamino) propionitrile
Alkylbenzenesulfonate
Allantoxanic acid, potassium salt
Alloxan
Allyl chloride
Allyl glucosinolate
Allyl isothiocyanate
6-Allyl-6,7-dihydro-5h-dibenz (c,e) azepine phosphate
Allylestrenol
(4-Allyloxy-3-chlorophenyl)acetic acid
Alternariol
Alternariol monomethyl ether and alternariol (1:1)
Alternariol-9-methyl ether
Aluminum aceglutamide
Aluminum chloride
Aluminum chloride hexahydrate
Aluminum lactate
Aluminium (III) nitrate, nonahydrate (1:3:9)
Aluminium potassium sulfate, dodecahydrate
Ambroxol hydrochloride
Ametycin
Amfenac sodium monohydrate
Amicardine
N1-Amidinosulfanilamide
Amidoline
5-((2-Aminoacetamido) methyl)-1-(4-chloro-2-(orthochlorobenzoyl) phenyl)-N,N-dimethyl-1H-S-triazole-3-carboxamide, hydrochloride, dihydrate
Aminoacetonitrile bisulfate
Aminoacetonitrile sulfate
2-Aminobenzimidazole
2-Amino-6-benzimidazolyl phenylketone
Aminobenzylpenicillin
5-Amino-1-bis (dimethylamide) phosphoryl-3-phenyl-1,2,4- triazole
2-Amino-5-bromo-6-phenyl-4 (1h)-pyrimidinone
4-Amino-2-(4-butanoylhexahydro-1h-1,4-diazepin-1-yl)-6,7-dimethoxyquinazoline hydrochloride
2-Amino-5-butylbenzimidazole
5-Amino-1,6-dihydro-7h-v-triazolo (4,5-d) pyrimidin-7-one
3-(2-aminoethyl) indol-5-ol
3-(2-aminoethyl) indol-5-ol creatinine sulfate
trans-4-Aminoethylcyclohexane-1-carboxylic acid
Aminoglutethimide
2-Amino-3-hydroxybenzoic acid
8-Amino-7-hydroxy-3,6-napthalenedisulfonic acid, sodium salt
4-Amino-N-(6-methoxy-3-pyridazinyl)-benzenesulfonamide
3-Amino-4-methylbenzenesulfonylcyclohexyl urea
2-Amino-6-(1,-methyl-4,-nitro-5,-imidazolyl) mercaptopurine
1-(4-Amino-2-methylpyrimidin-5-yl)methyl-3-(2-chloroethyl)-3-nitrosourea
2-Amino-4-(methylsulfinyl) butyric acid
5-Amino-2-napthalenesulfonic acid sodium salt
6-Aminonicotinamide

2-Amino-4-nitroaniline
4-Amino-2-nitroaniline
Aminonucleoside puromycin
meta-Aminophenol
2-Aminophenol
4-Aminophenol
meta-Aminophenol, chlorinated
7-(d-alpha-aminophenylacetamido) desacetoxycephalosporanic acid
3-Aminopropionitrile
beta-Aminopropionitrile fumarate
Aminopropyl aminoethylthiophosphate
3-(2-Aminopropyl) indole
Aminopteridine
2-Aminopurine-6-thiol
Aminopyrine sodium sulfonate
Aminopyrine-barbital
5-Amino-2-beta-d-ribofuranosyl-as-triazin-3- (2H) -one
4-Amino-2,2,5,5-tetrakis (trifluoromethyl) -3-imidazoline
2-Amino-1,3,4-thiadiazole
2-Amino-1,3,4-thiadiazolehydrochloride
2-Amino-1,3,4-thiadiazole-5-sulfonamide sodium salt
1-Amino-2-(4-thiazolyl)-5-benzimidazolecarbamic acid isopropyl ester
Amitriptyline-N-oxide
Amitrole
Ammonium vanadate
Amosulalol hydrochloride
Amoxicillin trihydrate
dl-Amphetamine sulfate
Ampicillin trihydrate
Amrinone
Amsacrine lactate
Amygdalin
Anabasine
Anatoxin I
Androctonus amoreuxi venom
Androfluorene
Androfurazanol
Androstanazol
Androstenediol dipropionate
Androstenedione
Androstenolone
Androstestone-M
Angel dust
Angiotonin
Anguidin
Aniline violet
6-(para-anilinosulfonyl) metanilamide
2-Anthracenamine
Antibiotic BB-K8
Antibiotic BB-K8 sulfate
Antibiotic BL-640
Antibiotic MA 144A1
Antimony oxide
Apholate
9-beta-d-Arabino furanosyl adenine
Arabinocytidine
Ara-C palmitate
Araten phosphate
Arathane
1-Arginine monohydrochloride
Aristocort
Aristocort acetonide
Aristocort diacetate
Aristolic acid
Aristospan
Aromatol
Arotinoic acid
Arotinoic methanol
Arotinoid ethyl ester
Arsenic
ortho-Arsenic acid
Arsenic acid, disodium salt, heptahydrate
Arsenic acid, sodium salt
Arsenic trioxide
Asalin
1-Ascorbic acid
1-Asparaginase
Atrazine
Atromid S
Atropine
Atropine sulfate (2:1)
Auranofin
Aureine
1-Aurothio-d-glucopyranose
Ayush-47
Azabicyclane citrate
Azactam
Azacytidine
Azaserine
Azathioprine
Azelastine hydrochloride
1-2-Azetidinecarboxylic acid
Azinphos methyl
Azo blue
Azo ethane
Azosemide
Azoxyethane
Azoxymethane
Baccidal
Bacmecillinam
Bal
Barbital sodium
Barium ferrite
Barium fluoride
Bayer 205
Baythion
Befunolol hydrochloride
Bendacort
Bendadryl hydrochloride
Benedectin
Benomyl
Benzarone
d-Benzedrine sulfate
Benzenamine hydrochloride
Benzene
Benzene hexachloride-g-isomer
1-Benzhydryl-4-(2-(2-hydroxyethoxy)ethyl)piperazine
Benzidamine hydrochloride
2-Benzimidazolecarbamic acid
1-(2-Benzimidazolyl)-3-methylurea
1,2-Benzisothiazol-3 (2H)-one-1,1-dioxide
1,2-Benzisoxazole-3-methanesulfonamide
Benzo (alpha) pyrene
Benzo (e) pyrene
Benzoctamine hydrochloride
para-Benzoquinone monoimine
Benzothiazole disulfide
2-Benzothiazolethiol
2-Benzothiazolyl-N-morpholinosulfide

2-(meta-Benzoylphenyl) propionic acid
2-Benzylbenzimidazole
Benzyl chloride
Benzyl penicillinic acid sodium salt
Beryllium chloride
Beryllium oxide
Bestrabucil
Betamethasone
Betamethasone acetate and betamethasone phosphate
Betamethasone benzoate
Betamethasone dipropionate
Betamethasone disodium phosphate
Betel nut
Betnelan phosphate
BHT (food grade)
Bindon ethyl ether
Binoside
4-Biphenylacetic acid
2-Biphenylol
2-Biphenylol, sodium salt
3-(4-Biphenylylcarbonyl) propionic acid
2,2-Bipyridine
Bis(para-acetoxyphenyl)-2-methylcylcophexylidenemethane
4,4-Bis(1-amino-8-hydroxy-2,4-disulfo-7-napthylazo)-3,3,-bitolyl,tetrasodium salt
1,4-Bis(3-bromopropionyl)-piperazine
1,3-Bis(carbamoylthio)-2-(N,N-dimethylamino)propane hydrochloride
trans-N,N,-Bis(2-chlorobenzyl)-1,4 cyclohexanebis (methylamine) dihydrochloride
Bis(2-chloroethyl) amine hydrochloride
4,-(Bis (2-chloroethyl) amino) acetanilide
4,-(Bis (2-chloroethyl) amino)-2-fluoro acetanilide
dl-3-(para-(Bis (2-chloroethyl) amino) phenyl)alanine
Bis(beta-chloroethyl) methylamine
Bis(2-chloroethyl) methylamine hydrochloride
Bis (2-chloroethyl) sulfide
N,N,-Bis (2-chloroethyl)-N-nitrosourea
N,N,-Bis (2-chloroethyl)-para-phenylenediamine
Bis (para-chlorophenyl) acetic acid
2,2-Bis (ortho, para-chlorophenyl)-1,1,1-trichloroethane
1,1-Bis (para-chlorophenyl)-2,2,2-trichloroethanol
Bis (beta-cyanoetyl) amine
Bis (dichloroacetyl) -1,8-diaminooctane
3,5-Bis-dimethylamino-1,2,4-dithiazolium chloride
Bis (dimethyldithiocarbamato) zinc
(((3,5-Bis(1,1-dimethylethyl)-4-hydroxyphenyl)methyl)thio)acetic acid 2-ethylhexyl ester
Bis (dimethylthiocarbamoyl) sulfate
2,4-Bis (ethylamino)-6-chloro-s-triazine
Bis (ethylmercuri) phosphate
Bis-HM-A-TDA
Bishydroxycoumarin
Bis (4-hydroxy-3-coumarin) acetic acid ethyl ester
1,4-Bis ((2- ((2-hydroxyethyl) amino) ethyl) amino)-9,10-athracenedione diacetate
Bis (isooctyloxycarbonylmethylthio) dioctyl stannane
Bis (2-methoxy ethyl) ether
Bisphenol A
1,4-Bis (phenyl amino) benzene
Bis (tributyl tin) oxide
2-(3,5-Bis (trifluoromethyl) phenyl) -N-methyl-hydrazinecarbothioamide (9CI)
Bladex
Bleomycin sulfate
Bomt
Bracken fern, dried
Bradykinin
Bredinin
Bremfol
Bromacil
Bromazepam
Bromocriptine
Bromocriptine mesilate
5-Bromo-2, -deoxyuridine
2-Bromo-d-lysergic acid diethylamide
6-Bromo-1,2-napththoquinone
Bromoperidol
Bromophenophos
4-Bromophenyl chloromethyl sulfone
Buclizine dihydrochloride
Budesonide
Bunitrolol hydrochloride
Buprenorphine hydrochloride
1,3-Butadiene
Butamirate citrate
1,4-Butanediamine
1,4-Butanediol dimethyl sulfonate
4-Butanolide
Butobarbital
Butoctamide semisuccinate
Butorphanol tartrate
Butoxybenzyl hyoscyamine bromide
2-Butoxyethanol
para-Butoxyphenylacetohydroxamic acid
Butriptyline
Bromoperidol
Bromophenophos
4-Bromophenyl chloromethyl sulfone
Buclizine dihydrochloride
Budesonide
Bunitrolol hydrochloride
Buprenorphine hydrochloride
1,3-Butadiene
Butamirate citrate
1,4-Butanediamine
1,4-Butanediol dimethyl sulfonate
4-Butanolide
Butobarbital
Butoctamide semisuccinate
Butorphanol tartrate
Butoxybenzyl hyoscyamine bromide
2-Butoxyethanol
para-Butoxyphenylacetohydroxamic acid

Butriptyline
n-Butyl acetate
n-Butyl alcohol
sec-Butyl alcohol
tert-Butyl alcohol
alpha,-((tert-Butyl amino) methyl) - 4-hydroxy-meta-xylene-alpha,alpha-diol
Butyl carbamate
Butyl carbobutoxymethyl phthalate
Butyl dichlorophenoxyacetate
Butyl ethyl acetic acid
Butyl flufenamate
n-Butyl glycidyl ether
n-Butyl mercaptan
n-Butyl-3,ortho-acetyl-12-b-13-alpha-dihydrojervine
1-(tert-Butylamino)-3-(2-chloro-5-methylphenoxy) -2-propanol hydrochloride
alpha-Butylbenzenemethanol
5-Butyl-2-benzimidazolecarbamic acid methyl ester
5-Butyl-1-cylcohexylbarbituric acid
2-sec-Butyl-4,6-dinitrophenol
4-Butyl-1,2-diphenyl-3,5-dioxo pyrazolidine
n-Butyl-N-nitroso-1-butamine
N-Butyl-N-nitroso ethyl carbamate
n-Butylnitrosourea
1-Butyl-2',6'-pipecoloxylidide
1-Butyl-3-sulfanilyl urea
1-Butyl-3-(para-tolyl sulfonyl) urea
1-Butyl-3-(para-tolylsulfonyl) urea, sodium salt
Butyl-2,4,5-trichlorophenoxyacetate
1-Butyryl-4-(phenylallyl) piperazine hydrochloride
Buzepide methiodide
Cadmium
Cadmium (II) acetate
Cadmium chloride
Cadmium chloride, dihydrate
Cadmium compounds
Cadmium oxide
Cadmium sulfate (1:1)
Cadmium sulfate (1:1) hydrate (3:8)
Cadralazine
Caffeic acid
Caffeine
Calcium ELIA complex
Calcium fluoride
Calcium phosphonomycin hydrate
Calcium trisodium diethylene triamine pentaacetate
Calcium valproate
Calcium-N-2-ethylhexyl-beta-oxybutyramide semisuccinate
Cambendazole
Camphorated oil
Candida albicans glycoproteins
Cannabidiol
Cannabinol
Cannabis
Cap
Caprolactam
Captafol
Captan
Carbamates
Carbaryl
Carbendazim and sodium nitrite (5:1)
Carbidopa
Carbinilic acid isopropyl ester
Carbofuran
Carbon dioxide
Carbon disulfide
Carbon monoxide
Carbon tetrachloride
Carboprost tromethamine
Cargutocin
Carmetizide
Carmofur
1-Carnitine hydrochloride
Carnosine
Carzinophilin
Cassava, manihot utilissima
Catatoxic steroid No. 1
d-Catechol
CAZ pentahydrate
Cefamandole sodium
Cefatoxime sodium
Cefazedone
Cefazolin sodium salt
Cefmetazole
Cefmetazole sodium
Cefroxadin
Cefuroxim
Celestan-depot
Cellryl
Cellulose acetatemonophthalate
Centbucridine hydrochloride
Centchroman
Cephalothin
Cervagem
Cesium arsenate
Cethylamine hydrofluoride
alpha-Chaconine
Chenodeoxycholic acid
Chlodithane
Chlorambucil
Chloramphenicol
Chloramphenicol monosuccinate sodium salt
Chloramphenicol palmitate
Chlorcyclizine hydrochloride
Chlorcyclizine hydrochloride A
Chlorcyclohexamide
Chlordane
Chlorimipramine
Chlorinated camphene
Chlorinated dibenzo dioxins
Chlorisopropamide
Chlormadinon
para-Chloro dimethylaminoazobenzene
2-Chloroadenosine
1-(3-Chloroallyl) -3,5,7-triaza-1-azoniaadamantane chloride
3-Chloro-4-aminoaniline
1-((para-(2-(Chloro-ortho-anisamido)ethyl)phenyl)sulfonyl)-3-cylcohexyl urea
Chlorobenzene
ortho-Chlorobenzylidene malononitrile
1-para-Chlorobenzyl-1H-indazole-3-carboxylic acid
7-Chloro-5-(ortho-chlorophenyl)-1,3-dihydro-3-hydroxy-2H-1,4-benzodiazepin-2-one
Chlorocylcine
6-Chloro-5-Cyclohexyl-1-indancarboxylic acid

6-Chloro-5-(2,3-dichlorophenoxy)-
2-methylthio-benzimidazole
5-Chloro-2-(2-
(diethylamino)ethoxy)benzanilide
7-Chloro-1,3-dihydro-5-phenyl,2H-
1,4-benzodiazepin-2-one
Chloroethyl mercury
1-(2-Chloroethyl)-3-cylcohexyl-1-
nitrosourea
1-Chloro-3-ethyl-1-penten-4-YN-3-
OL
Chloroform
4-Chloro-N-furfuryl-5-
sulfamoylanthranilic acid
Chlorogenic acid
endo-4-Chloro-N-(hexahydro-4,7-
methanoisoindol-2-YL)-3-
sulfamoylbenzamide
(-)-N-((5-Chloro-8-hydroxy-3-
methyl-1-OXO-7-isochromanyl)
carbonyl)-3-phenylalanine
5-Chloro-7-iodo-8-quinolinol
(4-Chloro-2-methylphenoxy) acetic
acid
2-(4-Chloro-2-methylphenoxy)
propanoic acid (R) (9CI)
4-Chloro-2-methylphenoxy-alpha-
propionic acid
7-Chloro-1-methyl-5-phenyl-1H-
1,5 benzodiazepine-2,4(3H,5H)-
dione
2-Chloro-11-(4-methylpiperazino)
dibenzo (b,f) (1,4) thiazepine
4-((5-Chloro-2-OXO-3(2H)-
benzothiazolyl)acetyl)-1-
piperazineethanol
4-(3-(2-Chlorophenothiazin-10-
YL)propyl)-1-piperazineethanol
4-Chlorophenylalanine
1-(para-Chloro-alpha-phenylbenzyl)-
4-(2-((2-hydroxyethoxy)
ethyl)piperazine)
1-(meta-Chlorophenyl)-3-N,N-
dimethylcarbamoyl-5-
methoxypyrazole
3-(para-Chlorophenyl)-
1,1,dimethylurea

5,(2-Chlorophenyl)-7-ethyl-1-
methyl-1,3-dihydro-2H-thieno (2,3-
e) (1,4) diazepin-2-one
N-3-
chlorophenylisopropylcarbamate
3-(4-Chlorophenyl)-1-methoxy-1-
methylurea
2-(ortho-Chlorophenyl)-2-
(methylamino)cyclohexanone
hydrochloride
3-(para-Chlorophenyl)-1-methyl-1-
(1-methyl-2-propynyl) urea
4-(para-Chlorophenyl)-2-phenyl-5-
thiazoleacetic acid
1-(para-Chlorophenylsulfonyl)-3-
propylurea
para-Chlorophenyl-2,4,5-
trichlorophenyl sulfone
4-Chlorophenyl-2,4,5-
trichlorophenylazosulfide mixed
with 1,1- bis(4-chlorophenyl)ethanol
Chloropromazine
Chloropromazine hydrochloride
Chloroquine
Chloroquine diphosphate
N-(3-Chloro-ortho-tolyl) anthranilic
acid
2-((4-Chloro-ortho-
tolyl)oxy)propionic acid potassium
salt
Chloro(triethylphosphine)gold
Chlorovinylarsine dichloride
4-Chloro-3,5-xylenol
Chlorphentermine
g-(4-(para-Chlorphenyl)-4-
hydroxiperidino)-para-
fluorbutyrophenone
Cholecalciferol
Cholesterol
Cholestyramine
Chorionic gonadotropin
Chromium chloride
Chromium (VI) oxide (1:3)
Chromium trichloride hexahydrate
Chromomycin A3
C.I. 45405
C.I. Direct blue 1, tetrasodium salt
C.I. Direct blue 6, tetrasodium salt

C.I. Direct blue 14, tetrasodium salt
C.I. Direct blue 15, tetrasodium salt
Cilostazol
Cinoxacin
Citreoviridin
Citrinin
Citrus hystrix DC., fruit peel extract
Clavacin
Clindamycin-2-palmitate
monohydrochloride
Clindamycin-2-phosphate
Cloazepam
Clobetasone butyrate
Cloconazole hydrochloride
Clofedanol hydrochloride
Clofexamide phenylbutazone
Clomiphene
racemic-Clomiphene citrate
trans-Clomiphene citrate
Clonidine hydrochloride
Clonixic acid
Cloxazolazepam
Clozapine
Coagulase
Cobalt (II) chloride
Corn oil
Corticosterone
Corticosterone acetate
Cortisol
Cortisone
Cortisone-21-acetate
Cottonseed oil (unhydrogenated)
Coumarin
Cravetin
meta-Cresol
Cumoesterol
S-1-Cyano-2-hydroxy-3-butene
Cyanotrimethylandrostenolone
Cycasin
Cyclocytidine hydrochloride
Cycloguanyl
Cyclohexanamine hydrochloride
Cycloheximide
Cyclohexylamine
Cyclohexylamine sulfate
2-(Cyclohexylamino)ethanol
N-Cyclohexyl-2-
benzothiazolesulfenamide

4-(4-Cyclohexyl-3-chlorophenyl)-4-oxobutyric acid
1-Cyclohexyl-3-para-tolysulfonylurea
Cyclonite
Cyclopamine
Cyclophosphamide hydrate
Cyclophosphoramide
alpha-Cyclopiazonic acid
5-(Cyclopropylcarbonyl)-2-benzimidazolecarbamic acid methyl ester
Cyprosterone acetate
Cysteine-germanic acid
Cytochalasin B
Cytochalasin E
Cytostasan
Cytoxal alcohol
Cytoxyl amine
Demeton-O + Demeton-S
Demeton-O-methyl
Demetrin
Denopamine
11-Deoxo-12-beta,13-alpha-dihydro-11-alpha-hydroxyjervine
11-Deoxojervine-4-EN-3-one
2,-Deoxy-5-fluorouridine
2-Deoxyglucose
2,-Deoxy-5-iodouridine
4-Deoxypyridoxol hydrochloride
Dephosphate bromofenofos
Depofemin
Depo-medrate
N-Desacetylthiocolchicine
Desoxymetasone
2-Desoxyphenobarbital
Detergents, Liquid containing AES
Detergents, Liquid containing LAS
Dexamethasone acetate
Dexamethasone 17,21-dipropionate
Dexamethasone palmitate
Dextran 1
Dextran 70
Dextropropoxyphene napsy
alpha-DFMO
Diabenor
Diacetylmorphine hydrochloride
Dialifor
Diamicron
2,4-Diamino-6-methyl-5-phenylpyrimidine
2,4-Diamino-5-phenyl-6-ethylpyrimidine
2,4-Diamino-5-phenyl-6-propylpyrimidine
2,4-Diamino-5-phenylpyrimidine
2,5-Diaminotoluene dihydrochloride
Diazepam
Diazinon
6-Diazo-5-oxonorleucine
Diazoxide
Dibekacin
5H-Dibenz (b,f) azepine-5-carboxamide
5H-Dibenz (b,f) azepine, 3-chloro-5-(3- (4-carbamoyl-4-piperidinopiperine
Dibenz (b,f) (1,4) oxazepine
Dibenzacepin
Dibenzyline hydrochloride
1,2-Dibromo-3-chloropropane
3,5-Dibromo-4-hydroxyphenyl-2 ethyl-3-benzofuranyl ketone
Dibromomaleinimide
1,6-Dibromomannitol
Dibutyl phthalate
N,N-Di-n-butylformamide
Dibutyryl cyclic amp
Dicarbadodecaboranylmethylethyl sulfide
Dicarbadodecaboranylmethylpropyl sulfide
1-(2,4-Dichlorbenzyl)indazole-3-carboxylic acid
Dichloroacetonitrile
(ortho-((2,6-Dichloroanilino)phenyl) acetic acid sodium salt
ortho-Dichlorobenzene
para-Dichlorobenzene
4,5-Dichloro-meta-benzenedisulfonamide
2,2,-Dichlorobiphenyl
Dichloro-1,3-butadiene
1,4-Dichloro-2-butene
2,2-Dichloro-1,1-difluorethyl methyl ether
5,5-Dichloro-2,2,-dihydroxy-3,3,-dinitrobiphenyl
Dichloroethane
1,1-Dichloroethane
2,3-Dichloro-N-ethylmaleinimide
Dichloromaleimide
Dichloro-N-methylmaleimide
2,4-Dichloro-4,-nitrodiphenyl ether
2,4-Dichlorophenol
(2,4-Dichlorophenoxy) acetic acid butoxyethyl ester
(2,4-Dichlorophenoxy) acetic acid dimethylamine
4-(2,4-Dichlorophenoxy) butyric acid
2-(2,4-Dichlorophenoxy) propionic acid
(+)-2-(2,4-Dichlorophenoxy) propionic acid
3,4-Dichlorophenoxyacetic acid
2,4-Dichlorophenoxyacetic acid propylene glycol butyl ether ester
2-(2,6-Dichlorophenylamino)-2-imidazoline
3,6-Dichloro-2-pyridinecarboxylic acid
Dichlorvos
Dicyclohexyl adipate
Dicyclohexyl-18-crown-6
Dicyclopentadienyldichlorotitanium
7,8-Didehydroretinoic acid
Dieldrin
Diethyl carbitol
Diethyl carbonate
Diethyl mercury
Diethyl phthalate
Diethyl sulfate
2-(Diethylamino)-2',6'-acetoxylidide
2-Diethylamino-2',6'-acetoxylidide hydrochloride
ortho-(Diethylaminoethoxy) benzanilide
2-(2-(Diethylamino)ethoxy)-5-bromobenzanilide
2-(2-(Diethylamino)ethoxy)-2,-chloro-benzanilide
2-(2-(Diethylamino)ethoxy)-3,-chloro-benzanilide

2-(2-(Diethylamino)ethoxy)-3,-
chloro-methylbenzanilide
(para-2-Diethylaminoethoxyphenyl)-
1-phenyl-2-para-anisylethanol
1-(2-(Diethylamino)ethyl)reserpine
7-Diethylamino-5-methyl-s-
triazolo(1,5-alpha) pyrimidine
N,N-Diethylbenzenesulfonamide
Diethylcarbamazine
Diethylcarbamazine acid citrate
Diethyldiphenyl dichloroethane
Diethylene glycol
Diethylene glycol monomethyl ether
1,2-Diethylhydrazine
1,2-Diethylhydrazine
dihydrochloride
N,N-Diethyllsergamide
N,N-Diethyl-4-methyl-3-oxo-5-
alpha-4-azaandrostane-17-beta-
carboxamide
3,3-Diethyl-1-(meta-pyridyl)triazene
a,a-Diethyl-(E)-4,4,-stilbenediol
bis(dihydrogen phosphate)
a,a-Diethyl-4,4,-stilbenediol
disodium salt
Diethylstilbesterol
Diethylstilbestrol dipalmitate
Diethylstilbestrol dipropionate
Diflorasone diacetate
Diflucortolone valerate
dl-alpha-Difluoromethylornithine
5-(2,4-Difluorophenyl) salicylic acid
Difluprednate
Digoxin
Dihydantoin
Dihydrocodeinone bitartrate
Dihydrodiethylstilbestrol
3,4-Dihydro-6-(4-(3,4-
dimethoxybenzoyl)-1-piperazinyl)-
2(1H)- quinolinone
5,6-Dihydro-N-(3-
(dimethylamino)propyl)-11H-
dibenz(b,e)azepine
10,11-Dihydro-5-(3-
(dimethylamino)propyl)-5H-
dibenz(b,f)azepine hydrochloride
5,6-Dihydro-para-dithiin-2,3-
dicarboximide
12,b,13,alpha-Dihydrojervine
10,11-Dihydro-5-(3-
(methylamino)propyl)-5H-
dibenz(b,f)azepine hydrochloride
1,7-Dihydro-6H-purin-6-one
7,8-Dihydroretinoic acid
Dihydrostreptomycin
4-Dihydrotestosterone
3-alpha,17-beta-Dihydroxy-5-alpha-
androstane
3-alpha,7-beta-Dihydroxy-6-beta-
cholan-24-OIC acid
1 alpha,25-Dihydroxycholecalciferol
3,4-Dihydroxy-alpha-
((isopropylamino)methyl)benzyl
alcohol
1-Dihydroxyphenyl-1-alanine
1-(-)-3-(3,4-Dihydroxyphenyl)-2-
methylanine
17R,21-alpha-Dihydroxy-4-
propylajmalanium hydrogen tartrate
DI(2-Hydroxy-n-propyl) amine
Diisobutyl adipate
Diisobutyl phthalate
alpha-(2-(Diisopropylamino)ethyl)-
alpha-phenyl-2- pyridineacetamide
Dilantin
Dilaudid
Diltiazem hydrochloride
Dimatif
Dimethoxy ethyl phthalate
1,2-Dimethoxyethane
3,6-Dimethoxy-4-
sulfanilamidopyridazine
Dimethyl adipate
O,O-Dimethyl
methylcarbamoylmethyl
phosphordithioate
Dimethyl phthalate
Dimethyl sulfate
Dimethyl sulfoxide
O,S-Dimethyl phosphoramidothioate
N,N-Dimethylacetamide
O,O-Dimethyl-S-(2-
(acetylamino)ethyl) dithiophosphate
4-(Dimethylamine)-3,5-XYLYL-N-
methylcarbamate
Dimethylaminoantipyrine
4-Dimethylaminoazobenzene
para-
Dimethylaminobenzenediazosodium
sulphonate
5-(3-(Dimethylamino)propyl)-2-
hydroxy-10,11-dihydro-5H-
dibenz(b,f)azephine
11-(3-Dimethylaminopropylidene-
6,11- dihydrodibenzo(b,e)thiepine
hydrochloride
10-(2-
(Dimethylamino)propyl)phenothiazi
ne
Dimethylbenzanthracene
1,1-Dimethylbiguanide
1-(2-(1,3-Dimethyl-2-
butenylidene)hydrazino)phthalazine
Dimethyldicetylammonium chloride
9,9-Dimethyl-10-
dimethylaminopropylacridan
hydrogen tartrate
6-alpha,21-Dimethylethisterone
N-(5-(((1,1-
Dimethylethyl)amino)sulfonyl)-
1,3,4-thiadiazol-2- YL)acetamide
monsodium salt
N,N-Dimethyl-para((para-
fluorophenyl)azo)aniline
Dimethylformamide
1,1-Dimethylhydrazine
1,2-Dimethylhydrazine
2,6-Dimethylhydroquinone
Dimethylimipramine
1,3-Dimethylisothiourea
1,3-Dimethylnitrosourea
3,3-Dimethyl-1-phenyltriazene
Dimethylthiomethylphosphate
N,N-Dimethyl-4-(para-
tolylazo)aniline
5-(3,3-Dimethyl-1-
triazeno)imidazole-4-carboxamide
citrate
2,6-Dimethyl-4-tridecylmorpholine
1,3-Dimethylurea
2,4-Dinitroaniline
4,6-Dinitro-ortho-cresol ammonium
salt

2,6-Dinitro-N,N-dipropyl-4-(trifluoromethyl)benzenamine
2,4-Dinitrophenol
2,4-Dinitrophenol sodium salt
Dinitrosopiperazine
Dinitrotoluene
Dinoprost methyl ester
Dinoprostone
n-Dioctyl phthalate
Dioxane
meta-Dioxane-4,4-dimethyl
1,4-Di-N-oxide of dihydroxymethylquinoxaline
1,3-Dioxolane-4-methanol
3-(2-(1,3-Dioxo-2-methylindanyl)) glutarimide
3-(2-(1,3-Dioxo-2-phenylindanyl)) glutarimide
3-(2-(1,3-Dioxo-2-phenyl-4,5,6,7-tetrahydro-4,7- dithiaindanyl)) glutarimide
2-(2,6-Dioxopiperiden-3YL)phthalimide
N-(2,6-Dioxo-3-piperidyl)phthalimidine
1,3-Dioxo-2-(3-pyridylmethylene)indan
Diphenylamine
Diphenylguanidine
Diphenylhydantoin
phenobarbital
3-(3,3-diphenylpropylamino)propyl-3',4',5'-trimethoxybenzoate hydrochloride
Dipropyl adipate
Diquat
DI-sec-octyl phthalate
Disodium ethylene-1,2-bisidithiocarbamate
Disodium etidronate
Disodium inosinate
Disodium methanearsenate
Disodium molybdate dihydrate
Disodium phosphonomycin
Disodium selenate
Disulfiram
Dithane M-45
2,2-Dithiobis(pyridine-1-oxide)magnesium sulfate trihydrate
2,2-Dithiodipyridine-1,1,-dioxide
Diuron
alpha-DFMO
Dobutamine hydrochloride
Domperidone
Dopamine
Dopamine hydrochloride
Doriden
Doxifluridine
Doxycycline
1-Dromoran tartrate
Duazomycin
Durabolin
Duricef
Dydrogesterone
Dye C
Econazole nitrate
Eflornithine hydrochloride
Elasiomycin
Elavil
Elavil hydrochloride
Elymoclavine
EM 255
Emoquil
Emorfazone
Enalapril maleate
Enavid
Endosulfan
Endrin
Enflurane
Enoxacin
Epe
Ephedrine
Epichlorohydrin
Epidehydrocholesterin
2-alpha,3-alpha-Epithio-5-alpha-androstan-17-beta-OL
4,5-Epithiovaleronitrile
EPN
Epocelin
1,2-Epoxyethylbenzene
Eraldin
Ergochrome AA (2,2)-5-beta,6-alpha,10-beta-5',6'-alpha,1-,-beta
Ergocornine methanesulfonate (salt)
Ergotamine tartrate
Ergoterm TGO
Erythromycin
Escherichia coli endotoxin
Escin
beta-Escin
Escin, sodium salt
Estradiol
Estradiol dipropionate
Estradiol polyester with phosphoric acid
Estradiol-17-valerate
Estradiol-3-benzoate
Estradiol-3-benzoate mixed with progesterone (1:14 moles)
Estradiol-17-caprylate
Estramustin phosphate sodium
Estra-1,3,5(10)-triene-17-beta-diol-17-tetrahydropyranyl ether
Estriol
Estrone
Ethanolamine
Ethinamate
Ethinyl estradiol
Ethinyl estradiol and norethindrone acetate
17-alpha-Ethinyl-5,10-estrenolone
dl-Ethionine
Ethisterone and diethylstilbestrol
6-Ethoxy-2-benzothiazolesulfonamide
2-Ethoxyethanol
2-Ethoxyethyl acetate
Ethyl alcohol
Ethyl all-trans-9-(4-methoxy-2,3,6-trimethylphenyl)-3,7-dimethyl-2,4,6,8-nonatetraenoate
Ethyl apovincaminate
Ethyl benzene
Ethyl (2,4-dichlorophenoxy) acetate
Ethyl fluclozepate
Ethyl hexylene glycol
Ethyl methacrylate
Ethyl methanesulfonate
Ethyl methyl 1,4-dihydro-2,6-dimethyl-4-(meta-nitrophenyl)-3,5-pyridinedicarboxylate
Ethyl morphine hydrochloride dihydrate

Ethyl thiourea
alpha-((Ethylamino)methyl)-meta-hydroxybenzyl alcohol
2-Ethylamino-1,3,4-thiadiazole
1-Ethyl-1,4-dihydro-7-methyl-4-oxo-1,8-naphthyridine-3-carboxylic acid
Ethyl-S-dimethylaminoethyl methylphosphonothiolate
Ethyl-N,N-dimethyl carbamate
Ethylene bis(dithiocarbamato)) zinc
Ethylene chlorohydrin
1,2-Ethylene dibromide
Ethylene dichloride
Ethylene glycol
Ethylene glycol diethyl ether
Ethylene glycol methyl ether
Ethylene oxide
Ethylenebis (dithiocarbamato) manganese and zinc acetate (50:1)
Ethylenediamine hydrochloride
Ethylenediaminetetraacetic acid
Ethylenediaminetetraacetic acid, disodium salt
Ethyleneimine
Ethylestrenol
2-Ethylhexanol
Ethyl-para-hydroxyphenyl ketone
Ethylmercuric phosphate
Ethyl-N-methyl carbamate
Ethyl-2-methyl-4-chlorophenoxyacetate
5-Ethyl-N-methyl-5-phenylbarbituric acid
2-Ethyl-2-methylsuccinimide
1-Ethyl-4-(2-morpholinoethyl)-3,3-diphenyl-2-pyrrolidinone
N-Ethyl-N-nitrosobiuret
1-Ethyl-1-nitrosourea
Ethylnorgestrienone
17-Ethyl-19-nortestosterone
N-Ethyl-para-(phenylazo) aniline
5-Ethyl-5-phenylbarbituric acid
1-5-Ethyl-5-phenylhydantoin
3-Ethyl-5-phenylhydantoin
5-(2-Ethylphenyl)-3-(3-methoxyphenyl)-s-triazole
2-Ethylthioisonicotinamide
Ethyltrichlorphon
Ethyl-3,7,11-trimethyldodeca-2,4-dienoate
Ethylurea and sodium nitrite (1:1)
Ethylurea and sodium nitrite (2:1)
Ethynodiol
Ethynylestradiol mixed with norethindrone
2-alpha-Ethynyl-alpha-nor-17-alpha-pregn-20-YNE-2-beta,17-beta- diol
Etizolam
Etoperidone
ETP
E. typhosa lipopolysaccharide
False hellebore
Famfos
Famotidine
FD&C red No. 2
FD&C yellow NO. 5
Feldene
Fencahlonine
Fenestrel
Fenoprofen calcium dihydrate
Fenoterol hydrobromide
Fenthion
Fenthiuram
Ferbam
Ferrous sulfate
Fertodur
Fiboran
Firemaster BP-6
Firemaster FF-1
Flavoxate hydrochloride
Flomoxef sodium
Floxapen sodium
Flubendazole
Flucortolone
Flunarizine dihydrochloride
Flunisolide
Flunitrazepam
Fluoracizine
N-Fluoren-2-YL acetamide
Fluorobutyrophenone
Fluorocortisone
5-Fluoro-2,-deoxycytidine
3-Fluoro-4-dimethylaminoazobenzene
Fluorohydroxyandrostenedione
2-Fluoro-alpha-methyl-(1,1,-biphenyl)-4-acetic acid 1-(acetyloxy) ethyl ester
4,-Fluoro-4-(4-methylpiperidino)butyrophenone hydrochloride
3-Fluoro-4-phenylhydratropic acid
5-Fluoro-1-(tetrahydrofuran-2-YL)uracil
Fluorouracil
Flutamide
Flutazolam
Flutoprazepam
Flutropium bromide hydrate
Folic acid
Fominoben hydrochloride
Fonazine mesylate
Formaldehyde
Formamide
Formhydroxamic acid
Formoterol fumarate dihydrate
N-Formyl-N-hydroxyglycine
N-Formyljervine
Forphenicinol
Fortimicin A
Fortimicin A sulfate
Fotrin
Fulvine
Fumidil
Furapyrimidone
Furazosin hydrochloride
2-(2-Furyl)-3-(5-nitro-2-furyl)acrylamide
Fusarenone X
Fusaric acid calcium salt
Fusariotoxin T 2
Fusidine
Fyrol FR 2
Gabexate mesylate
Galactose
Gastrozepin
Gentamycin
Gentamycin sulfate
Gentisic acid
Germanium dioxide
Gestoral
Gindarine hydrochloride
Glucagon

2-(beta-d-Glucopyranosyloxy) isobutyronitrile
d-Glucose
Gludiase
Glutaraldehyde
Glutril
Glycidol
Glycinonitrile
Glycinonitrile hydrochloride
Glycol ethers
Glycyrrhizic acid, ammonium salt
Gold sodium thiomalate
Gonadotropin releasing hormone agonist
Gossypol acetic acid
Grisofulvin
Guanabenz acetate
Guanazodine
Guanfacine hydrochloride
Guanine-3-N-oxide
Guanosine
HBK
Haloanisone
Halofantrine hydrochloride
Haloperidol decanoate
Halopredone acetate
Halothane
Haloxazolam
HCDD
Heliotrine
Hematoidin
Heptamethylphenylcyclotetrasiloxane
Heptyl phthalate
Heroin
Hexabromonaphthalene
Hexachlorobenzene
2,2',4,4',5'5'-Hexachloro-1,1,-biphenyl
3,3',4,4',5,5'-Hexachlorobiphenyl
Hexachlorobutadiene
Hexachlorocyclopentadiene
1,2,3,4,7,8-Hexachlorodibenzofuran
Hexachlorophene
4,5,6,7,8,8-Hexachlor-D1,5-tetrahydro-4,7-methanoinden
1-Hexadecanamine
Hexadecyltrimethylammonium bromide
Hexafluoroacetone
Hexafluoro acetone trihydrate
Hexamethonium bromide
Hexamethylmelamine
n-Hexane
1,6-Hexanediamine
2-Hexanone
Hexocyclium methylsulfate
Hexone
Hexoprenaline dihydrochloride
Hexoprenaline sulfate
n-Hexyl carborane
Histamethizine
Histamine diphosphate
Homofolate
Human immunoglobin COG-78
Hyaluronic acid, sodium salt
Hycanthone methanesulfonate
Hydantoin
Hydralazine
Hydralazine hydrochloride
Hydrazine
Hydrochlorbenzethylamine dimaleate
Hydrochloric acid
Hydrocortisone sodium succinate
Hydrocortisone-21-acetate
Hydrocortisone-17-butyrate
Hydrocortisone-17-butyrate-21-propionate
Hydrocortisone-21-phosphate
Hydrofluoric acid
10-beta-Hydroperoxy-17-alpha-ethynyl-4-estren-17-beta-OL-3-one
Hydroquinone-beta-d-glucopyranoside
N-Hydroxy ethyl carbamate
4,-Hydroxyacetanilide
N-Hydroxy-N-acetyl-2-aminofluorene
N-Hydroxyadenine
6-N-Hydroxyadenosine
3-alpha-Hydroxy-17-androston--one
17-beta-Hydroxy-5-beta-androstan-3-one
3-Hydroxybenzoic acid
para-Hydroxybenzoic acid ethyl ester
5-(alpha-Hydroxybenzyl)-2-benzimidazolecarbamic acid methyl ester
1-Hydroxycholecalciferol
Hydroxydimethylarsine oxide
Hydroxydimethylarsine oxide, sodium salt
9-Hydroxyellipticine
2-(2-Hydroxyethoxy)ethyl-N-(alpha,alpha,alpha-trifluoro-meta-tolyl)anthranilate
Hydroxyethyl starch
beta-Hydroxyethylcarbamate
1-Hydroxyethylidene-1,1-diphosphonic acid
17-beta-Hydroxy-7-alpha-methylandrost-5-ENE-3-one
7-Hydroxymethyl-12-methylbenz(alpha)anthracene
1-Hydroxymethyl-2-methylditmide-2-oxide
5-Hydroxymethyl-4-methyluracil
2-Hydroxymethylphenol
5-(1-Hydroxy-2-((1-methyl-3-phenylpropyl)amino)ethyl) salicyclamide hydrochloride
N-(Hydroxymethyl)phthalimide
3-(1-Hydroxy-2-piperidinoethyl)-5-phenylisoxazole citrate
2-Hydroxy-N-(3-(meta-(piperidinomethyl)phenoxy)propyl)acetamide acetate (ester hydrochloride)
Hydroxyprogesterone caproate
beta-(N-(3-Hydroxy-4-pyridone))-alpha-aminopropionic acid
4-Hydroxysalicylic acid
5-Hydroxytetracycline
5-Hydroxytetracycline hydrochloride
17-beta-Hydroxy-4,4,17-alpha-trimethyl-androst-5-ENE(2,3-d) isoxazole
Hydroxytriphenylstannane
dl-Hydroxytryptophan
5-Hydroxy-1-tryptophan
dl-Hydroxytryptophan
5-Hydroxy-1-tryptophan
Hydroxyurea
3-Hydroxyxanthine
Hydroxyzine pamoate
Hyoscine hydrobromide
Hypochlorous acid

Hypoglycine B
Ibuprofen piconol
Ifenprodil tartrate
IMET 3106
4-Imidazo (1,2-alpha) pyridin-2-yl-alpha-methylbenzeneacetic acid
Imidazole mustard
2-Imidazolidinethione
2-Imidazolidinethione mixed with sodium nitrite
2-Imino-5-phenyl-4-oxazolidinone
Improsulfan tosylate
Indacrinone
Indanazoline hydrochloride
1,3-Indandione
Indapamide
Indeloxazine hydrochloride
Inderal
Indium
Indium nitrate
1H-Indole-3-acetic acid
Indole-3-carbinol
Indomethacin
Inolin
Insulin
Insulin protamine zinc
Iocarmate meglumine
Iodoacetic acid
Iopramine hydrochloride
Iotroxate meglumine
Ipratropium bromide
Iron-dextran complex
Iron nickel zinc oxide
Iron-poly (sorbitol-gluconic acid)complex
Iron-sorbitol
Isoamygdalin
Isoamyl 5,6-dihydro-7,8-dimethyl-4,5-dioxo-4H-pyrano (3,2-c) quinoline-2-carboxylate
Isobutyl methacrylate
para-Isobutylhydratropic acid
Isocarboxazid
Isodecyl methacrylate
Isodonazole nitrate
Isoflurane
Isonicotinic acid hydrazide
Isonicotinic acid-2-isopropylhydrazide
Isooctyl-2,4-dichlorophenoxyacetate
Isophosphamide
Isoprenaline hydrochloride
Isoprenyl chalcone
Isopropyl alcohol
Isopropyl-2,4-D ester
Isopropylidine azastreptonigrin
4,4,-Isopropylidenediphenol, polymer with 1-chloro-2,3-epoxypropane
Isopropylmethanesulfonate
Isosafrole-n-octylsulfoxide
Isothiacyanic acid, ethylene ester
Isothiocyanic acid, phenyl ester
Isothiourea
Jervine
Jervine-3-acetate
Josamycin
Kanamycin
Kanamycin sulfate (1:1) salt
KAO 264
Karminomycin
Kepone
Kerlone
Ketamine
Ketoprofen sodium
Ketotifen fumarate
KΓ-868
Khat leaf extract
KM-1146
KPE
Lactose
Latamoxef sodium
Lead
Lead acetate (II), trihydrate
Lead chloride
Lead diacetate
Lead (II) nitrate (1:2)
Lecithin iodide
Lenampicillin hydrochloride
Lendormin
Lente insulin
Lentinan
Leptophos
1-Leucine
Leurocristine
Leurocristine sulfate (1:1)
Levamisole hydrochloride
Levorin
Levothyroxine sodium
Librium
d-Limonene
Linear alkylbenzenesulfonate,sodium salt
Linoleic acid (oxidized)
Liothyronine
Lipopolysaccharide, escherichia coli
Lipopolysaccharide, from B. Abortus Bang.
Lithium carbonate (2:1)
Lithium carmine
Lithium chloride
Lividomycin
Lobenzarit disodium
Locoweed
Lofetensin hydrochloride
Lucanthone metabolite
Luteinizing hormone antiserum
Luteinizing hormone-releasing hormone
Luteinizing hormone-releasing hormone, diacetate (salt)
Luteinizing hormone-releasing hormone, diacetate, tetrahydrate
Lyndiol
Lysenyl hydrogen maleate
d-Lysergic acid diethylamide tartrate
Lysergide tartrate
Lysine
Mafenide acetate
Magnesium glutamate hydrobromide
Magnesium sulfate (1:1)
Malathion
Maleimide
Malotilate
Maltose
Manganese (II) chloride (1:2)
Manganese (II) ethylenebis (dithiocarbamate)
Manganese (II) sulfate (1:1)
Maprotiline hydrochloride
Marezine hydrochloride
Maytansine
Mazindol

Mec
Meclizine dihydrochloride
Meclizine hydrochloride
Medemycin
Medrogestone
Medroxyprogesterone
Medroxyprogesterone acetate
Medullin
Melengestrol acetate
Mentha arvensis, oil
Mepiprazole dihydrochloride
Mepyrapone
Mequitazine
2-Mercapto-1-methylimidazole
1-(d-3-Mercapto-2-methyl-1-oxopropyl)-1proline (S,S)
N-(2-Mercapto-2-methylpropanoyl)-1-cysteine
6-Mercaptopurine monohydrate
6-Mercaptopurine 3-N-oxide
Mercaptopurine ribonucleoside
d,3-Mercaptovaline
Mercuric acetate
Mercuric oxide
Mercury
Mercury (II) chloride
Mercury (II) iodide
Mercury methylchloride
Merthiolate sodium
Mervan ethanolamine salt
Mescaline
Mesoxalylurea monohydrate
Mestranol mixed with norethindrone
Metalutin
Metaproterenol sulfate
Methadone
Methadone hydrochloride
dl-Methadone hydrochloride
Methallyl-19-nortestosterone
Methaminodiazepoxide hydrochloride
1-Methamphetamine hydrochloride
Methaqualone hydrochloride
Methedrine
dl-Methionine
l-Methionine
Methionine sulfoximine
Methofadin
Methophenazine difumarate
Methotrexate
Methotrexate sodium
Methoxyacetic acid
3-Methoxycarbonylaminophenyl-N-3,-methylphenylcarbamate
Methoxychlor
5-Methoxyindoleacetic acid
4-(6-Methoxy-2-naphthyl)-2-butanone
(+)-2-(Methoxy-2-naphthyl)propionic acid
2-(3-Methoxyphenyl)-5,6-dihydro-s-triazolo (5,1-alpha) isoquinoline
2-(para-(6-Methoxy-2-phenyl-3-indenyl)phenoxy)triethylamine hydrochloride
2-(para-(para-Methoxy-alpha-phenylphenethyl)phenoxy)triethylamine hydrochloride
N1-(3-Methoxy-2-pyrazinyl)sulfanilamide
Methyl alcohol
Methyl azoxymethyl acetate
Methyl benzimidazole-2-YL carbamate
2-Methyl butylacrylate
Methyl chloride
Methyl chloroform
Methyl (beta)-11-alpha-16-dihydroxy-16-methyl-9-oxoprost-13-EN-1- OATE
Methyl ethyl ketone
Methyl hydrazine
Methyl isocyanate
Methyl mesylate
Methyl methacrylate
Methyl (methylthio) mercury
Methyl parathion
Methyl pentachlorophenate
Methyl phenidyl acetate
Methyl salicylate
Methyl thiourea
Methyl urea and sodium nitrite
Methylacetamide
Methyl-5-benzoyl benzimidazole-2-carbamate
1-Methyl-2-benzylhydrazine
1-Methyl-5-chloroindoline methylbromide
Methylchlortetracycline
3-Methylcholanthrene
N-Methyl-4-cyclochexene-1,2-dicarboximide
N-Methyl-N-desacetylcolchicine
N-Methyl-dibromomaleinimide
beta-Methyldigoxin
17-alpha-Methyldihydrotestosterone
N-Methyl-3,6-dithia-3,4,5,6-tetrahydrophthalimide
Methylene chloride
Methylene dimethanesulfonate
N,N,-Methylenebis(2-amino-1,3,4-thiadiazole)
2-Methylenecyclopropanylalanine
Methylergonovine maleate
3-(1-Methylethyl)-1H-2,1,3-benzothiazain-4(3H)-one-2,2-dioxide
4-Methylethylenethiourea
3-Methyl-5-ethyl-5-phenylhydantoin
3-Methylethynylestradiol
x-Methylfolic acid
N-Methylformamide
Methylhesperidin
(alpha-(2-Methylhydrazino)-para-toluoyl)urea, monohydrobromide
4-Methyl-7-hydroxycoumarin
Methyl-ortho-(4-hydroxy-3-methoxycinnamoyl) reserpate
2-Methyl-1,3-indandione
N-Methyljervine
N-Methyllorazepam
Methylmercuric dicyandiamide
Methylmercuric phosphate
Methylmercury
Methylmercury hydroxide
1-Methyl-6-(1-methylallyl)-2,5-dithiobiurea
d-3-Methyl-N-methylmorphinan phosphate
N-Methyl-alpha-methyl-alpha-phenylsuccinimide
2-Methyl-1,4-naphthoquinone
2-Methyl-5-nitroimidazole-1-ethanol

N-Methyl-N,-nitro-N-nitrosoguanidine
4-(N-Methyl-N-nitrosamino)-1-(3-pyridyl)-1-butanone
N-Methyl-N-nitrosoaniline
N-Methyl-N-nitrosoethylcarbamate
N-Methyl-N-nitroso-1-propanamine
N-Methyl-N-nitrosourea
(3-Methyl-4-oxo-5-piperidino-2-thiazolidinylidene) acetic acid ethyl ester
10-Methylphenothiazine-2-acetic acid
N-Methyl-para-(phenylazo) aniline
3-Methyl-2-phenylmorpholine hydrochloride
N-Methyl-2-phenyl-succinimide
Methyl-4-phthalimido-dl-glutaramate
N-Methyl-2-phthalimidoglutarimide
N-Methylpyrrolidone
Methylsulfonyl chloramphenicol
17-Methyltestosterone
N-Methyl-3,4,5,6-tetrahydrophthalimide
Methylthioinosine
6-Methylthiouracil
6-Methyluracil
Metiapine
Meticrane
Metoprine
Metoprolol tartrate
Metrizamide
Mexiletine hydrochloride
Mezinium methyl sulfate
Mezlocillin
Mibolerone
Miconazole nitrate
Micromycin
Midodrine
Mikelan
Miloxacin
Miltown
Mineral oil
Mineral oil, petroleum extracts, heavy naphthenic distillate solvent
Mirex
Mithramycin
MN-1695
Mobilat
Molybdenum
Monoethylhexyl phthalate
Monoethylphenyltriazene
8-Monohydro mirex
Monosodium glutamate
Morphine hydrochloride
Morphine sulfate
Morphocycline
Moxestrol
Moxnidazole
Mucopolysaccharide, polysulfuric acid ester
Muldamine
Mycosporin
Nafoxidine hydrochloride
Naftidrofuryl oxalate
Naja nigricollis venom
Naloxone hydrochloride
Naphthalene
beta-Naphthoflavone
1-Naphthol
Navaron
Neem oil
Nembutal sodium
Neocarzinostatin
Neoprene
Neoproserine
Neosynephrine
Netilmicin sulfate
Nickel
Nickel carbonyl
Nickel compounds
Nickel subsulfide
Nickelous chloride
Nicotergoline
Nicotine
Nicotine tartrate (1:2)
N-Nicotinoyltryptamide
Nipradilol
Nisentil
Nitric acid
Nitrilotriacetic acid trisodium salt monohydrate
Nitrobenzene
Nitrofurantoin
Nitrofurazone
4-((5-Nitrofurfurylidene)amino)-3-methylthiomorpholine-1,1- dioxide
Nitrogen dioxide
Nitrogen oxide
Nitroglycerin
1-(2-Nitroimidazol-1-YL-3-methoxypropan-2-ol
Nitromifene citrate
2-Nitropropane
4-Nitroquinoline-N-oxide
Nitroso compounds
N-Nitroso compounds
N-Nitrosobis(2-oxopropyl)amine
Nitrosocimetidine
N-Nitrosodiethylamine
N-Nitrosodimethylamine
N-Nitrosodi-N-propylamine
N-Nitroso-N-ethyl aniline
N-Nitroso-N-ethylurethan
N-Nitroso-N-ethylvinylamine
N-Nitrosohexahydroazepine
N-Nitrosoimidazolidinethione
N-Nitrosopiperidine
1-(Nitrosopropylamino)-2-propanol
N-Nitroso-N-propylurea
Nizofenone fumarate
Norchlorcyclizine
Norchlorcyclizine hydrochloride
1-Norepinephrine
19-Norethisterone
Norethisterone enanthate
Norgestrel
1-Norgestrel
19-Norpregn-4-ene-3,20-dione
19-Nor-17-alpha-pregn-5(10)-en-20-yne-3-alpha,17-diol
19-Nor-17-alpha-pregn-5(10)-en-20-yne-3-beta,17-diol
19-Nor-17-alpha-pregn-4-en-20-yn-17-ol
Novadex
Nutmeg oil, East Indian
Nystatin
Ochratoxin
Ochratoxin A sodium salt
Octabromodiphenyl
Octachlorodibenzodioxin
Octoclothepine

Ofloxacin
Oleamine
Oleylamine hydrofluoride
Oncodazole
Ophthazin
Orgoteins
Orphenadrine hydrochloride
Oxaprozin
Oxatimide
Oxazolazepam
Oxepinac
Oxfendazole
Oxibendazole
Oxiranecarboxylic acid, 3-(((3-methyl-1-(((3- methylbutyl)amino) carbonyl)-,ethyl ester, (2S-(2-alpha-3- beta)R)))
N-(2-Oxo-3,5,7-cylcoheptatrien-1-yl)aminooxoacetic acid ethyl ester
2-(3-Oxo-1-indanylidene)-1,3-indandione
Oxolamine citrate
N-(2-Oxo-3-piperidyl)phthalimide
Oxybutynin chloride
Oxygen
Oxymorphinone hydrochloride
beta-Oxypropylpropylnitrosamine
Ozone
Padrin
Palm oil
Panoral
d-Pantethine
Pantocrin
Papain
Papaverine chlorohydrate
Paradione
Paramathasone acetate
Paraquat dichloride
Parathion
Paraxanthine
Pavisoid
PE-043
Penfluridol
Penicillic acid
Penitrem A
Pentachlorobenzene
2,3,4,7,8-Pentachlorodibenzofuran
Pentachloronitrobenzene
Pentachlorophenol
Pentafluorophenyl chloride
Pentazocine hydrochloride
Pentostatin
Pentothal
Pentothal sodium
Pentoxyphylline
Perchloroethylene
Perdipine
Perfluorodecanoic acid
Periactin hydrochloride
Periactinol
Perphenazine hydrochloride
Pharmagel A
1,10-Phenanthroline
Phenazin-5-oxide
Phenethyl alcohol
Phenfluoramine hydrochloride
Phenol
4-Phenoxy-3-(pyrrolidinyl)-5-sulfamoylbenzoic acid
Phenyl salicylate
Phenylacetic acid
(Phenylacetyl) urea
1-Phenylalanine
17-beta-Phenylaminocarbonyloxyoestra-1,3,5(10)-triene-3- methyl ether
para-(Phenylazo)aniline
2-Phenyl-5-benzothiazoleacetic acid
1-Phenyl-3,3-diethyltriazene
2-Phenyl-5,5-dimethyl-tetrahydro-1,4-oxazine hydrochloride
1-Phenyl-2-(1',1'-diphenylpropyl-3'-amino)propane
4-Phenyl-1,2-diphenyl-3,5-pyrazolidinedione
meta-Phenylenediamine
2-Phenylethylhydrazine
Phenylmethylcylosiloxane, mixed copolymer
N-Phenylphthalimidine
Phenyl-2-pyridylmethyl-beta-N,N-dimethylaminoethyl ether succinate
2-(Phenylsulfonylamino)-1,3,4-thiadiazole-5-sulfonamide
1-Phenyl-2-thiourea
Phomopsin
Phorbol myristate acetate
Phosphonacetyl-1-aspartic acid
Phosphoramide mustard cyclohexylamine salt
Phthalazinol
Phthalic anhydride
Phthalimide
Phthalimidomethyl-O,O-dimethyl phosphorodithioate
N-Phthaloly-1-aspartic acid
N-Phthalylisoglutamine
Physostigmine sulfate
Phytohemagglutinin
Picloram
Pilocarpine monohydrochloride
Pimozide
2,6-Piperazinedione-4,4,-propylene dioxopiperazine
Piperidine
3-Piperidine-1,1-diphenyl-propanol-(1) methanesulphonate
Piperin
Piperonyl butoxide
Pipethanate ethylbromide
Pipram
Pituitary growth hormone
Plafibride
cis-Platinous diammine dichloride
Platinum thymine blue
Podophyllin
Podophyllotoxin
Polybrominated biphenyls
Polychlorinated biphenyl (Aroclor 1248)
Polychlorinated biphenyl (Aroclor 1254)
Polychlorinated biphenyl (Kanechlor 300)
Polychlorinated biphenyl (Kanechlor 400)
Polychlorinated biphenyl (Kanechlor 500)
Polyoxyethylene sorbitan monolaurate
Potassium bichromate
Potassium canrenoate
Potassium chromate (VI)
Potassium clavulanate

Potassium cyanide
Potassium fluoride
Potassium iodide
Potassium nitrate
Potassium nitrite (1:1)
Potassium perchlorate
Potassium thiocyanate
Potato blossoms, glycoalkaloid extract
Potato, green parts
Pranoprofen
Prednisolone succinate
Prednisone 21-acetate
Predonin
9-beta,10-alpha-Pregna-4,6-diene-3,20-dione
17-alpha- hydroxypregn-4-ene-3,2 ortho-dione (9:10)
5-alpha-17-alpha-Pregna-2-en-20-yn-17-ol, acetate
Premarin
Primaquine phosphate
Primobolan
Prinadol hydrobromide
Procarbazine
Procarbazine hydrochloride
Procaterol hydrochloride
Prochlorpromazine
Progesterone
Prolinomethyltetracycline
Promethazine hydrochloride
Propadrine hydrochloride
Propane sultone
1,3-Propanediamine
1,2-Propanediol
Propanidide
3-Propanolamine
Proparthrin
Propazone
Propiononitrile
Propoxur
2-Propoxyethyl acetate
d-Propoxyphene hydrochloride
Propyl carbamate
Propyl cellosolve
n-Propyl gallate
Propylene glycol diacetate
Propylene glycol monomethyl ether
Propylene oxide
2-Propylpentanoic acid
2-Propylpiperidine
6-Propyl-2-thiouracil
Propylthiouracil and iodine
2-Propylvaleramide
2-Propylvaleric acid sodium salt
Prostaglandin A1
Prostaglandin E1
Prostaglandin E2 sodium salt
Prostaglandin F1-alpha
Prostaglandin F2-alpha
Prostaglandin F2-alpha-tham
Protizinic acid
Proxil
Pseudolaric acid A
Pseudolaric acid B
Purapuridine
Purine-6-thiol
Pyrantel pamoate
Pyrazine-2,3-dicarboxylic acid imide
Pyrazole
Pyrbuterol hydrochloride
Pyridinamine (9CI)
2,3-Pyridinedicarboximide
3,4-Pyridinedicarboximide
1-(Pyridyl-3)-3,3-dimethyl triazene
1-Pyridyl-3-methyl-3-ethyltriazene
5-(para-(2-Pyridylsulfamoyl)phenylazo)salicyclic acid
Pyrimidine-4,5-dicarboxylic acid imide
N1-2-Pyrimidinyl-sulfanilamide
Pyrogallol
Pyronaridine
N-(1-Pyrrolidinylmethyl)-tetracycline
Quaalude
Quercetin
Quinine
2-Quinoline thioacetamide hydrochloride
Ralgro
Refosporen
Reptilase
Reserpine
Retinoid etretin
all-trans-Retinylidene methyl nitrone
Rhodamine 6G extra base
2-beta-d-Ribofuranosyl-as-triazine-3,5(2H,4H)-dione
1-beta-d-Ribofuranosyl-1,2,4-triazole-3-carboxamide
Ricin
Rifamycin AMP
Rifamycin SV
Ripcord
Ritodrine hydrochloride
Rizaben
Robaveron
Ronnel
Rose bengal sodium
Rotenone
Rowachol
Rowatin
R Salt
Rubratoxin B
Rythmodan
Salicyclaldehyde
Salicyclamide
Salicyclic acid
Salicyclic acid, compounded with morpholine (1:1)
ortho-Salicylsalicylic acid
Salipran
Salmonella enteritidis endotoxin
Sarkomycin
SCH 20569
Scopolamine
Sefril
Selenium
Selenodiglutathione
Semicarbazide hydrochloride
Serum gonadotropin
Sfericase
Silicone 360
Sisomicin
S. Marcescens lipopolysaccharide
Smoke condensate, cigarette
Smokeless tobacco
Sodium para-aminosalicylate
Sodium arsenite
Sodium benzoate
Sodium bicarbonate
Sodium chloride

Sodium chlorite
Sodium chondroitin polysulfate
Sodium cobaltinitrite
Sodium colistinemethanesulfonate
Sodium cyanide
Sodium cyclamate
Sodium dehydroacetic acid
Sodium dichlorocyanurate
Sodium diethyldithiocarbamate
Sodium diphenyldiazo-bis(alpha-
naphthylaminesulfonate)
Sodium fluoride
Sodium (E)-3-(para-(1H-imidazol-1-
methyl)phenyl)-2-propenoate
Sodium iodide
Sodium lauryl sulfate
Sodium luminal
Sodium nigericin
Sodium nitrite
Sodium nitrite and carbendazime (1:1)
Sodium nitrite and 1-citrulline (1:2)
Sodium nitrite and 1-(methylethyl) urea
Sodium nitroferricyanide
Sodium pentachlorophenate
Sodium picosulfate
Sodium piperacillin
Sodium retinoate
Sodium saccharin
Sodium salicylate
Sodium selenite
Sodium selenite pentahydrate
Sodium sulfate (2:1)
Sodium d-thyroxine
Sodium tolmetin dihydrate
Sodium-2,4-dichlorophenoxyacetate
(22s,25r)-5-alpha-Solanidan-3-beta-OL
Solanid-5-ENE-3-beta, 12-alpha-diol
(22s,25r)-Solanid-5-EN-3-beta-OL
Solanine
Solcoseryl
Spectogard
Spiclomazine hydrochloride
Spiramycin
Spiroperidol
SRC-II, heavy distillate
1-ST-2121
Sterculia foetida oil
Steroids
Stimulexin
Streptomycin
Streptomycin
dihydrostreptomycin
Streptomycin sesquisulfate
Streptomycin sulphate
Streptonigran
Streptonigrin methyl ester
Streptozoticin
STS 557
Styrene
Subtigen
Succinic anhydride
Succinonitrile
Sucrose
Sulfadiazine silver salt
Sulfadimethoxypyrimidine
Sulfadimethyldiazine
Sulfamonomethoxin
Sulfamoxole-trimethoprim mixture
Sulfanilamide
6-Sulfanilamido-2,4-
dimethoxypyrimidine
5-Sulfanilamido-3,4-dimethyl-
isoxazole
Sulfanilylurea
N-Sulfanylacetamide
alpha-Sulfobenzylpenicillin
disodium
Sulfur dioxide
Sulfuric acid
Suloctidyl
Sultopride hydrochloride
Supercortyl
Superprednol
Surgam
Surital sodium
Surmontil maleate
Suxibuzone
Sweet pea seeds
Sygethin
meta-Synephrine hydrochloride
Synephrine tartrate
Synsac
2,4,5-T
T-1982
T-2588
Tagamet
Tarweed
TCDD
Tellurium
Tellurium dioxide
Temephos
Tenormin
Terbutaline sulphate
Terodiline hydrochloride
Testosterone
Testosterone heptanoate
Testosterone propionate
1,1,3,3-Tetrabutylurea
2,3,7,8-Tetrachlododibenzofuran
Tetrachloroacetone
1,1,3,3-Tetrachloroacetone
3,3',4,4'-Tetrachloroazoxbenzene
1,2,3,4-Tetrachlorobenzene
3,3',4,4'-Tetrachlorobiphenyl
2,4,5,6-Tetrachlorophenol
Tetracycline
Tetracycline hydrochloride
Tetraethyl lead
1-trans-D9-Tetrahydrocannabinol
2-(para-(1,2,3,4-Tetrahydro-2-(para-
chlorophenyl)naphthyl) phenoxy)
triethyl amine
2,3,4,5-Tetrahydro-2,8-dimethyl-5-
(2-(6-methyl-3-pyridyl)ethyl)- 1H-
pyrid 0-(4,3-beta) indole
Tetrahydro-3,5-dimethyl-4H,1,3,5-
oxadiazine-4-thione
5,6,7,8-Tetrahydrofolic acid
2-(1,2,3,4-Tetrahydro-1-
naphthylamino)-2-imidazoline
hydrochloride
4,-O-Tetrahydropyranyladriamycin
hydrochloride
para-(1,1,3,3-
Tetramethylbutyl)phenol, polymer
with ethylene oxide and
formaldehyde
2,2,9,9-Tetramethyl-1,10-decanediol
Tetramethyl lead
Tetramethylsuccinonitrile
Tetramethylthiourea
1,1,3,3-Tetramethylurea
Tetranicotylfructose
Tetrapotassium hexacyanoferrate

Tetrasodium fosfestrol
Tetrazosin hydrochloride dihydrate
Thalidomide
(+)-Thalidomide
(-)-Thalidomide
Thallium acetate
Thallium chloride
Thallium compounds
Thallium sulfate
Thebaine hydrochloride
para-(2-Thenoyl) hydratropic acid
Theobromine
Theobromine sodium salicylate
Theophylline
1-(Theophyllin-7-YL)ethyl-2-(2-(para-chlorophenoxy)-2-methylpropionate
Thiamine chloride
2-(Thiazol-4-YL) benzimidazole
2-(4-Thiazolyl)-5-benzimidazolecarbamic acid methyl ester
Thioacetamide
Thioinosine
Thiotriethylenephosphoramide
2-Thiouracil
Thiram
Thymidine
Thyroid
1-Thyroxin
Thyroxine
Tiapride hydrochloride
Ticarcillin sodium
Ticlodone
Timepidium bromide
Timiperone
Tinactin
Tindurin
Tinidazole
Tinoridine hydrochloride
Tiquizium bromide
2,4,5-T isooctyl ester
Titanium (wet powder)
Tizanidine hydrochloride
Tobacco
Tobacco leaf, nicotiana glauca
Tobramycin
Todralazine hydrochloride hydrate
Togal
Tolmetine
Toluene
para-Toluenediamine sulfate
ortho-Toluidine
Tormosyl
2,4,5-T propylene glycol butyl ether ester
Traxanox sodium pentahydrate
Triaminoguanidine nitrate
para,para,
Triazenylenedibenzenesulfonamide
Triazolam
Trichloroacetonitrile
1,2,4-Trichlorobenzene
Trichloroethylene
2,4,4,-Trichloro-2,-hydroxydiphenyl ether
(2,2,2-Trichloro-1-hydroxyethyl) dimethylphosphonate
N-Trichloromethylthio)phthalimide
4-(2,4,5-Trichlorophenoxy) butyric acid
alpha-(2,4,5-Trichlorophenoxy) propionic acid
Trichloropropionitrile
Triclopyr
Tricosanthin
Tridemorph
Tridiphane
Triethyl lead chloride
Triethylenetetramine
2,2,2-Trifluoroethyl vinyl ether
3,-Trifluoromethyl-4-dimethylaminoazobenzene
Trifluoromethylperazine
2-(8,-Trifluoromethyl-4,-quinolylamino)benzoic acid, 2,3-dihydroxy propyl ester
Trifluperidol
Triglyme
Trimebutine maleate
(beta)-Trimethoquinol
Trimethoxazine
5-(3,4,5-Trimethoxybenzyl)-2,4-diaminopyrimidine
Trimethyl lead chloride
Trimethyl phosphate
Trimethyl phosphite
3,3,5-Trimethyl-2,4-diketooxazolidine
Trimethylenedimethanesulfonate
exo-Trimethylenenorbornane
1,1,3-Trimethyl-3-nitrosourea
1,3,5-Trimethyl-2,4,6-tris(3,5-DI-tert-butyl-4-hydroxybenzyl) benzene
Triparanol
Tris (1-aziridinyl)-para-benzoquinone
Tris- (1-aziridinyl) phosphine oxide
Trisaziridinyltriazine
Tris (1-methylethylene) phosphoric triamide
Tritolyl phosphate
Tropacaine hydrochloride
1-Tryptophan
TSH-releasing hormone
Tungsten
dl-meta-Tyrosine
1-Tyrosine
Ubiquinone 10
Uracil
Uracil mixture with tegafur (4:1)
Uranyl acetate dihydrate
Urapidil
Urbacide
Urbason soluble
Urethane
Urfamicin hydrochloride
Uridion
Urokinase
Valbazen
Valison
Vanadium pentoxide (dust)
Vasodilan
Vasodilian
Vasodistal
Vasotonin
Venacil
Ventipulmin
Veratramine
Veratrine
Veratrylamine
Vincaleukoblastine
Vincaleukoblastine sulfate (1:1) (salt)

Vinyl chloride
Vinyl toluene
Vinylidene chloride
R-5-Vinyl-2-oxazolidinethione
Viomycin
Vipera berus venom
Viriditoxin
Visken
Vistaril hydrochloride
Vitamin A
Vitamin A acetate
Vitamin A acid
13-cis-Vitamin A acid
Vitamin A palmitate
Vitamin B7
Vitamin B12 complex
Vitamin B12, methyl
Vitamin D2
Vitamin K
Vitamin MK 4
Volidan
Vomitoxin
Wait's green mountain antihistamine
Warfarin
Warfarin sodium
White spirit
Xamoterolfumarate
Xanax
Xanthinol nicotinate
Xylene
meta-Xylene
ortho-Xylene
para-Xylene
Xylostatin
N-(2,3-Xylyl)anthranilic acid
Ytterbium chloride
Zaroxolyn
Zearalenone
Zimelidine dihydrochloride
Zinc carbonate (1:1)
Zinc chloride
Zinc (II) ELIA complex
Zinc oxide
Zinc (N,N,-propylene-1,2-bis(dithiocarbamate))
Zinc pyridine-2-thiol-1-oxide
Zinc sulfate
Zoapatle, crude leaf extract
Zoapatle, semi-purified leaf extract
Zotepine
Zygosporin A
Zyloprim

Appendix II. Acute Toxicity of Extremely and Highly Hazardous Pesticides – Classes IA and IB

Class Ia – Extremely hazardous		Class Ib – Highly hazardous			
Common name	LD_{50} (mg/kg) Oral, rat	Common name	LD_{50} (mg/kg) Oral, rat	Common name	LD_{50} (mg/kg) Oral, rat
Aldicarb	0.93	Acrolein	29	Mecarbam	36
Brodifacoum	0.3	Allyl alcohol	64	Mercuric	18
Bromadiolone	1.12	Azinphos-ethyl	12	Methamidop	30
Bromethalin	2	Azinphos-methyl	16	Methidathion	25
Calcium cyanide	39	Blasticidin-S	16	Methiocarb	20
Captafol	5000	Butocarboxim	158	Methomyl	17
Chloroetoxyfos	1.8	Butoxycarboxim	288 (dermal)	Monocrotoph	14
Chlormephos	7	Cadusafos	37	Nicotine	50 (dermal)
Chlorophacinone	3.1	Calcium arsenate	20	Omethoate	50
Difenacoum	1.8	Carbofuran	8	Oxamyl	6
Difethialone	0.56	Chlorfenvinphos	31	Oxydemeton-	65
Diphacinone	2.3	3-Chloro-1,2-	112	Paris green	22
Disulfoton	2.6	Coumaphos	7.1	Pentachloroph	80 (dermal)
EPN	14	Coumatetralyl	16	Propetamphos	10
Ethoprophos	26	Demeton-S-methyl	40	Sodium	10
Flocoumafen	0.25	Dichlorvos	56	Sodium	6
Hexachlorobenzene	10 000	Dicrotophos	22	Strychnine	16
Mercuric chloride	1	Dinoterb	25	Tefluthrin	22
Mevinphos	4	DNOC	25	Thallium	18
Parathion	13	Edifenphos	150	Thiofanox	83
Parathion-methyl	14	Ethiofencarb	200	Thiometon	120
Phenylmercury acetate	24	Famphur	48	Triazophos	82
Phorate	2	Fenamiphos	15	Vamidothion	103
Phosphamidon	7	Fluoroacetamide	13	Warfarin	10
Sodium fluoroacetate	0.2	Formetanate	21	Zinc	45
Sulfotep	5	Furathiocarb	42		
Tebupirimfos	1.3	Heptenophos	96		
Terbufos	2	Isoxathion	112		
		Lead arsenate	10		

Source: The WHO Recommended Classification of Pesticides by Hazard and Guidelines to Classification, 2004.

APPENDIX III. ACUTE TOXICITY OF MODERATELY HAZARDOUS PESTICIDES - CLASS II

Common name	LD_{50} (mg/kg) Oral, rat	Common name	LD_{50} (mg/kg) Oral, rat	Common name	LD_{50} (mg/kg) Oral, rat
Alanycarb	330	Gamma-HCH,	88	Profenofos	358
Anilofos	472	Guazatine	230	Propiconazole	1520
Azaconazole	308	Haloxyfop	393	Propoxur	95
Azocyclotin	80	HCH	100	Prosulfocarb	1820
Bendiocarb	55	Imazalil	320	Prothiofos	925
Benfuracarb	205	Imidacloprid	450	Pyraclofos	237
Bensulide	270	Iminoctadine	300	Pyrazophos	435
Bifenthrin	55	loxynil	110	Pyrethrins	500-1000
Bilanafos	268	loxynil octanoate	390	Pyroquilon	320
Bioallethrin	700	Isoprocarb	403	Quinalphos	62
Bromoxynil	190	Lambda-	56	Quizalofop-p-	1012
Bromuconazole	365	Mercurous	210	Rotenone	132-1500
Bronopol	254	Metaldehyde	227	Spiroxamine	500
Butamifos	630	Metam-sodium	285	TCA (acid)	400
Butylamine	380	Methacrifos	678	Terbumeton	483
Carbaryl	300	Methasulfocarb	112	Tetraconazole	1031
Carbosulfan	250	Methyl	72	Thiacloprid	444
Cartap	325	Metolcarb	268	Thiobencarb	1300
Chloralose	400	Metribuzin	322	Thiocyclam	310
Chlordane	460	Molinate	720	Thiodicarb	66
Chlorfenapyr	441	Nabam	395	Tralomethrin	85
Chlorphonium	178	Naled	430	Triazamate	50-100
Chlorpyrifos	135	Paraquat	150	Trichlorfon	250
Clomazone	1369	Pebulate	1120	Tricyclazole	305
Copper sulfate	300	Permethrin	500	Tridemorph	650
Cuprous oxide	470	Phenthoate	400	Xylylcarb	380
Cyanazine	288	Phosalone	120		
Cyanophos	610	Phosmet	113		
Cyfluthrin	250	Phoxim	1975 (dermal)		
Beta-cyfluthrin	450	Piperophos	324		
Cyhalothrin	144	Pirimicarb	147		
Cypermethrin	250	Prallethrin	460		

Source: The WHO Recommended Classification of Pesticides by Hazard and Guidelines to Classification, 2004.

Appendix IV. Acute Toxicity of Slightly Hazardous Pesticides – Class III. Part I

Common name	LD_{50} (mg/kg) Oral, rat	Common name	LD_{50} (mg/kg) Oral, rat
Acephate	945	Dicofol	690
Acetochlor	2950	Diethyltoluamide	2000
Acifluorfen	1370	Difenoconazole	1453
Alachlor	930	Dimepiperate	946
Allethrin	685	Dimethachlor	1600
Ametryn	1110	Dimethametryn	3000
Amitraz	800	Dimethipin	1180
Azamethiphos	1010	Dimethylarsinic acid	1350
Bensultap	1100	Diniconazole	639
Bentazone	1100	Dinocap	980
Butralin	1049	Diphenamid	970
Butroxydim	1635	Dithianon	640
Chinomethionat	2500	Dodine	1000
Chlormequat	670	Empenthrin [(1R)	>2280
Chloroacetic	650	Esprocarb	>2000
Copper	1000	Etridiazole	2000
Copper	1440	Fenothiocarb	1150
4-CPA	850	Ferimzone	725
Cycloate	>2000	Fluazifop-p-butyl	2451
Cyhexatin	540	Fluchloralin	1550
Cymoxanil	1196	Flufenacet	600
Cyproconazole	1020	Fluoroglycofen	1500
Dazomet	640	Flurprimidol	709
2,4-DB	700	Flusilazole	1110
Dicamba	1707	Flutriafol	1140
Dichlormid	2080	Fomesafen	1250
Dichlorobenzene	500-5000	Furalaxyl	940
Dichlorophen	1250	Glufosinate	1625
Dichlorprop	800	Hexazinone	1690
Diclofol	565	Hydramethylnon	1200

Source: The WHO Recommended Classification of Pesticides by Hazard and Guidelines to Classification, 2004.

Appendix IV. Acute Toxicity of Slightly Hazardous Pesticides – Class III. Part II

Common name	LD_{50} (mg/kg) Oral, rat	Common name	LD_{50} (mg/kg) Oral, rat
Iprobenfos	600	Propanil	1400
Isoprothiolane	1190	Propargite	2200
Isoproturon	1800	Pyrazoxyfen	1644
Isouron	630	Pyridaben	820
Malathion	2100	Pyridaphenthion	769
MCPA	700	Pyridate	2000
MCPA-thioethyl	790	Pyrifenox	2900
MCPB	680	Quinoclamine	1360
Mecoprop	930	Quizalofop	1670
Mecoprop-P	1050	Resmethrin	2000
Mefluidide	1920	Sethoxydim	3200
Mepiquat	1490	Simetryn	1830
Metalaxyl	670	Sodium chlorate	1200
Metamitron	1183	Sulfluramid	543
Metconazole	660	2,3,6-TBA	1500
Methylarsonic acid	1800	Tebuconazole	1700
Metolachlor	2780	Tebufenpyrad	595
Myclobutanil	1600	Tebuthiuron	644
2-Napthyloxyacetic acid	600	Thiram	560
Nitrapyrin	1072	Tralkoxydim	934
Nuarimol	1250	Triadimefon	602
Octhilinone	1470	Triadimenol	900
N-octylbicycloheptene	2800	Tri-allate	2165
Oxadixyl	1860	Triclopyr	710
Paclobutrazol	1300	Triflumizole	695
Pendimethalin	1050	Undecan-2-one	2500 (dogs, cats)
Pimaricin	2730	Uniconazole	1790
Pirimiphos-methyl	2018	XMC	542
Prochloraz	1600	Ziram	1400
Propachlor	1500		

Source: The WHO Recommended Classification of Pesticides by Hazard and Guidelines to Classification, 2004.

APPENDIX V. ACUTE TOXICITY OF ORGANOTIN COMPOUNDS

Acute oral toxicity of mono-organotin compounds

Compound	LD_{50} (mg/kg)	
Butylstannoic acid	> 6000	(mouse)
Butyltin trichloride	1400	(mouse)
Butyltin-S,S'S"-tris (isooethylmercaptoacetate)	1520	(mouse)
Otyltin trichloride	4600	(mouse)
Octyltin-S,S',S"-tris(2-ethylexylmercaptoacetate)	1500	(rat)

Acute toxicity of diorganotin compounds

Compound	LD_{50} (mg/kg)	
Dibutylin di(2-ethylhexoate)	200	(oral, rat, male)
Dibutyltin di(butyl maleate)	120	(oral, rat)
Dibutyltin di(nonylmaleate)	170	(oral, rat)
Dibutyltin dichlorlde	100-180	(oral, rat, male)
	112	(oral, rat, female)
	35	(oral, mouse)
	190	(oral, guineapig)
Dibutyltin-S,S'-bis (2- ethylhexylmercaptoacetate)	150	(oral, rat, male)
Dibutyltin dilaurate	243	(oral, rat)
Dibutyltin oxide	520	(oral, rat, male)
	39.9	(intraperitoneal, rat, female)
	24	(oral, mouse)
Dibutyltin sulfide	145	(oral, rat, male)
	150	(oral, rabbit)
	180	(oral, rat, female)
	800	(intraperiotoneal, rat, female)
	> 800	(intraperitoneal, rat, female)
Dioctyltin acetate	2030	(oral, rat)
Diocyltin dibutylmaleate	3750	(oral, mouse)
Dioctyltin-S,S'-bis(butylmercaptoacetate)	1140	(oral, white mouse)
Dioctyltin bis(dodecylmercaptide)	4000	(oral, white mouse)
Dioctyltin bis(2-ethylhexylmaleate)	2760	(oral, rat, male)
Dioctyltin-S,S'-bis (2-ethylhexylmercaptoacetate)	2010	(oral, white mouse)
Dioctyltin-S,S'-bis (laurylmercaptoacetate)	3700	(oral, rat, male)
Dioctyltin-S,S'-(1,4- butane-diol-bis-mercaptoacetate)	2950	(oral, rat, male)
Dioctyltin dichloride	8500	(oral, rat, male)
Dioctyltin dilaurate	6450	(oral, rat, male)
	800	(intraperitoneal, rat, female)
Dioctyltin di(1,2- propylene-glycolmaleate)	4775	(oral, rat, male)
Dioctyltin-S,S'-(ethyleneglycol-bis- dimercaptoacetate)	880	(oral, rat, male)
Dioctyltin maleate	4500	(oral, rat, male)

Dioctyltin ß-mercapto-propanoate	2050 (oral, rat, male)
Dioctyltin oxide	2500 (oral, rat, male)
Dioctyltin mercaptoacetate	945 (oral, rat, male)
Acute toxicity of triorganotin compounds	
Compound	**LD_{50} (mg/kg)**
2-trichloro-1-(butine-1'- oxide)-1-triethyl- stannyloxy)ethane	9.8 (intraperitoneal, rat)
	9.6 (intraperitoneal, mouse)
Tiethylstannylmethyl-(1-propynyl)formal	10.7 (intraperitoneal, rat)
Triethylstannylphenyl- acetylene	9.9 (intraperitoneal, mouse)
	9.1 (intraperitoneal, rat)
1-triethylstannyl-3- trimethylsiloxi-1- propyne	11.8 (intraperitoneal, mouse)
	11.4 (intraperitoneal, rat)
Triethyltin acetate	4 (oral, rat, female)
Triethyltin chloride	5 (intraperitoneal, rat, female)
Triethyltin sulfate	5.7 (intraperitoneal, rat, male)
Tributyltin acetate	46 (oral, white mouse)
	133 (oral, rat, male)
Tributyltin benzoate	108 (oral, white mouse)
	132 (oral, rat)
Tributyltin chloride	117 (oral, white mouse)
	129 (oral, rat)
Tributyltin laurate	180 (oral, white mouse)
Tributyltin oleate	195 (oral, rat)
Tributyltin salicylate	137 (ral, rat, male)
Bis(tributyltin oxide)	112-132 (oral, rat, male)
	180 (oral, rat)
	234 (oral, rat)
	194 (oral, rat, male) aqueous soloution
	148 (oral, rat, male) oil solution
	11.7 (dermal, albino rabbit)
	10.0 (oral, white mouse)
Trihexyltin acetate	1000 (oral, rat)
Trioctyltin chloride	10 000 (oral, rat, male)
Triphenyltin acetate	21 (oral, guineapig)
	24 (oral, guineapig)
	136 (oral, rat)
	81 (oral, mouse,male)
	7.9 (intraperitoneal, mouse,male)
	136 (oral, rat, male)
	491 (ral, rat, female)
	450 (dermal, rat, male)
	8.5 (intraperitoneal, rat, female)
	11.9 (intraperitoneal, rat, female)
	13.2 (intraperitoneal, rat, male)

	21 (oral, guineapig, male)
	3.7 (intraperitoneal, guineapig, male)
Triphenyltin acetate	5.3 (intraperitoneal, guineapig, male)
	30-50 (oral, rabbit, male)
Triphenyltin chloride	80 (oral, rat, male)
	135 (oral, rat, female)
Triphenyltin hydroxide	245 (oral, mouse, male)
	209 (oral, mouse, female)
	240 (oral, rat, male)
	360 (oral, rat, female)
	27.1 (oral, guineapig, male)
	31.1 (oral, guineapig, female)
	171 (oral, rat, male)
	268 (oral, rat, female)
Tricyclohexyltin hydroxide	710 (oral, mouse-peromiskus)
	1070 (oral, mouse-swiss white)
	540 (oral, rat)
	13 (intraperitoneal, rat)
	780 (oral, guineapig)
Tricyclohexyltin hydroxide	9 (intraperitoneal, guineapig)
	500-1000 (oral, rabbit)
	> 126 (intraperitoneal, rabbit)
	150 (oral, sheep)
	14 (intravenous, dog)

Acute oral toxicity of tetraorganotin compounds

Compound	**LD_{50} (mg/kg)**
Tetraethyltin	40.0 (mouse)
	15.0 (rat)
	40.0 (guinea pig)
	7.0 (rabbit)
	40 (mouse)
	9.0 (rat)
	40.0 (guinea pig)
	7.0 (rabbit)
Tetrabutyltin	6000 (rat)
Dioctyltin mercaptoacetate	945 (rat, male)

Source: World Health Organization. Environmental health criteria 15. *Tin and organotin compounds*; Geneva, Switzerland:1980.

APPENDIX VI. ACUTE TOXICITY OF SNAKE VENOMS IN MICE

Scientific name	Comon name	LD_{50} (mg/kg)
Intramuscular		
Hydrophis melanosoma	Black banded sea snake	0.08
Aipysurus laevi	Olive sea snake	0.09
Hydrophis ornatus	Reef sea snake	0.12
Hydrophis belcheri	sea snake species	0.16
Hydrophis elegans	Elegant sea snake	0.21
Crotalus mitchelli mitchelli	Panamint speckled rattlesnake	0.30
Naja nigricollis	Black spitting cobra	0.44
Bitis arientas	Puff adder	2.00
Crotalus enyo enyo	Orange rattlesnake	4.60
Vipera aspis	Asp viper	4.70
Bitis gabonica	Gaboon viper	5.20
Bitis nasicorais	Rhino viper	8.60
Crotalus exul	Rattlesnake species	9.20
Crotalus mitchelli pyrrhus	Southwestern speckled rattlesnake	9.60
Crotalus catalinensis	rattlesnake species	10.90
Crotalus pricei pricei	Twin-spotted rattlesnake	11.50
Crotalus polysticus	Mexican blotched rattlesnake	13.30
Crotalus atrox	Western diamondback rattlesnake	20.00
Agkistrodon contortrix laticinctus	Broad-banded copperhead	20.00
Intraperitoneal		
Oxyuranus scutellatus	Coastal taipan	0.009
Pseudocerastes fieldi	Field's horned viper	0.02
Bungarus multicinctus	Many banded krait	0.08
Bungarus caeruleus	Indian krait	0.09
Enhydrina schistosa	Beaked sea snake	0.11
Dendroaspis angusticeps	Eastern green mamba	0.12
Micrurus micrurus anomalus	coral snake species	0.12
Walterinnesia aegyptia	Desert cobra	0.13
Crotalus scutulatus scutulatus	Mojave green rattlesnake	0.16
Crotalus mitchelli mitchelli	Desert speckled rattlesnake	0.18
Naja haje	Egyptian cobra	0.19
Crotalus viridis concolor	Midget faded rattlesnake	0.20
Crotalus durissus terrificus	Cascabel	0.22
Naja kaouthia	Monocled cobra	0.23
Hydrophis cyanocinctus	Annulated sea snake	0.24
Cerastes vipera	Sahara sand viper	0.25
Lapemis hardwickii	Hardwicke's sea snake	0.26
Echis coloratus	Carpet viper	0.26
Austrelaps superbus	Lowland copperhead	0.28
Cerastes cerastes	Desert horned viper	0.30
Hydrophis spiralis	Yellow sea snake	0.32
Naja naja	Spectacled cobra	0.32
Naja melanoleuca	Black forest cobra	0.32
Dendroaspis viridis	Western green mamba	0.33
Hydrophis klossi	sea snake species	0.37
Daboia russelli russelli	Russell's viper species	0.40
Hydrophis melanosoma	Black banded sea snake	0.40
Naja flava	Cape cobra	0.40
Naja nigricollis	Black necked spitting cobra	0.40
Dendroaspis jamesoni	Jameson's mamba	0.41
Vipera ammodytes	Long-nosed viper	0.42
Naja nivea	Cape cobra	0.42
Daboia russelii formosensis	Russel's viper subspecies	0.49
Kerilia jerdoni	Jerdon's sea snake	0.53
Micrurus frontalis frontalis	Southern coral snake	0.59
Crotalus durissus durissus	Neotropical rattlesnake	0.67
Micrurus nigrocinctus	Black banded coral snake	0.70

Micrurus alleni	Coral snake species	0.71
Bitis arientas	Puff adder	0.72
Pseudechis colletti	Collett's snake	0.84
Dendroaspis polylepsis	Black mamba	0.94
Micrurus annulatus hertwigi	Coral snake species	0.95
Micrurus dumarillii	Coral snake species	0.96
Agkistrodon bilineatus	Cantil	1.16
Micrurus spixii obscurus	Amazonian coral snake subspecies	1.16
Bungarus fasciatus	Banded krait	1.55
Bitis gabonica	Gaboon viper	1.59
Ophiophagus hannah	King cobra	1.64
Vipera palaestinae	Palestine viper	1.90
Vipera bornmuelleri	True viper species	1.92
Crotalus viridis viridis	Prarie rattlesnake	2.00
Naja pallida	Sudan red spitting cobra	2.00
Vipera latifii	True viper species	2.07
Echis carinatus	Saw-scaled viper	2.15
Crotalus viridis lutosus	Great basin rattlesnake	2.20
Crotalus viridis caliginsis	rattlesnake species	2.26
Crotalus horridus horridus	Timber rattlesnake	2.27
Crotalus adamanteus	Eastern diamondback rattlesnake	2.29
Agkistrodon bilineatus taylori	Ornate cantil	2.30
Crotalus viridis helleri	Southern Pacific rattlesnake	2.44
Crotalus durissus totonacus	rattlesnake species	2.50
Crotalus mitchelli pyrrhus	Southwestern speckled rattlesnake	2.70
Bothrops asper	Terciopelo	2.84
Tropidolaemus wagleri	Wagler's viper	3.58
Bothrops atrox	Barba amarilla	3.80
Crotalus exul	rattlesnake species	3.9
Crotalus cerastes	Sidewinder	4.00
Crotalus catalinensis	rattlesnake species	4.10
Agkistrodon contortrix mokasen	Northern copperhead	4.22
Agkistrodon piscivorus leucostoma	Western cottonmouth	4.84
Calloselasma rhodostoma	Malayan pit viper	4.99
Trimeresurus flavoviridis	Okinawa habu	5.07
Agkistrodon piscovorus	Cottomouth/ Water moccasin	5.10
Crotalus atrox	Western diamondback rattlesnake	5.59
Lachesis muta	Bushmaster	6.17
Sistrurus milarius barbouri	Red pygmy rattlesnake	6.82
Agkistrodon contortrix contortrix	Southern copperhead	10.90
Trimeresurus okinavensis	Himehabu	15.00
Intravenous		
Pseudonaja textilis	Eastern brown snake	0.01
Oxyuranus scutellatus	Coastal taipan	0.01
Notechis scutulatus	Mainland tiger snake	0.04
Crotalus tigris	Tiger rattlesnake	0.06
Dispholidus typus	Boomslang	0.07
Crotalus viridis concolor	Midget faded rattlesnake	0.08
Bungarus multicinctus	Many banded krait	0.11
Daboia russelli russelli	Russels viper	0.13
Laticauda laticauda	Banded sea krait	0.16
Bungarus caeruleus	Indian krait	0.17
Hydrophis fasciatus	Banded small headed sea snake	0.18
Crotalus scutulatus scutulatus	Mojave green rattlesnake	0.19
Bitis caudalis	Horned puff adder	0.20
Pelamis platurus	Yellow bellied sea snake	0.22
Crotalus scutulatus salvini	Huamantlan rattlesnake	0.24
Acanthophis antarcticus	Common death adder	0.25
Atractaspis engaddensis	Mole viper	0.25
Dendroaspis polylepsis	Black mamba	0.25
Enhydrina schistosa	Beaked sea snake	0.26
Crotalus durissus terrificus	Cascabel	0.26
Pseudocerastes fieldi	Field's horned viper	0.28
Naja melanoleuca	Black forest cobra	0.29
Micrurus fulvius	coral snake species	0.30
Pseudechis australis	Mulga snake	0.30

Vipera palaestinae	Palestine viper	0.30
Walterinnesia aegyptia	Desert cobra	0.30
Laticauda semifasciata	sea snake species	0.34
Naja atra	Chinese cobra	0.35
Vipera latifii	True viper species	0.35
Naja naja	Spectacled cobra	0.35
Trimeresurus albolabris	White lipped tree viper	0.37
Naja kaouthia	Monocled cobra	0.37
Deinagkistrodon acutus	Sharp-nosed pit viper	0.38
Micrurus nigrocinctus	Black banded coral snake	0.39
Daboia russelii formosensis	Russell's viper subspecies	0.40
Laticauda colubrina	Banded sea krait	0.40
Naja haje	Egyptian cobra	0.42
Hydrophis cyanocinctus	Annulated sea snake	0.46
Bothrops jararacussu	Jararacussu	0.46
Pseudechis porphyriacus	Red-bellied black snake	0.54
Vipera berus	Adder	0.55
Naja nivea	Cape cobra	0.57
Vipera bornmuelleri	True viper species	0.61
Naja nigricollis	Black necked spitting cobra	0.63
Cerastes vipera	Sahara sand viper	0.64
Macrovipera lebetina	Levantine viper	0.64
Echis pyramidum (Egypt)	Saw-scaled viper	0.65
Dendroaspis jamesoni	Jameson's mamba	0.72
Micrurus alleni	Coral snake species	0.74
Tropidolaemus wagleri	Wagler's viper	0.78
Dendroaspis viridis	Western green mamba	0.80
Lapemis hardwickii	Hardwicke's sea snake	0.80
Vipera ammodytes	Long-nosed viper	0.80
Pseudocerastes persicus	Persian viper	0.83
Echis pyramidum (Kenya)	Saw-scaled viper	0.94
Naja oxiana	Central Asian cobra	0.96
Vipera aspis	Asp viper	1.00
Crotalus viridis viridis	Prarie rattlesnake	1.01
Micrurus dumarillii	coral snake species	1.02
Bothrops jararaca	Jararaca	1.10
Crotalus viridis helleri	Southern Pacific rattlesnake	1.14
Thelotornis kirtlandi	Twig snake	1.24
Bothrops asper	Terciopelo	1.24
Crotalus durissus durissus	Neotropical rattlesnake	1.24
Bungarus fasciatus	Banded krait	1.29
Rhapdophis subminatus	Red-necked keeled snake	1.29
Ophiophagus hannah	King cobra	1.31
Bitis arientas	Puff adder	1.32
Dendroaspis angusticeps	Eastern green mamba	1.50
Hemachatus haemachatus	Rinkhals	1.50
Crotalus intermedius	rattlesnake species	1.58
Atropoides picadoi	Central American pitviper species	1.60
Bothriechis schlegelii	Eyelash viper	1.60
Crotalus willardi willardi	Ridge-nosed rattlesnake	1.61
Crotalus adamanteus	Eastern diamondback rattlesnake	1.65
Echis pyramidum (Saudi Arabian)	Saw-scaled viper	1.67
Micrurus annulatus hertwigi	coral snake species	1.76
Bothrops alternatus	Urutu	1.96
Agkistrodon piscivorus leucostoma	Western cottonmouth	2.04
Vipera palaestinae	Palestine viper	2.09
Crotalus horridus horridus	Timber rattlesnake	2.11
Daboia russelli siamensis	Russell' viper subspecies	2.11
Crotalus exul	rattlesnake species	2.17
Hydrophis ornatus	Reef sea snake	2.20
Atractaspis dohomeyensis	Mole viper	2.24
Bothrops colombiensis	South American pitviper species	2.30
Bothrops neuwiedi	Jararaca pintada	2.30
Agkistrodon bilineatus	Cantil	2.40
Agkistrodon bilineatus taylori	Ornate cantil	2.40
Atropoides nummifer	Jumping viper	2.40

Agkistrodon contortrix mokasen	Northern copperhead	2.70
Echis ocellatus (Nigeria)	Saw-scaled viper	2.71
Crotalus atrox	Western diamondback rattlesnake	2.72
Bothrops atrox	Barba amarilla	2.84
Echis carinatus sochureki(Pakistan)	Saw-scaled viper	2.98
Trimeresurus okinavensis	Himehabu	3.00
Crotalus pricei pricei	Twin-spotted rattlesnake	3.07
Echis carinatus multisquamatus (Iran)	Saw-scaled viper	3.26
Crotalus polystictus	Mexican blotched rattlesnake	3.37
Trimeresurus flavoviridis	Okinawa habu	3.689
Agkistrodon piscovorus	Cottomouth/ Water moccasin	4.00
Aipysurus eydouxii	Spine-tailed sea snake	4.00
Cerastes cerastes	Desert horned viper	4.00
Echis coloratus	Carpet viper	4.38
Praescutata viperina	Viperine sea snake	4.50
Lachesis muta	Bushmaster	4.51
Porthidium nasuta	Hognosed viper	4.60
Cerastes cornutus	Desert horned viper	5.05
Calloselasma rhodostoma	Malayan pit viper	5.20
Trimeresurus elegans	Sakishima habu	5.23
Trimeresurus mucrosquamatus	Chinese habu	5.23
Trimeresurus stejnegeri	Chinese green tree viper	5.225
Vipera xanthina	Near east viepr	6.70
Bitis gabonica	Gaboon viper	6.72
Eristicophis macmohoni	Asian sand viper	7.5
Crotalus lepidus klauberi	Rock rattlesnake	9.00
Causus rhombeatus	Night adder	9.25
Gloydius halys	Pallas' viper	9.75
Agkistrodon contortrix contortrix	Southern copperhead	10.90
Sistrurus milarius barbouri	Red pygmy rattlesnake	12.59
Pseudocerastes persicus	Persian horned viper	15.26
Subcutaneous		
Oxyuranus microlepidotus	Inland taipan	0.025
Pseudonaja textilis	Eastern brown snake	0.0365
Aipysurus duboisii	Dubois's sea snake	0.044
Pelamis platurus	Yellow bellied sea snake	0.067
Acalyptophis peronii	Horned sea snake	0.079
Oxyuranus scutellatus	Coastal taipan	0.106
Bungarus multicinctus	Many banded krait	0.108
Hydrophis melanosoma	Black banded sea snake	0.111
Enhydrina schistosa	Beaked sea snake	0.1125
Boulengeria christyi	Congo water cobra	0.12
Notechis a. niger	Peninsula tiger snake	0.131
Boulengeria annulata	Banded water cobra	0.143
Echis carinatus	Saw-scaled viper	0.151
Hydrophis stricticollis	sea snake species	0.164
Hydrophis major	Olive-headed sea snake	0.193
Notechis a. occidentalis	Western tiger snake	0.194
Crotalus tigris	Tiger rattlesnake	0.21
Notechis scutulatus	Mainland tiger snake	0.214
Hydrophis elegans	Elegant sea snake	0.26
Aipysurus laevi	Olive sea snake	0.264
Laticauda semifasciata	Broad banded sea krait	0.273
Naja atra	Chinese cobra	0.29
Dendroaspis polylepsis	Black mamba	0.32
Notechis a. serventyi	Chapel island tiger snake	0.338
Hydrophis nigrocinctus	Duadin's sea snake	0.343
Crotalus scututatus	Mojave green rattlesnake	0.34
Bungarus caeruleus	Indian krait	0.365
Walterinnesia aegyptia	Desert cobra	0.4
Laticauda colubrina	Wide faced sea krait	0.435
Naja naja	Spectacled cobra	0.45
Hydrophis cyanocinctus	Annulated sea snake	0.464
Pseudonaja nuchalis	Gwardar/Western brown snake	0.473
Acanthophis antarcticus	Common death adder	0.5
Austrelaps superbus	Lowland copperhead	0.5

Lapemis hardwickii	Hardwicke's sea snake	0.541
Pseudonaja affinis	Dugite	0.66
Dendroaspis viridis	Western green mamba	0.7
Naja nivea	Cape cobra	0.72
Daboia russelli russelli	Russell's viper subspecies	0.75
Dendroaspis jamesoni	Jameson's mamba	1
Pseudechis papuanus	Papuan black snake	1.09
Naja haje	Egyptian cobra	1.15
Micrurus fulvius	coral snake sp.	1.3
Hoplocephalus stephensi	Stephen's banded snake	1.36
Daboia russelii formosensis	Russell's viper subspecies	1.37
Ophiophagus hannah	King cobra	1.7
Pseudechis australis	Mulga snake	1.94
Pseudechis porphyriacus	Red-bellied black snake	2
Pseudechis guttatus	Spotted black snake	2.13
Pseudechis colletti	Collett's snake	2.38
Hemachatus haemachatus	Rinkhals	2.65
Cryptophis nigrescens	Small eyed snake	2.67
Crotalus basciliscus	Mexican west-coast rattlesnake	2.8
Dendroaspis angusticeps	Eastern green mamba	3.05
Crotalus horridus horridus	Timber rattlesnake	3.1
Bungarus fasciatus	Banded krait	3.6
Vipera latifii	True viper species	4.61
Tropidolaemus wagleri	Wagler's viper	6.19
Vipera bornmuelleri	True viper species	6.25
Vipera berus	Adder	6.45
Vipera ammodytes	Long-nosed viper	6.59
Bothrops jararaca	Jararaca	7
Trimeresurus gramineus	Indian green tree viper	8.6
Deinagkistrodon acutus	Sharp-nosed pit viper	9.2
Vipera palaestinae	Palestine viper	9.4
Crotalus exul	rattlesnake species	9.92
Bitis gabonica	Gaboon viper	12.5
Trimeresurus albolabris	White lipped tree viper	12.75
Bothrops jararacussu	Jararacussu'	13
Crotalus polystictus	Mexican blotched rattlesnake	13.3
Bothrops neuwiedi	Jararaca pintada	14.2
Crotalus adamanteus	Eastern diamondback rattlesnake	14.6
Causus rhombeatus	Night adder	15
Cerastes cerastes	Desert horned viper	15
Bothrops alternatus	Urutu	15.8
Macrovipera lebetina	Levantine viper	16
Crotalus atrox	Western diamondback rattlesnake	18.5
Gloydius blomhoffi	Mamushi	20
Bothrops atrox	Terciopelo	22
Calloselasma rhodostoma	Malayan pit viper	23.4
Crotalus lepidus klauberi	Rock rattlesnake	23.95
Sistrurus milarius barbouri	Red pygmy rattlesnake	24.3
Emydocephalus annulatus	Turtle-headed sea snake	25
Agkistrodon contortrix contortrix	Southern copperhead	25.6
Agkistrodon piscovorus	Cottomouth/ Water moccasin	25.8
Bothriechis schlegelii	Eyelash viper	33.2
Lachesis muta	Bushmaster	36.9
Demansia olivacea	Olive whip snake	714.2

INDEX